GROUP THEORY:
AN INTUITIVE APPROACH

GROUP THEORY:
AN INTUITIVE APPROACH

GROUP THEORY: AN INTUITIVE APPROACH

Problem T.1: What is the transformation?

R. MIRMAN

Published by

World Scientific Publishing Co. Pte. Ltd.

P O Box 128, Farrer Road, Singapore 912805

USA office: Suite 1B, 1060 Main Street, River Edge, NJ 07661

UK office: 57 Shelton Street, Covent Garden, London WC2H 9HE

British Library Cataloguing-in-Publication Data
A catalogue record for this book is available from the British Library.

First published 1995
Reprinted 1998, 2000

GROUP THEORY: AN INTUITIVE APPROACH

ISBN 981-02-2183-5
ISBN 981-02-3365-5 (pbk)

This book is printed on acid-free paper.

Printed in Singapore by Regal Press (S) Pte. Ltd.

Preface

P.1 GROUPS, NATURE, UNDERSTANDING, UNREASONABLENESS

Groups as a tool — a computational tool — have become an essential part of mathematics, physics, chemistry, and now even other fields, and widely recognized as such. Groups as a tool — a tool for exploring, for inquiring, for understanding — are less widely recognized. There are reasons why groups are an important computational tool, reasons that come from the nature of what groups are, but perhaps more important, from what they probe. It is this relationship that is the foundation of their usefulness, and it is this that makes them so valuable, not merely in computation, but in illumination, that gives them the power to provide intuition, about groups themselves, but more, about the fields that we study with them.

The purpose of this book is the introduction — and exploration — of the theory of groups, to understand groups, to learn how to use them for computation and for proofs, but also to learn how to use them for the study and understanding of other fields — of all fields to which they can be applied. The purpose is to develop awareness, intuition, not merely about each of these areas, but to use what is achieved from one to develop it for others, to use groups to help master other subjects, and use these to help provide insight into groups.

Because this is a book about group theory, we emphasize using groups, and other subjects, as a tool for obtaining insight and judgment about groups, and for computation. Unfortunately this obscures the power — perhaps greater power — of groups to also provide understanding and intuition about other fields [Mirman (1995a,b)].

Although the main aim here is the understanding of groups and how to use them as tools, another is art appreciation, to help the reader realize — and wonder at — the beauty of mathematics, and its awesome applicability. That mathematics applies to the physical world is striking indeed, that it is consistent, that mathematics applies to mathematics, is more so. Eugene Wigner is said to have referred to the unreasonable effectiveness of mathematics in the natural world. But it is more than that, indeed not merely unreasonable but often stunning [Mirman (1995a)], and what is perhaps more surprising, is the unreasonable effectiveness of mathematics in mathematics. We hope that, after reading this book, this is something the reader will realize, not only from what is presented here, but also from what is noticed elsewhere, noticed, in part, because of the stimulation resulting from the study of the theory of groups.

P.2 WHO SHOULD BE INTERESTED IN THIS BOOK?

The obvious, but obviously unrealistic, answer, is everyone. Yet even those merely interested in group theory form a quite diverse set. Can a book appeal to a wide range of readers, from ones who plan to do advance mathematical research in group theory, or in other mathematical fields, to those who need, and care for, no more than a brief introduction to the terminology, plus everyone in between? The book should be considered an experiment to see if the answer can be yes. This requires some judgment by the reader. Not everyone will be interested in every word; it is likely that no one will.

An example of someone who might be least interested is an experimenter who needs a little understanding of the concepts and vocabulary to (make it appear that he can) read theoretical papers, but does not have the time or inclination to study the subject in depth. The book is written to be useful to, and appeal to, people such as these.

Who might be among those most interested? The flavor of the book is physical, and certainly theoretical physicists — and chemists — should be among the most interested. Because of this flavor it will hopefully appeal to applied mathematicians also. But should pure mathematicians be interested in it? In the author's (highly biased) opinion, very much so. To show part of the reason, we discuss below our educational philosophy, how group theory is presented elsewhere, how it should be presented, what is known, and what has been forgotten, about it.

The reasons for being interested in the book, especially for mathematicians, go beyond the material (although the material is important to much of mathematics). It is intended to be a text, but more, to be a developer of skills (including teaching skills), a reference, a source of material not available elsewhere, and particularly a stimulant to thought

— certainly something that mathematicians, especially those involved in group theory and related fields, should find useful. In accord with its pedagogical aim, problems are stressed, so the book should also be valuable to instructors as a resource for classes.

P.2.a Level

The book is designed to be used at the first-year graduate level (and of course, at higher levels). It can also be read by (relatively) advanced undergraduates, should any be interested. While it is written as a text for a course in group theory, as well as a supplementary text for other courses that involve group theory, it also has other purposes. In particular, it is designed to be useful — the most useful book — for anyone who wants to gain, by self-study, some introduction, or even a quick understanding of a definition, and at the other extreme, those working in group theory, or using it in their research.

P.2.b Prerequisites

Learning the material, but not necessarily just looking up something, requires some knowledge of matrices. A knowledge of the vocabulary of quantum mechanics would be helpful, but not necessary since quantum-mechanical terminology and examples are used for motivation and concreteness. Mathematicians being perhaps more motivated may need less explicit motivation, so if they are not familiar with quantum mechanics they can just skip those parts. Also some problems may require deeper knowledge, but those without it can skip them.

P.3 EDUCATIONAL PHILOSOPHY

Underlying every book is a set of aims, and a philosophy of how these are to be achieved, even if the author is unaware of them, even if confused and inchoate. It would help the reader in using a book if these were made explicit; here we mention some educational philosophy (and prejudices) that the author believes guided the aims, organization and writing of the book.

A book should be written to be used. Unfortunately few are. Most, if not almost all, texts at this level (not only in group theory) seem to take the attitude that the author presents the information to, even just throws it at, the reader, and then the reader is supposed to figure out what it means and how to learn and memorize it. This book is written in the belief, and hopefully in the manner, that it is the author's responsibility to figure out how to present the material in a way that the reader — every reader — can understand, learn, absorb and remember. Techniques to aid memory are crucial. The book is therefore somewhat, perhaps more than somewhat, repetitive. This may make it less graceful to read, but it makes the material easier to learn, thus to understand, and later to recall. And if that is true, it is a price worth paying.

One major aspect of the philosophy of this book is that it is better to consider each topic discussed in great depth, trying to achieve as deep an understanding as possible, rather than to try to cover as many topics as possible. As a result the subject matter is restricted. It is becoming more and more understood that this is the proper approach, certainly at elementary levels. But it is the proper approach at the level of this book also.

Thus this is aimed at being many things, perhaps too many, a reference certainly, a source of insight for those expert in the theory hopefully, but most of all a text.

P.3.a Training readers

A text is not only supposed to teach the particular material, but also help train the student to be a professional mathematician, physicist, chemist, ... Many problems are aimed at doing that, asking the reader to think and explain, to do things, at a simplified level, that would be done, or should be done, by a research scientist. At times the first sentence of a problem is completely trivial, the last completely impossible (at least for ordinary mortals like the author). One of the important skills of a researcher is choosing problems wisely. Those who choose trivial ones will produce trivial work. Those who look for grandeur will produce no (nontrivial) work. Many problems here are meant to provide useful exercises, helping students learn how far to go, and when to stop.

The book is designed to be easy to read and understand, with definitions of terms, upon which all else is based, emphasized, and with an attempt made to help the reader understand and remember them. However it is also designed to stimulate readers, from many different areas, to think. It tries to raise questions, even to irritate the reader into wondering about the material (but hopefully not about why the book was written the way it was). Thus each reader must decide for himself what he (or she) wants to get out of the book, what should be read, what briefly, and what carefully. Each must decide what problems to do, when to make a major effort, when to stop. And each must decide what he wants to think about, and learn about. This may, perhaps, annoy some readers. But still, it is good training. And that is a major purpose of this. The book is aimed at reaching a wide readership, but also affecting those who use it.

One of the things that a professional must learn is how to read. The book presents problems that try to stimulate the reader to be conscious of what he is reading, of being aware of it and its implications, and to learn that he is (usually) not so conscious, that he reads without full awareness of what is being said (perhaps something that he shares with the writer, who may not always be fully aware of the implications of what he is writing). And it tries to make him understand the importance of such awareness.

More generally, the book raises questions about the material, hopefully forcing, or at least stimulating, the reader to think, and not only about the material. This helps him learn the subject, but it is beyond that, a valuable exercise.

In proving and explaining results, we try to use the simplest and most obvious method possible, not the most esoteric, nor the one that shows how much unusual mathematics we know. It is best if the proofs and discussions follow clearly and obviously from the assumptions underlying a particular system, for this not only makes it easier to follow and understand, but leads to greater understanding of the system and the assumptions. It better prepares the reader to consider other cases with different assumptions and postulates, and helps him to develop proofs and analyses for these. The closer to the foundations the better — it is safer and less likely to lead to falls. And this makes the reader aware of foundations, strengthening the foundation of his skills.

Many readers will go on to teach, this and other topics. It is hoped that the style of the book is irritating enough to make the reader aware of his own reactions, of how well he is learning and why, of how he would like the material to be presented to make it easy for him to learn, understand and remember, of how he thinks the presentation can be improved. Having learned something about how he himself learns, he will be better prepared to teach others. And, perhaps also, he will have better learned the material himself.

P.3.b What is motivation, and why is it important?

Motivation is central, and here emphasized; the book states the reasons the material is being considered, and the questions (or at least some) that are to be studied and hopefully answered. It is often felt that professional mathematicians and physicists are so motivated that they will respond with great eagerness to a set of disconnected arguments, and even equations. Perhaps. But this misses an important point. To understand a subject fully it is necessary to place it in context, to see what it is related to, why, and how can it help answer questions of interest. To what is it relevant? What is its structure, how can it be seen as, not a series of individual equations and topics, but as a coherent whole, what questions are important, and how does the discussion help the reader to understand and answer them? These, context, coherence, form the essence of motivation, and these are stressed here.

Motivating the reader by emphasizing the coherence of the subject is essential to learning, but it may also make the reader see, and so wonder at, the coherence, not only of group theory, but of mathematics as a whole.

Not only motivation, but explanation, examples, concreteness, are the crux of the approach. We try to explain not only facts, but reasons, not only what is happening, but why. What properties of groups, and the (physical) objects to which they are applied, give the results being derived? Why should these hold? How do they arise from the assumptions and definitions about groups and representations, and about nature? What do they imply about the mathematics underlying the material, mathematics in general, about physics (and other fields)? Why are these properties important?

P.3.c Vocabulary is fundamental

The most important part of learning a subject is mastering and understanding the vocabulary. The book stresses the meaning and content of terms and concepts, trying to make these not only as clear, but as familiar, and easy to remember, as possible. It may be unique in its emphasis not merely on vocabulary and the meaning, but also the origin, of words, giving, where possible, Greek and Latin roots and cognates, and relating these to not only the definition, but the mathematical and physical significance, of the terms. Illustrations, examples, even counter-examples, are provided to help the reader see what the definitions mean.

This approach makes it easier to understand, learn, and remember the material, and, if wanted, just quickly look it up. And it, and the related problems plus other aspects, help develop general mathematical and physical skills, so that hopefully after working through the book a student will not only know, and understand, group theory, but will be better trained as a physicist and mathematician — and be better able to explain mathematics and physics to others.

P.3.d Failures of the writer

Here we have set standards (perhaps too explicitly), with no expectation of being able to meet them. That the book does not satisfy the required standards is no reason to ignore or lower them. It is rather a reason to notice and be more aware of them, so to be able to see and understand the weaknesses, to be able to overcome them, here and elsewhere, thus to come closer to the standards set, and thereby raise them. We will always fail to meet standards. But it is better to fail to meet high standards than fail to meet low ones.

And failure here to meet proper standards is not an excuse for the reader also to fail, in his study of this book, or in his own writing. Besides if the reader, (as a result of) having thought about this discussion, notices failures here (and elsewhere) it will help him think about, and decide, what is proper. And so, and because he will then be more aware of them, his own standards and quality of work will improve, and he will help enforce and raise standards in general — including those of his, or her, students.

In this book, as with any, there may be errors, typographical ones certainly, possibly even statements that are wrong. The reader should go through the book carefully, and with some skepticism, checking everything. There is no reason to believe that the writer (definitely not this writer) is perfect, or can produce a perfect book.

P.3.e How to react to this book

The ultimate purpose of this book is to present group theory in such a way that even the author can understand it. If it is clear, as it should be, that this has not been completely achieved, that should not be taken by the reader as a reason for discouragement, but as a challenge.

If the book seems very long — and it certainly seems very long to the author who had to write it but found the material too interesting to leave any out — it should be remembered that it is not necessary to read it all (or even most of it). The book is not an attempt to tell everything you wanted to know about group theory but did not realize all these topics were there to be learned about, and far too much more in addition.

P.3.f Please read the problems (but don't do them all)

Mathematics (and physics) books consist of statements presented as if they are immediately obvious. However, perhaps not all statements in these books are quite so obvious. But constraints of reality make it impossible to explain and prove everything. Thus this book also consists of statements that are not proven. To show that they are not as obvious as they are often regarded, some are explicitly given as problems. This does not mean that less is explained here than elsewhere;

the material is usually as fully explained as in other books, often (too) much more so. But this use of problems is designed to make the reader aware of what is not explained, rather than hiding it, as is usually done. The reader can ignore this, taking the statements as given (as seems to be done with other books) — although knowing that a statement may not be fully obvious can make it easier to read a proof (the reader will at least know that he does not understand something, not because he is confused, but because it is not explained). But all problems should be read, for much of the material is in them. And even if a problem is not attempted, just reading it, and especially thinking about (reasons for not doing) it, is useful.

One of our purposes is consciousness raising, and having much of the discussion in the form of problems might help raise the reader's consciousness of how much is usually unexplained. Perhaps that will make him read more carefully and productively in general.

What we have sometimes done is to copy proofs, almost line for line (and given the source if the reader does not believe this) but changed each sentence to a problem. Thus instead of "It then follows ..." we have written "Show that it then follows ...". Presenting the proof this way stimulates thought about the material more than a list of statements does. And it should show how much is usually left out. If the proof is clear, as originally given, the problems are trivial. If not, and if not for most or all of the proofs so presented, then this has implications about the way material on mathematics and physics is usually presented, which the reader might want to ponder.

Another reason for many problems is that only a small part of any subject can be discussed. Often problems raise questions that are not discussed, or not discussed explicitly. While solving them will greatly enhance the reader's understanding, just reading them will, at least, make the reader aware of these topics. Often references are given so those interested can learn about the existence of material that could not be included, and where to obtain further information (unfortunately these are at times to books too little known, and now out-of-print; but hopefully the references here will, at least, help preserve some memory of them, and perhaps for some, more than memory).

There are a large number of problems, and many are quite long. Readers who look at these might get discouraged and not try any — or even not try to read the book. There is no reason to get discouraged, and no reason to try all problems. The book is designed to appeal to, and be useful to, many different types of readers; hopefully the number of problems will provide sufficient ones for each set of interests. This requires that each reader make decisions about what is of greatest interest and use. But there is no reason to put undue effort into ones of lesser interest — except, of course, the introductory ones.

In some cases the same problem, especially for the more advanced ones, is given in several different ways. It is difficult to know which attack will prove most successful (for a particular reader), so it helps to suggest several. And thinking about different views of the same point, the same problem, should raise questions and ideas, so encourage breadth and depth.

A few problems are really research projects. So what are they doing in an introductory text? Because a book starts at an elementary level does not mean that the reader cannot be ready to do advanced work after studying it. It would be nice if some, at least, were to learn enough to be able to do research problems. A few such problems have been included to test whether that happens.

However there are also pedagogical reasons. One of the most important things we have to learn in studying a subject is what we want to know. What questions should we ask, what is important? Knowing what to ask is a sign of understanding, and thinking about the relevant questions forces understanding. Any text, besides trying to teach a topic should also help the reader to learn how to do work in the field. Thus a list of questions, of problems, some trivial, some impossible, and hopefully some in between, read carefully, and thought about, adds to comprehension. And a list for one subject will, hopefully, guide and stimulate the reader to construct lists for others. Training someone to see what the issues are, what questions to ask, will lead to a habit of doing so, a habit of asking questions and of thinking about what should be asked, here and in general.

Of course any reader who can solve one of the advanced problems will not only develop his own knowledge and understanding, but that of everyone else — and in such cases the solutions should definitely be published.

P.4 STYLE

It might seem that the way a book is written is not really important (apparently many authors think so). But style not only reveals much about us, it greatly influences our effectiveness [Mirman (1995a), appendix]. Understanding the stylistic decisions of the author, and the reasons (he thinks) he made them, can help in interpreting what he (thinks he) is saying. Here, to help the reader, we try to clarify some points about how language is used in this book.

The writing style may at times seem somewhat unusual. Most books claiming to be about mathematics are written in a very formalistic style, as if they were dressed in a tuxedo — elegant perhaps, but constraining.

This book is dressed in work clothes. Its style is not elegant, and will not impress people as the tuxedo-dressed books do. But work clothes are often more practical, and someone dressed in them is more likely to get the job done.

P.4.a In defense of passivity

The grammar checker used to check the manuscript complains that too many sentences are in the passive voice. It seems that our use of language has made the active voice more comfortable, unfortunately. Nevertheless it is often misleading, not so much in what it says, as in what it implies. It is correct to say "representations are given by ...", and somewhat misleading to say "we give representations by ...", since this is a property of the representations, not of the way we handle them. And its emphasis is wrong. What is important is not what we are doing, as is suggested by the ordering of the words, but the property of the representations (although egotistical writers may disagree).

Indeed, in order to use the active voice, a mysterious new class of agents often has to be brought in: "workers have found that the mass of the proton is 1836 times that of the electron". Who are these workers, and why are they so interested in the masses of the particles? And why is their role being emphasized by mentioning them first? But of course, we are not concerned with the workers, whoever they may be, or whatever their motivations. Why not say what we want to say, and what we are interested in: "the mass of the proton has been found to be 1836 times that of the electron"? (This is different from "the mass is ...", a more definitive statement.)

This desire to avoid the passive voice at times leads to absurdity: "Workers have accelerated the proton ...". These workers must have highly charged personalities, quite magnetic too, if they, rather than accelerators, are used to accelerate particles.

Here we have tried to compromise between accuracy and (apparent) readability or perhaps just bad habit. At the beginning the active voice is used more than later (when our style may be more familiar), with the reader hopefully understanding the meaning, even if it is at times slightly confused by the wrong voice.

P.4.b That and which

There is much confusion about the use of these words. There are definite rules about which to choose, but apparently no one knows what they are. If the reader realizes what the writer believes, perhaps wrongly, to be correct, he or she is more likely to understand what the author is trying to say.

The difference, as it is understood by the author, between that and which is: *that* defines, *which* expands. A word that is followed by a *that* is specified, in part, by the words following the *that*. However a term that is followed by *which* is given completely by its name, or other information. The words following the *which* add information about an additional property of the thing the term names.

Thus a *which*, being a modifier, is preceded by a comma. The problem is that there are cases in which it should not be. Hence if the reader comes across a *which* not so preceded, he should realize that there are three possibilities: the wrong word was used, a comma was left out (perhaps merely a typographical error), or the term is defined before the *which* is reached, the material following merely adding information about it, but even so a comma does not fit. Knowing what the author believes is correct increases the chance that the reader will be able to guess correctly what the author wants to say.

P.4.c Numbers

Another thing for which rules are confusing, and not consistent, is the proper form of numbers: should they be written in numerical form, or spelled out? Since the rules cannot always be applied consistently, we try a different style here, hopefully with some consistency. However we do not rule out deviations for typographical reasons (and unfortunately perhaps because of simple errors).

When a number is a noun or an adjective, it is a word, and is spelled out, "... one way ..."; when it is a numerical value, "the matrix element is 1", it is written as a numeral. It is impossible to be completely consistent, because sometimes there is ambiguity (or at least the author is confused enough to think so) about which category the number fits in. But we try to do this carefully.

P.5 RIGOR

Any discussion of mathematics should be rigorous, whatever that means. This book makes no pretense of rigor, although hopefully it will attain it. In particular this is a mathematical physics book, and an introduction, so we have aimed for the level of rigor standard for such works. We would like to attain rigor, though hopefully, not rigor mortis. But it is not rigorously clear what rigor is, or how to achieve it, or indeed whether it has any meaning, or is achievable at all. Thus not wishing to try to attempt to achieve something that is not defined, probably not definable, and possibly nonexistent, we have just aimed at being careful, clear and understandable. That we have not succeeded should not obscure what the goal is.

Since we are interested in applications, we consider the most restricted mathematical structures, so that everything is well-behaved, differentiable as often as required, and so on. This being understood in books like this, we do not state explicitly the restrictions (nor do we know explicitly what they all are). Any reader who wishes a discussion of these restrictions can look at any of the books on the subject. In fact it would be well to look at several, since the necessary restrictions listed differ among them — though they are all, presumably, equally rigorous. Here it can be assumed that all restrictions apply, and if it is wished to extend these results, then they should be carefully checked to see which, if any, can be removed, and if not how the results have to be modified.

Proofs elsewhere should be looked at, if for no other reason than that it is useful to think about proofs and decide which, if any, are rigorous, and if not why not and how they can be made so.

P.5.a What is rigor?

Rigor — the term denoting the combination of clarity and logic, with clarity emphasized — is essential. But what is it? It helps in studying a book to understand the views of the author. What does this author mean by rigor? And why?

A basic requirement of rigor is the careful, explicit, precise, simple, and fully understandable definition of all terms, symbols and concepts. Vocabulary is basic. Language, terminology, nomenclature, should be as standard and as widely accepted and understood as possible. All words should be familiar to (at least, almost) all readers, not just to some friends of the author. The words, symbols, terms, used must express exactly the meaning that the author intends, and also be understood by the reader in such a way that he understands exactly what the author means. Language is a bridge. That the precise, correct meaning enters the bridge is, alone, useless. What is essential is that it cross the bridge (and that the author take great care to see that the language is used and understood, and understood in the same way, on both sides of the bridge). Language must be comfortable, and be understood precisely and with the same meaning, on both sides of the bridge, by both the author and the reader. Rigor means choosing language to inform, not to impress. After all it is the purpose of a book or paper to communicate (although many are written solely because they look good on the author's publication list, and read like it).

Standard, simple language makes it more likely that the reader will fully comprehend (correctly) what the author is trying to say, and so make the material widely accessible. If the reader does not know the language, even though that ignorance is subtle and hidden, or if the writer and reader do not use the same language, especially if they are not aware that their languages differ, then no matter how logical the discussion, it is lacking in both rigor and value. A discussion is rigorous if it can easily be, not only followed, but understood, both formally and intuitively.

The importance of language can be seen especially in quantum mechanics, where its misuse has caused much confusion [Mirman (1995a)].

Logic is a technique for using language to convey precise meaning. Rigorous derivations are completely logical, with the logic explicit and correct, completely clear, with all assumptions listed (especially those that most people do not realize when reading). If the meaning is not precise, and both comprehensible and comprehended, then the argument is illogical. Logic that does not *convey* meaning, exact, accurate, correct meaning, has no logic.

If the reader cannot understand and easily follow the arguments and proofs, and thoroughly check them, then no matter how generally they are stated or how brilliant the author is in flaunting his knowledge of mathematics, they are not only not rigorous but if not wrong, at least useless. Logic is meaningless if it cannot be followed (correctly). A proof is not rigorous if it cannot be understood, learned from and verified. Rigorous proofs do not consist of vague statements with an attached list of names and an implied "these experts have checked that the result is correct so it is unnecessary to show someone like you a meaningful derivation of it".

Likewise in discussions of physics, all assumptions and terminology should have well-defined and interpretable physical meaning. That a term means something mathematically is irrelevant if it does not physically. And theories must be based on (clearly stated) physical assumptions, not merely mathematical ones. There is no point in a mathematically rigorous derivation of a physically incorrect or vacuous result. Elegant mathematics and long rigorous discussions that are used, not to clarify, but to cover up the lack of physical content may impress and fool some people. They are unlikely to fool nature.

We have emphasized — not attained — clarity (without, we hope, skimping the logic). We believe that this is the proper standard for rigor. If the reader can follow, understand and check, if all assumptions, implicit or explicit, are clear, if the terms are not only memorized but understood, if the limitations of the results come thorough, then, no matter how simple and obvious the terminology or the proof may be, it is rigorous. And if not, that lack can be seen and corrected.

P.5.b How many assumptions does rigor require?

Rigor means stating all definitions and results as generally as possible, and stating whatever restrictions make these less than completely general. However this requirement is probably impossible to meet. No matter how generally something is stated it is likely that someone can invent a broader structure such that the stated result is, in that context, limited. It might be suspected that in analogy to Gödel's theorem which states (loosely speaking) that it is impossible to prove everything within a system, it may be that it is impossible to state everything (or anything) in complete generality, or even completely state the restrictions that prevent it from being completely general.

For one thing there is the problem of language, which is always ambiguous and approximate. The meaning of words, and the symbols for words, and the grammar, require many postulates. This is certainly true for physics. Even the number of assumptions in a statement is not clear. But understanding what assumptions are made, and how many, can help in learning a subject. Here then we list some questions the reader might think about in studying, and not only this book.

Problem P.5.b-1: How many assumptions are there in saying that Poincaré invariance holds for a space? One, or an infinite number? If the curvature at each point differs, then giving it at every point requires an infinite number of assumptions. Is the statement that space is flat, that the curvature at each point is zero, only a single assumption? If every direction were different (as if space had an inherent gravitational field), giving the directions would than require how many assumptions? Does the statement that all directions are the same, require only one? Is there a difference for abstract mathematical spaces?

Problem P.5.b-2: Is the observation that the Dirac equation holds at each time, and space, point one assumption or an infinite number? The equation holds for each electron — is this one statement or many? Of course an electron can be defined as a particle, with given charge and mass, for which the Dirac equation holds. But then the question is shifted to whether a particular particle, a particular physical object, say one in a specific trap, is an electron. If there are an infinite number of different types of particles (say with different masses but all with spin-$\frac{1}{2}$), how many assumptions is it to say that the Dirac equation holds for all? And what is the cardinality of that number?

Problem P.5.b-3: How many assumptions are there in the quantum theory of measurement, or in the analysis of the Stern-Gerlach experiment? The reader might find it interesting to see how many can be listed — even organizing a competition to see who can list the most.

Problem P.5.b-4: For group theory we assume the existence of an inverse for each operator. But physically to go from one point to another, and then back to exactly the same point, is impossible. What physical assumptions are needed to define the concept of a group, especially a continuous one? What assumptions are needed to define a particular group? And how many? Perhaps mathematically it is possible to define a group by saying that each operator has an inverse — that is by applying a word to the set of operators — and then define the word "inverse" or "product", or any of the others that enter, with no further assumptions. Or are these definitions assumptions? Is there a difference between mathematics and physics? And what assumptions are needed to make the assumptions needed to define the words needed to define the concept of a group?

Problem P.5.b-5: It might seem that only few postulates are needed for group theory — certainly very few are stated. But there is a requirement that the elements (presumably enumerable, distinguishable and susceptible to being labeled) have a product. But what is a product? Just a word or does it have properties that all products, no matter what, must have? Are there any assumptions that these requirements, on products and groups, are realizable and compatible? Thus in considering factor groups, we need products of sets. Do all sets have products? Or else what are we assuming about the sets of group elements? The same questions can be asked about the other words (or are they more than just words?) used in group postulates.

Problem P.5.b-6: And then there are topological groups. To give a topology we give a list of points — and there are a lot of them — that form the neighborhoods of each point — of which there are a lot. Is this list one postulate or many? Indeed to say that a set of points is in the neighborhood of some given one requires many assumptions about the definition of the words (or are they just words?): point, neighborhood, nearness and so on. Thus to say that a topology can be imposed on a set implies that there are many assumptions made about that set, and so about the definition of topology. Consider a space made up of points whose properties vary, say each point is itself a set, and not all the sets have the same cardinality (if that is possible). Would the imposition of a topology and the definition of topology itself require one assumption or many, and of what cardinality would this set of assumptions have, if a cardinality can be defined for it? Can the properties vary; is this an assumption?

Problem P.5.b-7: How many postulates are needed for the definition of Lie groups? Have all ever been given? Is it possible? Again a competition to see who can think of the most postulates would be interesting.

Problem P.5.b–8: Suppose we consider the set consisting of all real numbers between 0 and 1, all rational numbers between 1 and 2, and all integers between 2 and 100. Is it possible to define a group over this set? What assumptions are made in doing so? Or in not being able to do so? And how many?

Problem P.5.b–9: How many assumptions are needed for group theory or for the definition of a particular group, a finite number, $\aleph_o, \aleph_1, \aleph_\aleph$, or more? Does it matter whether the group is finite, infinite, topological, or other? What is the cardinality of the (smallest) set of assumptions needed for group theory? Can this smallest set — if there can be different cardinalities — be found? How? What is it? Would it be useful? If it is not clear that these are real questions then the lists of assumptions given, or implied, in the different books on group theory can be checked, and compared. The total number of (different) ones given by all these together makes an impressive, and probably incomplete, enumeration.

Problem P.5.b–10: What theorems about group theory can be proven within the theory itself — any? To what extent is it necessary to go to larger structures (Gödel's dilemma?)? What is the smallest of these systems? Is there any, and does it depend on the type of group? Can embedding group theory in different structures give different theorems? What would be the physical implications of such results? What is the cardinality of the set of assumptions of the smallest structure in which all the results of group theory can be proven?

Problem P.5.b–11: If not all theorems can be proven within a given system, can all assumptions about it be stated using terms defined for that system? Can all assumptions about group theory, finite groups, topological groups, other groups, be stated using only a finite, well-defined (whatever that is) set of terms? If not, how rigorous can a subject be made? Are the answers different if the subject is considered as mathematics, then if it is considered as physics?

Problem P.5.b–12: Do the (obvious) answers to these questions differ before and after reading the book?

ACKNOWLEDGEMENTS: For much help, this is to thank the Department of Natural Sciences of Baruch College (City University of New York), and particularly the Physics Division, Dr. C. J. H. Lee and Dr. Susan Hezlet of World Scientific Publishing Company.

Table of Contents

List of Tables

List of Figures

Chapter I

The Physical Principles of Group Theory

I.1 CHANGE IN THE WORLD AROUND US

Change is the essence of life, and of physics. Things change, the world changes, we change — else there would be no physics, and no life. And we change our view of the world. Motion, change, and their description, are fundamental [Mirman (1995a)]. How things change, move, are transformed, and how our view of them changes, and does not, as they, and we, change — this is the essence of our discussion. Change, transformation, what it is, when and how it occurs, and when it does not, change or lack of it in our view of a system, these help us understand not merely the system, but nature. And it is the foundation on which nature exists.

If we step away from an object, or turn ourselves around, or move or rotate the object, it looks different. But not always; views of some things remain the same. A tire, or donut (plain), rotated around a perpendicular axis through its center, is unchanged (at least visually). And this invariance tells us much about it, about ourselves, about nature.

Physics, the science of the description of motion, of the most essential of changes, of transformation, and the mathematics that we study here, group theory, the language describing change, its occurrence and its nonoccurrence, provide the most intrinsic of tools for understanding nature. Why? How can we develop these, and use them to study physics and the universe, and the objects in it?

The manufacture and understanding of the mathematical tools, and the use of them, particularly, but not only, to understand nature, is our aim. The mathematics is useful for — but not only for — reasons of physics; it is built on a foundation of physical laws, a machine resting on experimental observations. So, though our subject is mathematics, it is with physics that we begin.

First, to conform to common usage, and because it is more accurate, we change the word we use from "change" to "transformation", and we transform, at least for a while, to a more technical topic of discussion (we put in "for a while" to emphasize that inverse transformations play an important part in our considerations).

Much of group theory, especially its applications, involves symmetry, so changes that produce no change. But it is important to understand that transformations can provide information though they may not leave a system (an object, or set of objects) invariant, and also that the symbols need not be interpreted as transformations. While it is often necessary to discuss specific interpretations and special, and important, cases, the reader should be careful to realize that these limitations need not be inherent in the subject, but in the way it is presented, and keep an open mind.

I.2 FOUNDATIONS

Since this is to a large extent a book about physics (we chauvinistically regard chemistry as a branch of physics), despite the view that group theory is a branch of mathematics, we start by setting the context, summarizing how physics, and observers, describe physical systems. It is important, more than is generally realized, that the assumptions, the restrictions, the meanings, be explicit and clear. So to understand the physics, we try to discuss these in depth. This is also a book about mathematics and (we hope) can be read, understood, and appreciated, without any knowledge of physics. But an understanding of the technology of art, of what an artist is trying to do and why, helps in understanding and appreciating his creations. So a sense of the fields in which mathematics is applied, and how and why it is used, and why it is applied to these fields, and used in these ways, helps in the understanding and appreciation of the mathematics. For this reason, among others, we use and emphasize physics, and physical applications. Here we try to provide motivation, context, for the material that follows. There will be no loss in mathematical understanding if it is skipped (except for the sections on the group axioms). Whether there will be a loss to the reader's physical (and mathematical) intuition, and ability to do physics and mathematics, must be decided by the reader, who takes the risk of that loss.

Also this chapter, and the next two, indicate some of the problems considered in group theory, and reasons for them, and a sample of the questions we ask, and try to answer, in this book.

We consider the concept that underlies the physical application of group theory, physical transformations, and study how different observers describe a physical system, and what restrictions on these systems we are thus lead to. The application of group theory to the solution of a physical problem, the state-labeling problem, is considered as part of the introduction. Then we consider examples of groups by studying transformations and symmetries of some objects.

After that there is no further escape — we come to the mathematics, although we shall try to disguise it as physics.

I.2.a The need for observers

The universe (that we explore) is governed by quantum mechanics [Mirman (1995a)]. We, like nature, use quantum mechanical terminology, and quantum mechanical examples — to provide motivation, and make abstract terms concrete; those who do not know, or do not enjoy, quantum mechanics can slide over it without losing any of the mathematics, or better, translate these terms into a language they enjoy more. So a system, an electron, an atom, a molecule, a crystal, a person, the universe itself, is described by a wavefunction, also called statefunction or statevector, which are better terms; for variety we use these interchangeably. But a single system alone (the universe?) is meaningless. For physics there must be (at least) two, called the object and the observer (which is not a human being, since we are studying physics, not sociology or psychology, but another physical system); it is the relationship between systems which has physical meaning. We can ask only how the behavior of one influences that of the other (say by affecting the positions of pointers in those parts of the observer that we call gauges). This relationship is described by the statefunction. The statefunction of an object is defined with respect to another physical one. In common usage we say that it is given relative to a coordinate system — a physical object [Mirman (1979)].

There must be observers, and there must be the possibility of many (say, differently placed or oriented); these see different, though, in some way, related, statefunctions. And the system, being resident in space, is constrained by its properties and by the laws of physics (a redundancy?). These limit the possible statefunctions (and possible observers). Indeed much about physical systems, statefunctions, and experimental results, is determined by the nature of space and of physical laws, and not by the systems.

This then is the foundation of the (experimental) science of physics — the observer and the observed, physical systems, their relationships and their description. Though the depiction of physical objects and how they behave is dependent on the observer, all see the same system, so their findings are related, and this relationship, so their observations, must be restricted. The "important" properties of a system are independent of (invariant) who (really what) studies it, though their description, as with the components of a vector, may be observer dependent. The most fundamental premise on which science, and life, is based is that there are properties of systems that are independent of observers.

Equivalently, a fixed observer can look at a transformed physical system; these are two ways of saying the same thing. Both the observed and the observer are physical systems. They are related in some way — an angle might specify their relative orientation, this different for different observers. To change the observer, it, or the system must be rotated; both cases need a physical action, like a torque on one of the objects. (The requirements that the properties of the system be independent of this angle, and that observations by relatively-rotated observers be related — that one observer can calculate from his observations those of any other observer, given their relative angle — may be more subtle than it appears. In each situation the meaning of this statement must be defined.)

I.2.b Observations must be related

The idea of transformation is central to group theory, and to physics. Usually this means transformations between observers relating the observations made by one to those made by others (these are different observers of course; we might consider several experimenters at one point, but then they are a single observer — only if they are, say, relatively displaced are they different). But relationships between observers are greatly restricted, as are the possible experimental results that an observer can obtain (observations). Part of these restrictions comes from the nature of the system, but a large part is the result of the properties of space (that is, presumably, the properties of nature, the laws of physics).

Observations made by one observer — in one coordinate system — must be related to those of any other. That relationship must not make any observer (of a class, say inertial ones) preferred — all such observers (coordinate systems) are physically equivalent; the laws of physics found must be the same for all.

This is the physics, but how does it give constraints, and tools for understanding? What are these fundamental properties, these limitations, and their effects? How can we determine them? How can we use them to understand as much as possible from the most essential properties of nature and of physical laws (definitely a redundancy) and the basic properties of the object? And how can we then go beyond this, to obtain the complete statefunction — using these general requirements as a tool to simplify, even make possible, its computation and the extraction of its predictions?

In summary, that is how we (start to) apply group theory.

I.3 WHAT ARE TRANSFORMATIONS?

To see how invariance and the properties of transformations restrict, and determine, the nature and behavior of physical systems, we first examine how observers are related: given the description provided by one observer how do we find that for another? So we analyze the concept of transformation, a relationship between statefunctions describing a system, as seen by different observers.

Transformations (in physics) are (seemingly) strongly restricted; they (all) have common properties, so strongly constraining physical laws — the nature of the universe. Besides their physical consequences these restrictions have mathematical ones as well.

Given a physical object what can we do to it while leaving it unchanged? Unless it is extremely delicate, we can rotate it or move it to a different position. Or, we can leave the object unchanged and transform the observer who can turn his head, or step away. So we have examples of physical transformations, rotations and translations. Of course if the object is quite delicate we might wind up with other examples of physical transformations; however since we are concerned only with transformations that leave the important properties of the system invariant, we do not consider these. Also if we transform the observer we must do so in a way that its rules of measurement are not altered, as in cases for which turning a neck blurs vision, or parts of an instrument are transformed, say by stretching a meter stick.

These, rotations and translations, are important transformations and for a well-know reason, space (that is, physics?) is invariant under them. Physical laws do not depend on positions or orientations of objects, or observers — a prime reason why some transformations are fundamental. We discuss others later. Perhaps there are further ones that we (or at least this author) do not know about. There are phenomena that provide hints; of what, is unfortunately not clear [Mirman (1995b)].

I.3.a Transformations of real objects

Objects and transformations can be abstract; it is better to start with familiar things. Since many readers like to eat while they study we consider first some food, and it is generally accepted that a reasonable food to start with is an orange. If we take an orange and rotate it about an obvious axis, or we walk around it, we find that it remains (essentially) the same. Observers at different positions all see the same thing. And all observers are the same; there is no way of using their reports of what the orange is like to distinguish one observer from another. However if we rotate around a perpendicular axis, so that one observer looks at the equator of the orange, the other at a pole, then the observers have different views, and, based on their reports of what the orange looks like, we can tell which observer is which.

So we have examples of transformations, some leaving the (view of the) object unchanged, others changing it.

For a ball — a sphere — any rotation (keeping its center fixed) leaves it unchanged; it looks the same no matter how it is turned, no matter how an observer revolves around it. It has more symmetry than an orange. And knowing that a sphere is spherically symmetric (however this might be determined) we know what any observer sees, based on what one sees (for example the color); for an orange the observer looking at the equator cannot tell what the one looking at a pole sees. This is an interesting point, well-worth thinking about. An apple gives another example having transformations that leave the system somewhat changed, but not greatly. If it has a stem, the pole would look very different from the equator. But even at the equator rotations produce slightly different views. And eating an apple, or an orange, transforms it to an object very different from the original.

Of course while a ball is unchanged by rotations there are transformations that do change it, like squashing or cutting. For some objects and some transformations there is no change. For others there are small changes by some transformations, perhaps larger ones by others. And there are transformations producing such drastic change that we cannot consider the final object to be the same, in any sense, as the original.

Problem I.3.a-1: There are limitations on what an observer can do to a sphere, or itself, without changing the view. State precisely what transformations leave invariant — the view of — a sphere.

Problem I.3.a-2: Analyze the transformations of an egg — with care or we shall wind up with unwanted transformations.

Problem I.3.a-3: What are the transformations of a banana, a pear, a starfruit?

Problem I.3.a-4: Next we come to an important question in botany. Why are the transformations that leave the object invariant (unchanged) different for a grapefruit than for an ugli fruit? There is a related question in economics. Does this difference in symmetry have anything to do with an ugli fruit generally being more expensive than a grapefruit?

Problem I.3.a-5: A closer look at the orange will reveal something about transformations that leave it invariant. Actually a tangerine, peeled, is visually clearer. It is not left invariant by a rotation through any angle, but only by rotations through a finite set of angles. Thus the transformations are not specified by a continuous parameter (the angle) but are discrete and finite. What are the angles of rotations that leave a tangerine invariant? Why are the symmetry groups (the set of transformations leaving the object invariant) of the peeled and unpeeled tangerines different?

Problem I.3.a-6: Of course fruit is not the only food with symmetry. Compare the transformations of a donut and a bagel.

Problem I.3.a-7: Food is not all that we can use to study transformations. These, and symmetry under them, have esthetic implications and have found an important place in art. The reader might wish to look at his wall (if it has wallpaper) [Schattschneider, (1990)] or his floor (if tiled) and study how the designs change, and remain the same, under various transformations. Symmetry and transformations appear in many other aspects of life; perhaps the reader can even notice these in his clothes. It may never have occurred to him that he can study physics and mathematics by going to the supermarket (and not only the vegetable section) or by staring at the floor. But the richness of mathematics and the laws of physics, and group theory, have only been slightly tasted by these examples.

I.3.b Transformations and invariance

Why do transformations tell us anything about physics? The laws of physics, Newton's Laws, Schrödinger's equation, the Dirac equation, for example, are rotationally invariant — they do not depend on an angle (direction). It is always surprising to see how readily this idea is accepted since it is obviously not true — the vertical direction is clearly different from a horizontal one. Only a more sophisticated analysis, not immediate intuition, reveals that the vertical is different because it is the line joining two material bodies. But there is no line in space, independent of objects, that is distinguished.

But this pseudo-example shows the difference between the laws depending on a direction and those, as in reality, for which all directions are the same. If the reader were to stand up and rotate, keeping his head the same distance from the floor, and then rotate by laying down, it would be clear that there are important differences between the two types of rotation. We can simulate a law that has a built-in direction by putting the gravitational force in Newton's second law. These two rotations suggest how transformations would be different if the laws were not invariant under the transformations.

Problem I.3.b-1: But objects also would appear different. A wrecker's ball, a heavy object that can be taken as a perfect sphere, is used to demolish buildings (and here, we hope, preconceptions). Considering gravity, how do transformations of it show that the laws of physics are (apparently) not rotationally invariant? With gravity, is it really symmetrical? This would be visually clearer if the ball were made of a soft, so easily deformable, material.

I.3.c Transformations that almost are symmetries

These examples should indicate how invariance of space affects (determines?) the laws of nature and properties of physical objects, the sets of transformations useful in their study, and what happens if invariance fades. And this last has important implications.

Another, rather strange, kind of invariance (sort of) involves protons. We can transform one into a neutron — wherever a proton (state-function) appears in a physical law we replace it by a neutron (state-function). Is the law invariant? The two particles have different mass and different charge. But we ignore charge and notice that the mass difference is very small compared to either mass so ignore it. Then we find that the (type of interaction called the) strong interaction is invariant under this transformation, a rotation of isospin ("rotating" neutrons and protons into each other, and similarly for other objects having these interactions).

So we get isospin invariance: under isospin rotations, laws involving only strong interactions are almost invariant, almost, but not quite. This can be extended to SU(3) transformations (and so on): the laws are somewhat invariant, but less than for isospin transformations.

We use group theory to study systems invariant under transformations, but it is not limited to that. The transformations need only have physical meaning, and both invariance and partial invariance (often known as broken symmetry) provide that. But generally the closer the system is to completely invariance the more group theory describes its behavior and restricts it. Or perhaps group theory completely describes such partially invariant systems as the nucleon, but we just do not understand things well enough to see how.

Of course we can find "almost" (more technically "broken") symmetry in classical physics. Rotate this book in its plane. It is almost the same, but not quite. The difference between classical physics and quantum mechanics is that for the latter, symmetry, exact or broken, provides strong restrictions on, and detailed information about, the behavior of objects that are symmetric, while for classical physics it adds to our esthetic enjoyment of these objects. Perhaps this difference might be significant.

Problem I.3.c-1: Give examples to show that the difference between the applications of group theory in quantum mechanics and classical physics has been somewhat overstated.

It is useful to keep in mind the questions of how the results depend on quantum mechanics, and why, and to what extent similar results might hold classically.

The appearance of isospin emphasizes another point. There are operators that change one physical state, perhaps one physical system, to another, and these form groups, often giving symmetry. Group theory is introduced, certainly here, using symmetry. But there can be many types of operations on physical, and mathematical, systems. These need not all be symmetry operations. But they may form groups. The definition of a group is so broad, and the theory so wide, that there can be, and are, myriad kinds of objects to which it applies, and for which it is useful. If there is a group describing a physical system there may be symmetry, perhaps subtle or hidden, and that should be considered. But there may not be.

It is interesting that the way our language has developed, an open mind has become the antithesis of an empty one.

I.3.d What transformations should we consider?

These examples show the appearance, and importance, of transformations in many areas of life. But what transformations? For a sphere there is a set of transformations given by its properties, leaving these unchanged, rotations. In other cases we can consider various sets of transformations, those that leave an object unchanged, those that change it only slightly as for an apple or this book, those changing it more drastically as for a carelessly handled egg, or food that is cooked or eaten. The transformations that we are concerned with are those that help us gain useful information, and those leaving a system unchanged, or almost so, are likely to be included. But which transformations are useful depends on the system, on what we want to know, and perhaps on our own cleverness in seeing how to get information from them.

Here we assume (only?) that we have a set of transformations, picked for whatever reasons, and that the set has the properties given below. What we want is to find how we can get information just by knowing that a set has these properties — that it forms a group. For illustration we often consider particular sets, but what these are, and how to pick them and why, is really another topic. Still knowing what information we can obtain, and how, is itself an important guide to deciding what sets of transformations can be useful.

I.3.e Why we must examine the meaning of transformations

Consider a system of two electrons; all experimental results must be invariant under their interchange, so the statefunction must be either symmetric or (actually) antisymmetric under this interchange. For rotations we can transform either the observer or the observed. Can we consider transformations between observers for this interchange? There is a difference between the two types of transformations. We can actually rotate an atom (with a torque) but cannot interchange electrons — there is no way of distinguishing them. What a transformation means in a specific case must be carefully thought about for there may be subtleties that could mislead. (The meaning of antisymmetrization has been considered elsewhere [Mirman (1973)]).

Thus the physical meaning of the requirements placed on transformations, and that the description is the same for all (a word that must be defined in each case) observers, and the reasons they hold, have to be evaluated and stated for every system. These transformations are not mathematical operators on functions but statements about the laws of physics (the nature of the universe) and about particular systems, and operations on them. They are summaries of experiment — whose experimental meaning and support must be understood and justified.

I.3.f The centrality of transformations

Physical objects, both the observed and the observer, can be transformed (fortunately). Transformations, those leaving an object unchanged, and those changing it, tell us much about such objects (food is a physical object). There are causes for the effects transformations have; knowledge of these effects and of why they occur, and the way we react to them (look at wallpaper) is fascinating and useful. They inform us, not only about objects but about the universe — for good reasons.

Though they are operations, mathematical and physical, that we use as a tool to study systems and nature, and that is the way they are usually presented, transformations are more fundamental: they are physics, they underlie nature — they likely determine all properties of the universe [Mirman (1995a,b)]. Since this view underlies the whole book, it should be, if not fully justified, at least explained.

Why are transformations fundamental? There must be interactions, and observers. But when objects interact they change each other, they cause transformations, translations, rotations, boosts (changes of velocity). Motion, without which there would be no universe, no physics, no life, is just the set of these transformations.

And for physics to make sense different observers — distinct material objects — must be able to relate their observations, to communicate, that is to transform their observations into those of the other objects in the universe. It is not merely that transformations are fundamental, but there is nothing else.

Functions describing physical objects depend on coordinates and are labeled by numbers supplied by the group of these transformations. These labels have physical meaning: they can be measured, and they describe properties of statefunctions (like wavelength). The quantum mechanical statefunctions actually come from the transformation groups allowed by nature, and by the system. But more, the need for, and properties of, quantum mechanics itself, even probability and uncertainty, come from the requirement that objects interact [Mirman (1992)] (as does the dimension of space [Mirman (1986, 1988a,b, 1995a)], and, in a perhaps more subtle way, so does quantum gravity [Mirman (1995b)]), so that objects must be described by functions, and these have to be transformable.

While we do not — yet — fully understand all implications of the properties of the groups of transformations, the information they provide about (say) physics is vast, and it is not unreasonable to believe that they are physics.

I.3.g The example of orbital angular momentum

An example of the relationship, in quantum mechanics, between observations, and how physical laws and properties of systems are restricted and explained, is given by the orbital angular momentum states of, say, an atom. An observer sees, for each integer l, a set of states with $l_z = -l, \ldots, l$; l_z is the component of the angular momentum with respect to that observer's z axis. Another observer finds the same set of states now labeled by $l_{z'}$, the component with respect to its (different) z axis. The functions describing the system are the same, but the angles they depend on are different, being measured from different axes; if the angular momentum is "up" in one frame (the wavefunction gives an "up" state), in another the wavefunction describes the angular momentum at some angle to the z' axis (the state is a superposition of states, these giving different angular momentum components, along z'). The atom is unchanged, as are the functions describing it and the energies of the states. But the wavefunction is different for different axes.

The functions of one set of angles are (linear) combinations of those of the other set; these combinations relate the two observers — they tell how the description of the system given by one is related to that of the other. The coefficients of the functions in the sums depend only on the angles, and are the same for any two observers who are rotated through the same angle. They do not depend on the orientation of z, which in fact cannot be determined, so has no meaning.

This illustrates the basic points. One observer finds a set of statefunctions describing a system. Another, who is related in some well-determined way, here by being rotated through a given angle around a given axis, also determines the statefunctions. These must be related; the two sets are the same functions (of different angles) and the functions of one set are linear combinations of those of the other, with the coefficients depending on the angle. And there are properties of the system, for example the energy, that are invariant; all observers find the same set of values. These requirements are enough to determine how the statefunctions depend on the angles.

I.3.h If symmetry starts to break down

To emphasize — and clarify? — these points consider conditions for which they are not true.

If the atom is placed in a weak magnetic field the energy of a state "up" along the axis parallel to the field is different from one "up" along a different axis. However for all observers (all coordinate systems differing only by being relatively rotated) the number of states with given angular momentum, but different components along their z axes, is the same, $2l + 1$. The functions describing these states are the same. It is only the Hamiltonian expectation values for corresponding states (those with the same l_z — the same value of the angular momentum along each observer's own z axis) which differ. And the functions describing a state seen by one observer are sums of those of a different observer. Thus the description, and these properties, of the system are the same for all observers.

However if the magnetic field is very strong then the difference in energies between the states with different l_z is greater than the difference between states of different l (orbital angular momentum is no longer a good quantum number). It is no longer useful (but still possible) to say the system is described by sets of states, each labeled by l, and that the number in each set is the same for all observers. In one coordinate system, that with z axis along the field, the functions describing the system have a simple form and each has a definite energy (Hamiltonian eigenvalue). But for other observers the functions are complicated, depending on the angle between the z axis and the field. And they do not have definite energy, but are linear combinations of states that do. The energy eigenstates are not eigenstates of l and $l_{z'}$; the Hamiltonian, and the operators for orbital angular momentum, and its z' component, do not commute and cannot be simultaneously diagonalized (unlike the case of no field). A statefunction, an energy eigenstate, now a sum of states of different l, is a linear combination, not of a few states in another system, all with the same angular momentum, but of many; these with different values of l_z, and now of l.

Thus the description of the system is different for different observers. And there is one observer who is special, the one with z axis along the field; its description is simpler than that of others. If there is no magnetic field then there is no way of picking a direction of space — it is isotropic. This is a property of space (or is it better to say that this is a property of the way the laws of physics describe space?). With the field, there is one direction that is different (really only because there is a physical object, the field, in that direction). We can see the difference in the properties of the system, the way that they are described, and the relationship between different observers when space is, and is not, isotropic.

Obviously if space were not isotropic, physical systems, their description, and the relationship between different observers, would be very different. But it is isotropic (which is nice in many ways), among its other properties. It is important to find the consequences of this for physics. This is a role of group theory.

Spin provides another example. An observer sees two states (for an electron); these have the same energy. Another observer finds the same two states, and the same energy, having the same functional form, but now referring to its z axis. Again a magnetic field picks out one observer and gives the energy of the "up" and "down" states different. What happens if the field is very strong? There are always two states and they are not mixed with states from different sets (a strange statement because for electrons there are no other sets; but consider a nucleus, which might have different spin states between which a magnetic field, but not a rotation, could cause transitions). However in a strong field, one of the states has so much more energy that the other decays into it; experimentally it (almost) appears as if there is only a single state, so it would be impossible to quantize along any axis but that parallel to the field. For no field there is an equal probability of finding an electron with spin up and with spin down, along any axis, but with a strong field the probability is different for up and down, and each depends on the angle between z and the field. There is a distinct difference between observers.

I.4 WHAT WE THINK PHYSICAL TRANSFORMATIONS ARE LIKE

It is a remarkable fact (perhaps not often enough remarked on) that very many physical transformations all have certain characteristics, and these include the ones that seem to give the most fundamental information about physics and the world it describes. Limiting our study to such transformations we develop a theory, a machine, to extract their implications and the restrictions these place on nature. This is why group theory, the study of transformations with these properties, being so inclusive, and so restrictive, is of such use.

I.4.a The group axioms

A group consists of symbols (which we here regard as, so for concreteness call, transformations) and a rule, the product, for combining them, giving the transformation that is the product (the combination) of every two. The transformations are labeled by a set of parameters θ, schematically $T(\theta)$; they might be indices, or rotations labeled by angle θ, or the interchange of two electrons with the label their coordinates. The product is $T(\theta)T(\theta')$, with the transformation on the right taken (here) to act first; $T(\theta')$ is applied to a state, and then $T(\theta)$ acts on the result. These sets of transformations have the following properties (the group axioms):

A) There exists a transformation (labeled $T(0)$) that leaves the relationship between the object and the observer unchanged. This is the unit transformation (the identity); every group has an identity.

B) For every transformation, $T(\theta)$, there exists another, $T(-\theta)$, its inverse, which brings the system back to its original state. Thus the action of these two carried out in succession, the product of these two, leaves the system unchanged, so is equivalent to the unit transformation:

$$T(\theta)T(-\theta) = T(-\theta)T(\theta) = T(0). \qquad \text{(I.4.a-1)}$$

C) There is a single transformation $T(\phi)$ equivalent to every product of transformations (every set of transformations applied in succession); the transformed state obtained using $T(\theta)T(\theta')$ is the same as that obtained from the same original system using $T(\phi) = T(\theta)T(\theta')$. This property is clear for rotations. The order of the transformations is often important — they give different states; generally,

$$T(\theta)T(\theta') \neq T(\theta')T(\theta). \tag{I.4.a-2}$$

D) Transformations are associative,

$$\{T(\theta_1)T(\theta_2)\}T(\theta_3) = T(\theta_1)\{T(\theta_2)T(\theta_3)\}. \tag{I.4.a-3}$$

(To be extremely rigorous further axioms can be added [Lomont (1961), p. 17], but it is difficult to think of objects, mathematical or physical, that do not satisfy them. However if the reader can, then it is important to add these axioms.)

Problem I.4.a-1: The definition of the inverse is stated here using two equations; is this one requirement, or two?

Problem I.4.a-2: From the axioms show that [Tung (1985), p. 25; Aivazis (1991), p. 1]

$$T(0) = T(0)^{-1}, \quad T(0)T(\theta) = T(\theta). \tag{I.4.a-4}$$

Problem I.4.a-3: Does it matter in which order the axioms are given; do any depend on previous (or perhaps erroneously listed, later) ones?

Problem I.4.a-4: Although we talk about transformations — we are interested in physics — a group is a set of abstract symbols with a product satisfying these axioms. We are interested in transformation groups, for which the symbols stand for transformations, but generally the results are independent of the interpretation given the group symbols. Restate the axioms abstractly, using symbols instead of transformations, and with no interpretation.

I.4.b Why groups of transformations are important

These properties are important because physical transformations have them, and this is a major reason we are interested in them, but not the only one. These requirements are so general that objects in various fields, including many in mathematics, also form groups. The mathematical theory of such objects is thus a tool for the study of many systems, providing knowledge about a vast array of things and fields, in a comprehensive and simple way. By aiding the extraction of the consequences of the group requirements, it provides understanding of much about mathematics, physics, and beyond.

In physics there are many systems, very different, that have similar properties, obeying the same differential equations, for example. Why? They have the same symmetries, the same time development —

often we need just say they have the same symmetries — so are described by the same groups. But the number of groups is limited. So descriptions, representations, must be the same. The behaviors, the differential equations, the states, are determined by the symmetries, so by the groups — and these are so few in number that there is not much choice for physics. Group theory gives lists of possible behaviors, and objects must obey.

A large class of physical transformations has these properties; it may seem obvious that physical transformations should (that it is obvious is not surprising, since we are so familiar with transformations that do). It was once obvious that physical functions should always have derivatives. The understanding of fractals changed that. Thus, however obvious these restrictions on physical transformations may (or may not) be, all we state is that we limit ourselves here to such transformations.

Problem I.4.b-1: Nevertheless the most fundamental physical transformations do have these properties — they do form groups, and the most fundamental properties of the universe (seem) closely related to, and likely follow from, transformations with these properties. The reader might find it interesting to consider whether there is any way of understanding why all these should — must — be so.

Problem I.4.b-2: There are various physical reasons why groups are of interest. Consider the set of all linear operators that commute with a Hamiltonian. (What is a Hamiltonian? What does it mean to say that an operator commutes with a Hamiltonian?) Show that these form a group (which need not be finite dimensional) [Heine (1993), p. 18]. Explain why they have this property.

I.4.c Pseudo-counterexamples

The requirements on the transformations given by the axioms form the foundation of the subject, therefore it is worthwhile to understand them. One of the best ways of doing so is by dreaming up operators that do not obey the axioms — this helps clarify their meaning. We try to do this here. Some examples (indeed all) may seem artificial, forced and certainly not natural. This is regrettable but hardly surprising. Physical transformations do obey the axioms (at least the ones we are most familiar and comfortable with), so counter-examples are necessarily contrived.

First, the product of two transformations of the set must be a transformation of the set.

Displacements can be added so the product of two (here addition is the product of transformations) is in the set. But with a wall it would be difficult to add two displacements that are close to it. Consider walking along the edge of a building then turning and walking along another edge. The product of the two transformations, the vector sum of the displacements, would require walking through the walls of the building. Mathematically this can be done, physically certainly not. And if there really were an end of the world, group theory (and much else) would be in serious trouble at that point.

Also for the edge of an irregular cliff, while two displacements may both individually be possible their sum may not be. The sum of displacements is commutative; the order of addition does not matter. This is true mathematically, but it is easy to draw a cliff for which addition in one order is possible, addition in the other is not. This emphasizes that because operators can be defined mathematically does not mean they have physical meaning or are physically possible, or are the ones that we physically want.

Another example is a system of three atoms, A, B and C, and transformations that combine two objects to form a third. The product of transformations $A+B$ and $(A+B)+C$ is not in the set for there are no transformations combining three objects at once. While a set may not form a group it may be possible to extend it to one. On the other hand if we cook something this may require a series of transformations of the food. It would often be difficult to do the preparation and cooking in one step.

If we put an atom in a magnetic field and then hit it with a charged, spinning particle in one step, the result is different from the effect of the two transformations in succession because the field affects the particle. Here we might consider the "product" in the set but it would be different from the effect of two transformations in succession.

These are artificial because the transformations whose product we consider are of different types, in distinction to, say, the product of two rotations.

The second requirement is that the inverse of each transformation be in the set. Growing old is a good, though unfortunate, counter-example. Statistical mechanics provides transformations that can go only one way. Of course these involve large numbers of objects rather than a single, simple system. This point may be significant.

Again in combining two objects, there is no inverse unless we add the transformations of breaking up objects into their component parts. Of course if we consider the well-known transformation of mixing coffee and cream it is (physically) difficult to add to the set the inverse transformation. Cooking gives other examples. We can also consider placing an atom in a magnetic field strong enough to tear it apart. It is difficult to think of a magnetic field (being turned on or off) that would put the atom back together again. We might call these Humpty-Dumpty transformations — they have no inverses.

The author must admit that he has been unable to think of any set of transformations that does not include an identity, and which cannot be extended to include one.

Problem I.4.c-1: Find sets of transformations that do not contain identities. Is it possible to imagine a set with more than one identity?

How about the associative law? Returning to our trip around the building, we now further climb up the wall. The product is vector addition. Suppose that we can walk through the building, that is carry out the product of the first two transformations, and there we find stairs. The product of the three transformations is possible. But carrying out the first transformation we arrive at a corner where there are no stairs so that the product of the last two transformations is impossible. For these, the associative law does not hold.

One can also imagine various chemical transformations, say adding to a substance a mixture of chemicals A and B and then to this mixture adding C which could give a very different result from adding A to a mixture of B and C.

Problem I.4.c-2: Are there concrete situations in which the product of three transformations is possible, but is not associative?

Problem I.4.c-3: Add to vegetables a mixture of soup and water and then store for a while at a temperature below 0^oC (the third transformation). Then, starting with the original ingredients, add soup to the vegetables and to this add frozen water. Are these transformations associative?

Problem I.4.c-4: The edge of a cliff provides us with another (somewhat dangerous) example; the operators are displacements and the product is the vector sum. Draw a diagram showing a case where carrying out the first displacement allows the performance of the sum of the second two, but carrying out the sum of the first two does not allow the performance of the third. Might it be possible that all three can be carried out (physically), but the product is not associative?

Problem I.4.c-5: Suppose we performed transformations on a system, an electron, an atom, a molecule, say by inserting it into an electric field, a magnetic field, and an oscillating electromagnetic field. Would these transformations always (ever) be associative? Would it matter which two fields were turned on first?

Problem I.4.c-6: Hysteresis leads to other examples. It is worthwhile considering a piece of iron in magnetic fields, which transform it. Construct a set of transformations and determine under what conditions, if any, they are associative. There is an important point here. The effect of one transformation here depends on what previous transformations have been carried out, that is the history of the object. For transformations for which this is true, would associativity be expected in general? Why? Is this true for the transformations that form groups?

Problem I.4.c-7: Consider materials with nonlinear optical activity, and see whether, and when, transformations by them are associative.

Problem I.4.c-8: Is there a rule or algorithm that distinguishes these transformations from, say, rotations? Is it possible to tell when, and why, a set of transformations is associative?

Not only can there be physical transformations that are not associative, but there can be mathematical objects. Later we mention a set of numbers, octonions, with a rule of combination that is not associative (pb. III.2.e-5, p. 75) and more important, Lie algebras (chap. XIII, p. 367). The requirement of associativity really means something.

Problem I.4.c-9: For the set of positive rational numbers, show that division as the operation of combination is not associative [Grossman and Magnus (1992), p. 11]. Yet multiplication is associative; explain how its apparent inverse can not be. For this set, is there any operation that is associative? Are there sets for which this operation is associative?

Certain properties of physical transformations may seem obvious, and these lead them to form groups. But that they do, follows from (too often unnoticed) properties of physical laws. The description of an object given by Newton's laws, Schrödinger's equation, Dirac's equation, is independent of the history of the object. This is a reason that transformations form groups, or perhaps it is better to say: these laws must hold because geometry requires transformations to form groups.

Problem I.4.c-10: One reason for the importance of groups is that a set of transformations that commute with a Hamiltonian forms a group (pb. I.4.b-2, p. 16), thus when such a set occurs, as it often does, the states of the system described by the Hamiltonian are restricted by the group. The analyses of these restrictions, and how they arise, is a fundamental purpose of this book. Prove that a set of operators commuting with a Hamiltonian (or any operator?) form a group [Lax (1974), p. 6]. (Obviously this is true if the operators do not act on the states that the Hamiltonian does; assume that they do.) The proof is simple that the product of two such operators is in the set. More significant are the questions of the existence of inverses and of associativity. To rotate an object requires that it be accelerated, at least twice. But if the object is charged, or constructed of charged particles, such acceleration produces radiation. (And if not so constructed then other forces must be applied giving equivalent results). Thus, producing the exact inverse of a *physical* rotation becomes problematical. Likewise, it is not obvious exactly how to define the product of two rotations to give the same result as each of the individual ones applied in succession, if we really consider radiation. This raises questions, as the reader has undoubtedly noticed, with associativity. However this is not what we mean by a rotation. For one thing we can define it as a mathematical operation. But then we cannot say, for example, that it is physically required to commute with a certain set of Hamiltonians, for it has no physical meaning. This point should be clarified and made stronger. We can think of it as a comparison of the measurements of two observers, relatively rotated, that is a rotation of the observer (except that this also is a physical process, whose exact meaning has to be clarified). But this reference to the observer has a less clear meaning for other transformations, rotations in isospin space, say, or for time reversal or gauge transformations [Mirman (1995b)], or the interchange of identical particles, and perhaps the meaning and properties of transformations become even more subtle in other cases (some perhaps not now known). Thus we consider in this book those transformations, in particular ones commuting with Hamiltonians, that have inverses and obey the rule of associativity. They thus form groups perhaps because it is a property of nature that physical transformations form groups, or perhaps because we are only aware of transformations that form groups, or perhaps because we are limiting ourselves to consider only transformations that form groups, or perhaps because we apply the name transformation to operations that have properties to form groups, or perhaps .... Why do they form groups?

Problem I.4.c–11: The reader should develop other counter-examples of transformations that do not obey group axioms, and from them find the assumptions, too often hidden, about physics and nature that underlie those aspects of physical laws that lead to, or follow from, the group properties of physical transformations.

I.5 GROUPS

Physical transformations (at least those we are most comfortable with) have the properties required by the group axioms. Physicists think of them as transformations on physical systems, mathematicians (often) regard them as abstract symbols, and have developed a detailed theory of such collections. We wish to use the results of this theory. So following mathematicians we call a set of symbols or objects (such as transformations) with a combination rule, a group, if it has the properties (the group axioms):

A) the existence of an identity,

B) the existence (more properly, definition or inclusion) of an inverse for each symbol,

C) the existence of a product symbol for each two symbols, and

D) the product is associative.

The things in the set we call the group elements , or transformations, or operations or symbols, or sometimes other things.

A group then is a set (of symbols, transformations or whatever we wish) together with a rule (their product) for combining them (in pairs) such that, using the defined product, they obey the group axioms. Stated this way, a group is an abstract mathematical object. In physics we are (usually) interested in operations on physical objects or (thus) on statefunctions describing them. We regard a group as a set of such operations, and talk about transformation groups, the abstract symbols being thought of as physical transformations.

Of course given a physical object we can conceive of various transformations, no matter how unphysical or absurd, say the transformation of an electron into a computer (or perhaps this is not that unphysical or absurd). But what we are really interested in, and what we limit ourselves to, are groups of transformations that leave the system invariant, or almost so — symmetry groups — or in some other way have a "reasonable" relationship to the system. If a set of transformations leaves a system invariant it limits its properties, giving us information about it. But there may be other transformations that can be useful and eventually we consider some.

Problem I.5-1: In this definition of a group the phrase "in pairs" is used. Is this necessary, always, ever? Is it sufficient?

I.5.a Uniqueness

One of the most dangerous parts of speech are the articles. They are so simple and natural that we read them without perceiving, as if they were too small to be noticed. But they can carry some big assumptions, well hidden by the small size of the words of this part of speech. And they can be wrong.

We speak in the axioms of "an" identity and "an" inverse for an element, or "the" product. Of course to a mathematician the word "the" is not a problem. He always assumes that the product is well-defined and unique. But since the axioms here refer to physics we must be sure in each situation that this is really true. But even for a mathematician the other articles must be shown to be correct.

Problem I.5.a-1: Thus show that a set obeying the group axioms has "an" identity, a single one, also that each element has a unique inverse. Does the order of multiplication matter: does

$$aa^{-1} = a^{-1}a? \qquad \text{(I.5.a-1)}$$

Problem I.5.a-2: Another place where there might be trouble is in strings of products. Thus ABC has a unique meaning by the assumption that multiplication of operators A, B, C is associative. But is this true for longer strings? Show that $ABCDE\ldots$ is unambiguous, for all operators and all lengths of strings.

Problem I.5.a-3: How is the inverse of the product AB related to the inverses of A and B; what is $(AB)^{-1}$?

Problem I.5.a-4: Given elements A and B of a group, and $AC = B$. Is the element C unique? How about D, with $DA = B$? Show that both C and D always exist. Does $C = D$?

Problem I.5.a-5: Take the product of a fixed A with each B_i of group G. Show that the elements AB_i (and B_iA) are distinct ($AB_i \neq AB_j$, $i \neq j$), and that all elements of G can be written in this form.

Problem I.5.a-6: Consider a set of elements A_i obeying all the group axioms, except associativity. Let

$$A_1 A_2 \ldots (A_k \ldots) \ldots A_n = A_1 A_2 \ldots (A_l \ldots) \ldots A_n, \qquad \text{(I.5.a-2)}$$

that is, the products are equal for two different placements of the parentheses. Then show that the product is the same for all placements of the parentheses; the associative law holds [Hall (1959), p. 25]. Is it necessary to include all group elements in this product?

Problem I.5.a-7: Undoubtedly there are many other assumptions hidden away, some that should be checked physically, others mathematically, and perhaps others that do not follow from the group axioms, but which no one has ever noticed. The reader will undoubtedly enjoy searching for these, both in the group axioms and in other parts of the discussion of group theory and its applications, and also seeing which of these assumptions are about mathematics and which about physics, and why, and why they are necessary, and correct, if they are.

I.5.b Some related concepts

For reference we list some related terms. A set is a collection of symbols (or objects, or whatever we enjoy collecting; although this begs the question of what a collection is). A groupoid is a set with a product defined for each pair of symbols (and the product is in the set; as seen above there are sets for which the product is not in the set). A semigroup is, as the name implies, half a group. Here the product of the groupoid is associative. If we require in addition an inverse (necessitating an identity) for all elements, we have a group.

Adding more requirements gives other algebraic structures. However we shall first consider the consequences of the requirements we have imposed to get a group.

Problem I.5.b-1: Growing old gives an example of a semigroup, unfortunately. Why?

Problem I.5.b-2: In the above problems at what stage from set to groupoid to semigroup to group are each of the properties required? This is a question to keep in mind in reading the rest of this book.

I.5.c Transformation groups and symmetry groups

While groups are often taken as sets of symmetry operations, they are more basic then that. For example the rotation group (which we define more carefully in chap. X, p. 269), the set of rotations of space, is the set of transformations that leave magnitudes and angles unchanged. Take the x axis horizontal, y vertical. Suppose that we rotate by angle θ, giving axes x' and y'. These are related by

$$x' = x cos\theta + y sin\theta, \quad y' = -x sin\theta + y cos\theta. \qquad \text{(I.5.c-1)}$$

In Newton's second law we should include a force that depends on the distance between a particle and the center of the Earth, and a rotation changes the coordinates of both. However suppose we ignore the latter and just take gravity along the y axis. Then (it appears that) space is not invariant under rotations in the xy plane: in the law, gravity gives a term in the y equation, but not in the x, while giving a term for both x' and y'. However the $x'y'$ axes are still related to the xy axes by this equation (a group transformation). This rotation group is a transformation group over space — it is the set of operations that transform one coordinate system to another (rotated) one. But it is not (apparently) a symmetry group.

Groups of transformations are more fundamental than symmetry groups. A symmetry group of a system is a transformation group of it, though a transformation group need not be a symmetry group. Some properties of objects obtained using group theory come from the relevant group being a symmetry group, others only from it being a transformation group. This should be understood, for the results from the latter can hold more widely, for other objects also (and these may be of interest). The rotation group is (a subgroup of) the transformation group of (Euclidean) space, though it need not be a symmetry group (especially of physical space). It is one of those remarkable facts, again not often enough remarked on, that it is actually a symmetry group.

Problem I.5.c-1: Write Newton's second law with gravity in the $x'y'$ system and notice that it is (apparently) not the same as in the xy system. Also check that these transformations do form a group; this is the two-dimensional rotation group.

I.5.d The value of group theory

To many it may seem obvious that if there is a mathematical formalism describing physical quantities it is useful. But for skeptics it is worthwhile summarizing why this is true for group theory. One use for the formalism is to provide a systematic method for studying systems with symmetry (for which there are transformations, having meaning for the system, that leave it unchanged). In particular many systems, though vastly different, have underlying similarities because they all permit the same transformations (like symmetry under rotations). Group theory allows us to study all these at once — it is unnecessary to solve the same problem for each.

It also allows us to see relationships between systems that may look different, and to see (certainly more clearly) how the properties of a system follow from its symmetries, and those of space and time, thus to better understand physics and nature. It helps to develop intuition, inherently valuable, and often useful in applications.

Essentially a group is important in studying a system because the transformations are — in some way — related to its properties. This is clear if they are symmetries, if the system is unchanged by them (or almost so), which may be a property of the system, as for a sphere, or of the laws of physics — nature — as for rotational invariance. The symmetries of an object depend also on the laws of physics: a perfect sphere, for visual clarity made of cloth, in a gravitational field, would be quite strange. In other cases, and there are other cases, the reasons that there is a group of transformations related to the system is less clear; trying to find the reasons is an important (but often overlooked) step in understanding it.

There are many questions that we might ask and that the theory can help us study, and can especially suggest what physical assumptions are needed to answer the questions. Given two systems (the observed and the observer) can one be replaced by another system (the transformed one) such that there are simple relationships between the experimental results obtained in the two cases? What properties of nature, and the system, require (allow) this? The system is described by a set of statefunctions. What are the possible numbers of members of each set? How are they related by the transformations? What are their functional forms? What are the relevant operators of the system? What are their expectation values between these functions? Which of these operators are diagonal; what is their maximum number? What properties of the laws of physics and of the system determine the answers to these questions?

We aim then to develop the mathematical structure to study and answer these questions, to understand the physics underlying the structure, and the answers it gives, and to help the reader translate between from one language, that of group theory, to another, that used in describing physical systems — often the language of quantum mechanics. As the reader will see, quantum mechanics and group theory often differ in no more than vocabulary [Mirman (1988a,b, 1991a-d, 1992, 1995a,b)]. And this is not surprising, for quantum mechanics is group theory, but with physical interpretation of the terms. To illustrate, we consider a quantum mechanical problem, and see why it is group theoretical, in disguise.

I.6 GROUPS AND QUANTUM-MECHANICAL STATES

In quantum mechanics the fundamental object in the study of a system is its state, described by a (state)function from which experimental quantities are found. Thus the first step is to specify the state of a system, giving enough information to allow the statefunction to be determined, then from it find the experimental predictions.

To give the state, we must first name it, to distinguish it from others. This is the state-labeling problem, that of finding a complete set of labels, quantum numbers (perhaps with continuous values), that fix the state vector — wavefunction — of the system. This assigns parameters to the system, such that anyone who knows their values can find the statefunction (often an explicit function of coordinates) and the experimental predictions, probabilities for positions, or for transitions between states, say. These parameters distinguish one state from another. And different observers, though viewing the same state, look at it differently, seeing for example the object's spin at different angles to their z axes. Some parameters are observer dependent; they tell what an individual observer sees.

What are these parameters, and what determines the set of numbers that label states? How are they related to the properties of the system, to its symmetry group (if they are)? Why might groups be relevant to state labels, and their values?

We wish to understand how group theory labels states and see how to determine these labels and their values, and obtain statefunctions from them; given a system we wish to find the restrictions on it imposed by general requirements of physics, isotropy and homogeneity of space say, and then find its wavefunctions and their labels. From these we calculate the experimentally obtained quantities, and we can see how the quantities are determined and limited by the physical laws, using the group restricting and describing the system. However we do not consider at this point the extent to which the state of an arbitrary system is simply determined by general laws and by group theory, but limit the discussion to a particular system.

I.6.a Labeling the states of the hydrogen atom

For the hydrogen atom, these labeling parameters are the principal quantum number, n, the orbital angular momentum quantum number, l, the spin of the electron (always $\hbar/2$, which we take as $\frac{1}{2}$) and their z components. Knowing these values we know the wavefunction.

Total angular momentum quantum numbers, and n, are the same for all (relatively rotated) observers — the total angular momentum operator commutes with rotations. Angular momentum z-components are observer dependent (they are changed by rotations) and give the information about what a specific observer sees.

I.6.b The reason for the labels

Why are these the labels? They exist, and are relevant, because of properties of the atom, but more because of general properties of physical laws and of space, and the properties of transformations that are determined by the nature of space.

For the three-dimensional simple harmonic oscillator, and for any spherically-symmetric system, the angular-momentum quantum numbers are the same as for the hydrogen atom, and have the same values. The quantum numbers, their number and their values, are given by group theory, by the rotation group in three dimensions (for the hydrogen atom in four dimensions, although this last dimension is not physical). For both the hydrogen atom and this oscillator, the energy is independent of all quantum numbers except the principal one; for all spherically symmetric systems it is independent of the z component. This follows not from the properties of the system, but of space.

The labels (aside from the principal quantum number) of the hydrogen atom, or the harmonic oscillator — the orbital angular momentum, spin (angular momentum), and their z components — are the eigenvalues of operators describing physical quantities (these operators are generators of the rotation group — what this means we have to see).

I.6.c The procedure for labeling states

Generally (though perhaps not always) we label states by using operators, preferably with physical significance, just large enough in number to give a set of eigenvalues that distinguish states — different states differ in at least one eigenvalue of the set. This provides a complete set of labeling operators; their eigenvalues label the states, completely specifying them. The statefunctions are eigenfunctions of these operators.

For the statefunction to be an eigenfunction of all, the labeling operators must commute so they can be simultaneously diagonalized. What are these operators? For the rotation group, they are a mutually commuting subset of the (still incompletely defined) group generators plus algebraic functions (polynomials) in them. They are so chosen that no other operator (except functions of these; $G_1^2 + G_2$ is a function of G_1 and G_2) commutes with every member of the set.

Problem I.6.c-1: Why is this last condition necessary?

So given a group describing a system, this set of operators, a property of the group, not of the system (or correctly, therefore of the system), plus others needed to make a complete set, must be determined, and their eigenfunctions and eigenvalues, the states of the system and their labels, found. Then the operators and their eigenvalues must be physically interpreted. Thus, why do the eigenvalues of the rotation-group commuting-operators give the total angular momentum and its z component? What do these have to do with angular momentum? This is an example of what must be explained.

Finding these operators, and their eigenvalues and eigenvectors, from the symmetries of the system — the set of transformations leaving the system invariant (a term which must be defined in each case) is one of the essential tasks of group theory.

I.6.d Why is angular momentum connected with group theory?

The commuting operators coming from the rotation group are L^2 and L_z, giving the total angular momentum and its z component. Why? What do we mean by angular momentum, why has this quantity been chosen to receive a name and so much attention? (When we study Lie algebras we shall explain the connection between the group and these operators, and why we use such an evasive term like "coming from".)

The quantities that we introduce so casually in elementary physics are not arbitrary, there are good reasons for them. The importance of many comes from their being conserved — they are constant in time. And there is a reason why they are conserved, and so are flattered by a name and a law stating their conservation. Thus we have to relate these names from elementary physics to the operators of groups. (That quantities are conserved is both important, and useful, in other ways [Oyibo (1993)]).

The eigenvalues of these angular momentum operators are constant in time, for any system that is rotationally invariant — one in which the Hamiltonian of the system commutes with generators of rotations (L_x, L_y, L_z), therefore also

$$L^2 = L_x^2 + L_y^2 + L_z^2. \tag{I.6.d-1}$$

We need only multiply the eigenvalues by a constant (so they have the units previously picked) and give them names. They are physically significant because they are conserved; this results from their commutation with the Hamiltonian, which follows from the rotational invariance of the system. We should not ask why these are the angular momentum operators for that is the name we give them — angular momentum has meaning (only) because it is given by the eigenvalues of these operators. The concept of angular momentum appears in physics because the eigenvalues of these operators label states and are conserved.

We see later how these rotation operators are related to the linear momentum operators and why angular momentum and linear momentum are related in the way that we learned in elementary physics.

What is significant is not the names angular momentum and z component, but the set of operators — the group — commuting with the Hamiltonian, and their eigenvalues and eigenfunctions.

This illustrates how group theory leads to the introduction of physically significant quantities, though we may not be aware of the group theoretical, or the physical, reasons these quantities are introduced — but we should be.

Chapter II

Examples of Groups

II.1 GROUPS IN THE REAL WORLD

The best way to study group theory is to study actual groups — actual physical, and mathematical, objects and their transformations and symmetries. Here we use real objects to gain understanding, not only of how groups give us information about them, but also about the nature of groups and their structure, and how such entities illuminate what would otherwise be abstract mathematics.

The objects are geometric shapes having transformations that leave them invariant. But these are shapes of physical bodies and it is often useful to think of them this way. There can be many bodies that have these shapes: molecules, crystals, macroscopic objects (including works of art), and so on. To be concrete we often discuss examples, say molecules; the reader should try to think of others, especially ones he has had some experience with, be these organic molecules, food, tinkertoys, buildings, or just lines on a piece of paper. A vast number of examples of things with various symmetries is given by Hargittai and Hargittai (1987). It is a good source which should be referred to while reading this book. Other good sources include Lockwood and Macmillan (1978), and Shubnikov and Koptsik (1974); see also Fejes Toth (1964), Ghyka (1977), Senechal (1990), Weyl (1989) and Yale (1988). (Although we discuss physical interpretations of the mathematics it is really more accurate to speak of geometrical interpretations — though the reader should think about whether he agrees with this. The physics, though partly in the geometry, and partly in the determination of the symmetry, resides perhaps more in the statefunctions, and when we discuss them the relationship between physics and these aspects of group theory might be clearer.)

The purpose of this chapter is not only to introduce the groups, and relate them to physical objects, but also to raise some of the questions that are interesting and important, and suggest why. This is often done in problems. Many of these can easily be answered, others are more difficult, and for some the material presented so far may be insufficient. But if the reader puzzles over them — for a little while — he will be better prepared for the discussions of the later chapters, where many are considered, though not necessarily with reference to this one. It is worthwhile to try, at least some of, these again, after going through the rest of the book.

Abstract group theory regards all objects with the same group as identical, more accurately as having isomorphic groups; when mathematicians cannot distinguish between two groups we say they are isomorphic — the groups, not the mathematicians (the closest thing to isomorphic mathematicians is identical twins). Another example of isomorphism is that between groups of symbols and of matrices (which even mathematicians would recognize as distinct); these forming a representation, something of major importance. If groups are isomorphic (literally have equal shapes), there is a one-to-one correspondence between their elements, and the product of each pair of elements of one group corresponds to the product of the corresponding elements of the other. We can thus consider any interpretation, and usually the most helpful in gaining understanding are comfortable things, thus we start with these.

II.2 GROUPS OF PHYSICAL OBJECTS

The most familiar groups of transformations are of physical objects. We usually think of transformations as physical actions, not as actions on mathematical symbols; thus we do not recognize that we are dealing with advanced mathematics when we do something like turn around. So here we consider simple examples to relate the mathematical concepts to (hopefully) our physical intuition.

II.2.a The group table

The discussion in this chapter is skimpy; we mention these groups to give some idea of the types of possible groups, how their transformations are related to the objects whose symmetry they describe, and to introduce them. Later we turn to a more detailed study. In many problems the symmetry group of the object is requested. To supply this, make a table giving the product transformation for every pair of transformations. On the top row list the group elements, and likewise in the leftmost column list these in the same order, both starting with the unit element. At each intersection, write the element (found perhaps experimentally) which is the product of the one at the left of the row, with the one at the head of the column, this acting first (using the given order of multiplication; sec. I.4.a, p. 14). This is the group (multiplication) table; it specifies the group. The table should be checked to show that it does give a group [Maxfield and Maxfield (1992), p. 12].

II.2.b Two-element groups

The simplest groups (beyond the trivial one of a single element, the identity) have two elements, the identity, E, and one other R, with $R^2 =$ E. There are many realizations of these (isomorphic) groups.

II.2.b.i *The group of a pair of hands*

A pair of hands, one left, one right, mirror images of each other,

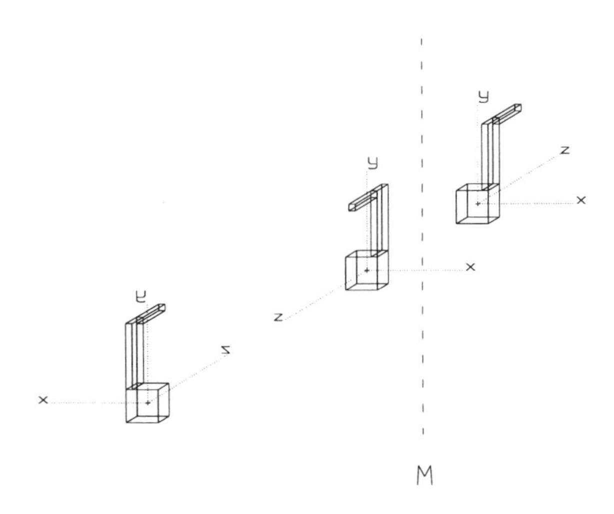

Figure II.2.b.i-1: HANDS AS AXES

is invariant under the reflection in the mirror, thus under the two-element group, where R is the reflection. This, using a highly simplified hand, is shown, with axes, and the mirror M; the figure also shows a second right hand obtained by a rotation of π about the y axis. The pair of right hands (on the left side) is thus invariant under this two-element group of the identity and the rotation; the pair of hands (on the right), one left and one right, is invariant under a two-element group, with

R the reflection. These two-element groups, mathematically identical, have different physical (and geometrical) meaning.

II.2.b.ii *The group of a held football*

What can be done to a football such that it is impossible to tell whether anything was done at all? Its long axis cannot be (arbitrarily) rotated, and it cannot be rotated about this axis for it has a seam, and the position of that would change. The ball can be rotated by π about a line perpendicular to the axis, through its center, and the center of its seam. Thus the group leaving it invariant has two transformations, one the identity — doing nothing will leave it invariant — the other a rotation of π around this single line. The product of two rotations of π brings the football back to its original orientation — it is equivalent to the identity. Using circles to indicate stitches, and labeling the rotation axis M, we get a very schematic football and pair of hands:

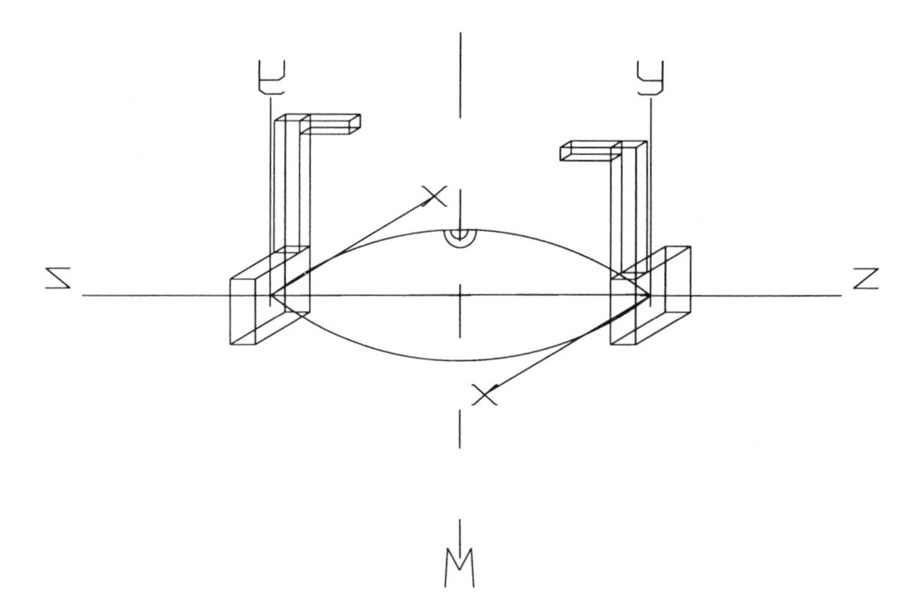

Figure II.2.b.ii-1: SYMMETRY OF A HELD FOOTBALL

This illustrates that the symmetry group depends on the transformations that we wish to allow. We could also put a mirror along the short axis, through the center of the football, going through the line labeled by (the reflection-invariant) M, and the ball has symmetry — reflection symmetry — through this mirror. This transformation cannot be carried out physically (without dissolving, and reassembling, the ball), but the system has symmetry in the sense that if we replaced every little piece of ball by the corresponding little piece that is at the reflected position, the resultant object would be identical to the original. So if we also wished to consider reflections in a mirror there would be other transformations, the reflection and the product of the reflection and rotation. We do not want to do this now so we have the football held at its two ends by identical right hands (supplied by identical-twin football players). Replacing the stipples by identical current loops is another way of eliminating the reflection symmetry. These all rotate in one direction and their images rotate in the opposite direction — right and left-handed loops are interchanged — so the current-carrying football has symmetry only under rotations of π about the axis, but not under reflection.

Problem II.2.b.ii-1: Find the group table experimentally (with a real football), and also using the product of rotations. Also check that the two hands do go into each other under the rotation, but not under the reflection. This is a two-element group. Show that all two-element groups have the same group table — in an abstract sense they are all the same.

Problem II.2.b.ii-2: Note that with the ball held by two right hands there is no reflection symmetry. Suppose that the hands are removed, giving reflection symmetry (in the planes marked M, through the center and also, for a line of footballs, through the ends, as can be seen with a drawing). How many elements does the resultant group have? What is its group table? What would the group be, if in addition, the stitches were removed?

II.2.c The group of a triangle

The invariance group of an equilateral triangle [Falicov (1966), p. 64; Grossman and Magnus (1992) p. 16; Stephenson (1986), p. 141] consists of those transformations that change it into an identical one, at the same position; we cannot tell by looking at the triangle whether such a transformation has been carried out. Which transformations are these? Take the triangle as a solid, say with arrows on its edges, giving it a sense of rotation (these might be three vectors adding to zero), or with the two faces colored differently (or both). Every transformation leaving the triangle invariant is one of three operators, the rotations of 0, $2\pi/3$ and $(2\pi/3)^2$ about a perpendicular through the center of the triangle, or is the product of these (the result of carrying them out in succession).

Problem II.2.c-1: Is the last phrase redundant? Construct a triangle from pencils (with points and erasers) and check that these three transformations form its (complete) invariance group — there are no others that leave it invariant. What point is the center? For each pair of transformations find, experimentally, the single transformation having the same effect. Does the order in which the transformations are carried out matter? Repeat using a drawing. Construct the group table, giving the equivalent transformation for each of the 3×3 products; the list of the three operators plus this table of their products defines the (abstract) group. Is part of this sentence redundant? Pick coordinates for each vertex and find, algebraically, all transformations of these keeping the triangle equilateral, with the same position and orientation, and all products, and the group table; verify that it is the same as found experimentally. Check using the table, and experimentally, that this set of transformations forms a group — the group axioms are obeyed.

Problem II.2.c-2: Now remove the arrows and color the two faces the same. This gives six symmetry operators, the three rotations, plus now three $\pi/2$ rotations, these about each of the three lines through a vertex, bisecting (so perpendicular to) the opposite side. Check this with a drawing. Again every invariance transformation is one of these six, so any product leaves the triangle invariant, thus must be one of these. However, explain how the product of a transformation which flips the triangle, and a rotation which does not, can do only one or the other. Check that these six transformations form the (complete) invariance group; it is called D_3. Are there transformations leaving the triangle invariant that cannot be performed physically (without taking it apart) — is the invariance groups of three sticks, and of pencils, larger (or smaller) than that of three lines on paper? If so, are these transformations related to the six? How? For each pair of transformations find, experimentally, the single one that has the same effect. Does the order matter? Why? Write down the 6×6 group table, and also that for the triangles of pencils, and of sticks (which have no direction, as do pencils), if these are different. Using coordinates for the vertices find all transformations leaving the triangle invariant, and the group table, and verify that it is the same as found experimentally. Check both experimentally, and from the table, that these transformations form a group. How are the tables of the different triangles related? Another interpretation of these transformations, whose implications we will explore in depth, is obtained by considering three objects, here the three vertices, although that is not really relevant. These six transformations produce all permutations of them. Write down every permutation of three symbols, counting their number, all products, and the resultant table, and compare with the table for the triangle. Identify each permutation with a transformation.

Problem II.2.c–3: Do these problems (experimentally and mathematically) if the vertex points are replaced by objects. First take these as circles two of one color, the other different. Then take them as squares, rectangles, triangles, of different orientations and color combinations. These give examples of (what could be called) "broken symmetry" (sec. I.3.c, p. 8). Suppose the triangle were only isosceles. What would its invariance group be? How are these groups related to that of the equilateral triangle? This is another example of "broken symmetry". Explain the relationship between these types.

Problem II.2.c–4: Think of the triangle as a molecule consisting of three atoms at the vertices. The atoms can be taken to have spin (all in the same direction), or be spherically symmetric. Find the symmetry groups. What are the groups for the equilateral and isosceles triangles if two of the atoms are identical and the third different? What groups are obtained if not all spins are parallel? How are these related to the ones for identical atoms? If all atoms are distinct, would there be symmetry? Which groups apply to these cases, that for the triangle with differently colored sides, or the one for which the two sides are the same, or neither?

Problem II.2.c–5: Suppose the triangle were only a set of lines in a plane (instead of three dimensions). What is the group now? Answer the above questions, especially the ones about broken symmetry and about molecules, for two dimensions. Is there any simple way of getting these answers from the ones above?

Problem II.2.c–6: These are perhaps the simplest systems to start with, after the (trivial?) line. Is a line (segment) trivial? What is the group of transformations that leave it invariant? It can be considered as a cylinder, of finite radius, or of zero radius. Does it matter? What would the effect of putting an arrow at one end be?

II.2.d The water group

The well-known molecule H_2O provides another group of some interest [Lomont (1961), p. 117]. The molecule has an O atom and two hydrogen atoms which we take as a triangle (in a plane). There are four transformations, the identity, reflection in the plane in which the molecule lies, a rotation of $\pi/2$ about a line through O perpendicular to the line joining the two H's, and a reflection through this line. The last two have the same effect, while the second may not seem to do anything. If the molecule existed in a two-dimensional space, its plane, there would be no distinction between the last two, and no meaning to the second. But it is actually in three dimensions and physically these are all meaningful and distinctive transformations. And this is relevant to its properties.

Problem II.2.d-1: Draw the molecule, and check these transformations. What is — experimentally — the group table? Using coordinates for the atoms find the transformations and group table. Check that these are the same, and are group tables. What is the group if all atoms are the same? How is it related to this one? Check that all elements of this group commute; since all (except the identity) give the same result, they do not have much choice. We use the nickname "water group" for this.

II.2.e The pyramidal group

The pyramid, with an equilateral triangle as its base, is another physical object with the same group as that of the triangle with differently colored sides.

Problem II.2.e-1: Build from cards, or sticks, such a pyramid, find the transformations leaving it invariant, and their group table; also do this using coordinates. Show that the axioms are obeyed. Is there a difference if the edges are pencils? A molecule with atoms at the vertices is another interpretation. Construct a model using balls and sticks and check that the transformations found these different ways agree. Compare the group table with that of the various triangles. Would you have expected these comparisons? Does it matter if the atoms have spin?

Problem II.2.e-2: Verify that with the z axis passing through the vertex and perpendicular to the base, the symmetry group consists of the identity E, and three rotations of $2\pi/3$ about the z axis (C_3, C_3^2, $C_3^3 = E$); C_n is the symbol for a rotation of $2\pi/n$. These give as symmetries reflections (σ) in the three planes passing through the corners and the z axis (σ_v, $\sigma_{v'}$, $\sigma_{v''}$) corresponding to the transformations of the triangle that interchange two corners with the third fixed; however, while symmetry transformations, they are not rigid-body ones: the molecule must be distorted to carry them out. For a molecule with a base of three identical atoms, and a different atom at the vertex, how can these transformations be carried out? The rotations are clear. To obtain the effect of the reflections, rotate the vertex by π about an axis through one corner of the base bisecting the opposite side, bringing it from above the base to below. This interchanges two corners leaving the third unchanged. Then push the vertex atom through the base back to its original position. Thus we can physically perform these six transformations without breaking the molecule (though distorting it). We can perform no further ones leaving the molecule invariant, except by removing atoms and interchanging them when separated from what remains of the molecule. Are there any? Check this experimentally, and that these are all the symmetry transformations so this is the largest symmetry group. What is the subgroup of symmetry transformations — the subset that itself forms a group — that treats the pyramid as a rigid body? Show that if two such pyramids are attached at their bases giving an object invariant under π rotations about any of three axes through the base, there is a (rigid-body) symmetry group, which we can call the double-pyramidal group; it is labeled D_3. Which axes are these?

II.2.f The groups of the square

The group of the square [Baumslag and Chandler (1968), p. 77; Falicov (1966), p. 60; Joshi (1982), p. 5; Maxfield and Maxfield (1992), p. 15, 59, 199, 200; Tung (1985), p. 26; Aivazis (1991), p. 9] consists of those transformations, treating the square as a *rigid body*, that take it into one at the same position with the same orientation — that leave it invariant. These permute the corners; they can be carried out physically (without cutting the edges). However this definition is ambiguous, leading to more than one group. When we mention later, as we shall, the group of the square, the name will be taken to refer to each (or at least the relevant ones) of these.

Problem II.2.f-1: Why is there ambiguity?

Problem II.2.f-2: First we take the square to have arrows on its edges, the sum of four vectors adding to zero for example, with the two faces colored differently. The only symmetry transformations are rotations about the center, of integer powers of $\pi/2$, giving a four-element group (called C_4). Carry out all symmetry transformations on a square card, with differently colored sides. Consider how each changes the set of vertices into a different set and show that the product of every pair of transformations can be replaced by a single transformation. Find the products for all pairs and the group table (and show that this is a group).

Problem II.2.f-3: With no arrows and the same colors, there are also two rotations of π about lines perpendicular to pairs of opposite edges, through their centers, and two further π rotations about lines through opposite vertices. This group (of rigid-body transformations) has eight elements. Its order, the number of its elements, is therefore 8. Verify this with a drawing, and a model. Repeat the analysis and find the group table (with $8 \times 8 = 64$ entries), and show that it is a group. This group is D_4. If the square is only a rectangle its group is D_2. Why does the group of the square have eight elements? Label the vertices in order and take vertex 1 in, say, the upper-right corner. Now vertex 2 can be at either the upper left or lower right. Once this is decided, the positions of all corners are fixed. There are four choices for vertex 1 so there are eight arrangements of the vertices, so eight transformations (including the identity) from any initial one. Assign coordinates to the vertices, and find all symmetry transformations, and all products; write equations giving the coordinates after each transformation in terms of the ones before. Find the group table and check that it is that found experimentally. How is this table related to that for the square with differently colored sides? Taking the square as a quantum mechanical object (say a molecule) its statevector depends on coordinates; what are they? With the origin at the center of the square draw the x and y axes parallel to the sides, the z axis perpendicular to the plane. This gives coordinates for each atom, and it is these that the statevector is a function of. Find the symmetry transformations on them. There are further transformations, reflections in mirrors through the symmetry axes (which are, however, not rigid-body transformations). Different group theory books give different sets of eight transformations, some including reflections, some not. Why can they do this? List all sets. If we include reflections in the plane of the square (treating it as a three-dimensional object) the group is D_{4h}. An example of a molecule with this group is cyclobutane, $(CH_2)_4$ [Hargittai and Hargittai (1987), p. 295].

Problem II.2.f–4: The order of transformations is important; not all commute: transformations T_1T_2 and T_2T_1 give different arrangements of the vertices. Verify this experimentally (and from the group table). A group all of whose elements commute is an Abelian group. This is an example of a non-Abelian group.

Problem II.2.f–5: For a molecule the corners are not points; they have shapes. Does this matter? How? Vary their orientations, keeping them similarly oriented, and then change their relative orientation. Find their groups. Answer the above questions.

Problem II.2.f–6: Instead of a square take a rectangle. What is its invariance group? Relate this group to that of the square. Repeat this for a trapezoid. What are the group tables? Can these be obtained from the group tables of the square? How? These are subgroups. Why?

Problem II.2.f–7: Let a square (and similarly a rectangle) be the shape of a molecule with four atoms placed at the vertices; what are the groups if three (and if two) atoms are identical and the other(s) different? How many groups are there for the rectangle (trapezoid)? What is the relationship of these to the groups for four identical atoms? They are called subgroups of the group of the square. Are they?

Problem II.2.f–8: For all these subgroups verify from the group tables, and experimentally, that they are groups. Is this necessary, or does it follow from above?

Problem II.2.f–9: As with the triangle, we can consider the transformations as permutations, now on four symbols. Write all these permutations, their products, and the group table, and verify that the (abstract) group is the same of that of the square (that is, the groups are isomorphic). Identify each permutation with a transformation of the square. Does this correspondence seem reasonable? Why?

Problem II.2.f-10: Using these group tables we can check an important point; each element of any group must appear in each row, and column, of the group table and only once. If it appeared more than once, $AB = AC$, so $B = C$. Why must every element appear in each row? Does the same hold for columns?

II.2.g The double-rectangle group

That different objects (may) have different groups is obvious; but in what way do they differ? They could have unequal orders (numbers of elements). If groups have the same order though, could they be different? Why would there be a difference between objects that allow the same number of transformations?

Consider two identical (and identically colored) parallel rectangles, in different planes, with the center of one under that of the other. The covers of this book makes a reasonable, though not perfect, model. A rectangular sandwich is a better, but less familiar, one. A sheet of blank paper, considering the two sides as the rectangles, is a good example.

Problem II.2.g-1: Find all transformations, including reflections, that leave this figure invariant, both experimentally and by inserting coordinates (where?) and find the (which?) transformations on them. Compute the group table. Show that the group (D_{2h}) has eight elements. Thus this, like the group of the square (which?), is of order eight. However they are different (non-isomorphic). Prove that there is no way to relabel elements of one to make its group table correspond to that of the other. Take a double rectangle and perform on it the operations that leave invariant the square, and vice-versa (a computer-graphics program would be nice). Why are the groups not the same? Thus there can be, and usually are, more than one group for each order. We shall see, using these examples, various ways in which groups differ.

Problem II.2.g-2: What is the symmetry group if the two rectangles are squares? Is it (isomorphic to) any group we have met? Why?

Problem II.2.g-3: What are the symmetry groups for the double rectangle, and double square, if the two faces are relatively rotated?

Problem II.2.g-4: The symmetry group depends on the transformations that we wish to consider. We can obtain the effect of a reflection by moving atoms of a molecule past each other. These are not rigid-body transformations, but they can be carried out without tearing the figure (or removing an atom from a double-rectangular molecule). Not including these reduces the symmetry group. What is the resultant group? How about for the double-square? Are these related to any previously-met groups? Should they be? Is there a relationship between this set of transformations, and those including mirrors?

II.2.h The group of the cube

The obvious generalization of the square to three dimensions is the cube [Armstrong (1988), p. 37, 55; Burn (1991), p. 58, 61; Hall (1959), p. 19, 304].

Problem II.2.h-1: Find experimentally the invariance group of the cube; repeat using coordinates for each vertex, and verify that it is the same, however found. Check that the group axioms are obeyed. How is the group of the square related to it? Do this both with, and without, mirrors allowed. Give ways of eliminating mirror symmetry. Check that it includes these symmetry axes:

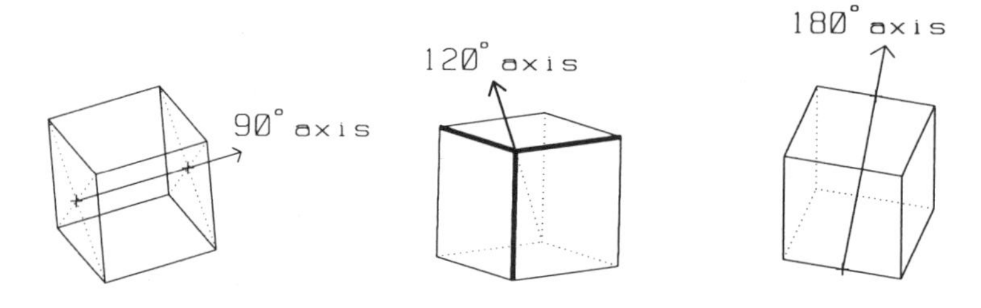

Figure II.2.h.-1: SYMMETRY AXES OF A CUBE

Problem II.2.h-2: Suppose a three-dimensional object has two faces square, and the others rectangular. What is its group? How is it related to the groups of the cube, square, rectangle?

Problem II.2.h-3: Although harder to visualize, there is a generalization of the cube to any dimension. Find the group for the four-dimensional hypercube, checking that it is a group. Repeat for five and six dimensions. Can this be done for arbitrary dimension, or only for specific values? A computer program to do it would be nice.

II.2.i Groups of given order

The order of a group does not determine it; there can be more than one. However, there are limitations. There are only a finite number of distinct groups of any order (number of transformations) since there are only a finite number of ways of inserting a finite set of symbols in a group table (and not all give groups). These groups determine the possible numbers and types of objects with symmetry, such as the square and the double rectangle. Thus, group theory has the important consequence of restricting, and enumerating, the kinds of physical objects with symmetry. It is difficult to think how such limitations can be found without group theory. The nature of physical transformations, and the associated mathematics, leads to constraints on physical objects, and provides methods for finding these and their consequences.

Problem II.2.i-1: For $n = 2, 3, 4$, give all ways of inserting n symbols into an $n \times n$ table, and show that not all, indeed few, give groups. Can you easily recognize which do, and which do not? Why?

Problem II.2.i-2: For a group with two elements, give all group tables. Do the same for groups with 3,4,5,6,7, and 8 elements [Baumslag and Chandler (1968), p. 148; Joshi (1982), p. 22; Ledermann (1953), p. 54; Lomont (1961), p. 34]. How many groups are there for each order? Notice that there is no simple pattern. This is quite time consuming. So develop a (computer-implemented) algorithm to do it. Are there general rules that follow from the output of such programs?

II.3 CONTINUOUS GROUPS

The groups considered so far are finite groups — the number of their transformations is finite. But these are not all. Some of the most important groups are continuous groups. Their operators are functions of continuous parameters — the number of their transformations is (of course) infinite, and these depend on parameters which are continuous.

II.3.a Rotation (orthogonal) groups

The most familiar groups are probably the rotation groups, in two dimensions called SO(2), in three dimensions, SO(3). The (generalization of the) rotation group in a space of dimension n is SO(n). Rotations keep the lengths of vectors and the angle between them (their dot product) invariant. Distances do not change if you turn around. Thus the group in n dimensions that keeps dot products of real vectors (like displacement or momentum) invariant is the orthogonal group (with a name to remind us that it keeps perpendicular vectors perpendicular), SO(n), or if we include inversions, O(n) — the "S" means special, not including inversions (and thus no reflections). O(n) is also the group of all $n \times n$ orthogonal matrices; for SO(n) the matrices are limited to those with unit determinant. Eventually we discuss why these two definitions give groups that are the same (isomorphic). For three space plus one time dimension, the corresponding (pseudo-)rotation group is the Lorentz group, SO(3,1). The complete Lorentz group is O(3,1). The operators of these groups are not finite in number, but depend on continuous parameters, angles, and for the Lorentz group, velocities. They are examples of continuous groups.

The rotation group is fundamental because space is isotropic ("equal turning", that is equal — the same — in all directions, no matter how an observer turns).

The group of rotations in a plane, the two-dimensional rotation group, SO(2), is a function of a single angle, θ, so has only a single operator (function), $R(\theta)$. Obviously, the product of two elements is

$$R(\theta_1)R(\theta_2) = R(\theta_1 + \theta_2). \qquad \text{(II.3.a–1)}$$

This group is Abelian; the order of rotations about the same axis is unimportant.

For three dimensions the rotation group, SO(3), is a function of three parameters; a rotation can be written as a product of a rotation by θ_x, around x, θ_y, around y, and θ_z around z (although the rotations usually chosen are somewhat different). The group is noncommutative; the order in which a series of rotations — around different axes — is carried out on an object affects the final orientation of the object.

We shall consider, besides finite rotations, infinitesimal ones — for which the θ's are infinitesimal. This gives, for SO(3), three operators, L_x, L_y, L_z, generating rotations around the three axes, and in addition the operator

$$L^2 = L_x^2 + L_y^2 + L_z^2. \qquad \text{(II.3.a–2)}$$

These are also (up to a constant necessary only because of the choice of units), the three components of angular momentum, and the operator giving the square of the angular momentum (properly, their eigenvalues give the values for these). The question why these operators, which transform one function to another, should give physical properties of an object is interesting (sec. I.6.d, p. 27); we consider them later.

Problem II.3.a–1: Take this book and rotate it about an axis along a side, then about an axis through the center (of the side) perpendicular to the cover. Notice the orientation. Now perform these two rotations in the opposite order. Compare the two final orientations. Do this for axes along the bottom and one of the sides. Do rotations about different axes commute? Rotate about a single axis by $\pi/12$ then by $\pi/6$, then starting from the original orientation around the same axis in reverse order. Do rotations about the same axis commute?

Problem II.3.a–2: Why are the groups that keep orthogonal vectors orthogonal and of the same length, and of $n \times n$ orthogonal matrices, the same? Verify that the latter do preserve lengths and angles. How do the groups of matrices isomorphic to O(n), and that isomorphic to SO(n), differ?

Problem II.3.a-3: It is always useful to prove that the sets that we have been calling groups, are actually so. Do this with the rotation groups, experimentally, and using the groups of orthogonal matrices, first for SO(2), and O(2), then SO(3), and O(3), then, if possible, in general.

II.3.b The SU(2) group

While the orbital-angular-momentum quantum number has only integer values, there is another kind of angular momentum, spin, with half-integer values. This implies, correctly, that there is a group closely related to SO(3) — SU(2), the special unitary group in two (complex) dimensions. We discuss eventually what this relationship is (chap. X, p. 269).

II.3.c The isospin group

A nucleon can have its spin rotated, but in addition, neutrons and protons can be turned into each other. Another quantum number is needed to describe them, isotopic spin (properly, isobaric spin; we avoid the name problem by calling it isospin). The nucleon has, besides rotational invariance, another group of (almost) symmetries, the isospin group. Isospin quantum numbers have half-integer values. This suggests, correctly, that there is a relationship between this group and SU(2). Physically they have (very?) different meanings, mathematically they are identical (except that we use different symbols for their operators); the isospin and SU(2) groups are isomorphic. The rotation group and SU(2) are related, but are not quite the same; they are homomorphic (have like shapes).

II.3.d Unitary groups

In quantum mechanics the probability is given by the absolute square of the statefunction, and the transition amplitude by the dot product (also called, quite reasonably, scalar product) of two statefunctions. These must remain invariant under symmetry transformations and the set of transformations which keeps the dot product of two complex n-dimensional vectors (like statefunctions) invariant is the unitary group (because it keeps the unit, the norm, invariant) SU(n), or if an overall phase factor is allowed, U(n).

A unitary matrix is one whose inverse equals its transposed complex conjugate:

$$U_{ij}^{-1} = U_{ji}^{*}. \tag{II.3.d-1}$$

The group U(n) is the group of all $n \times n$ unitary matrices; if the matrices are limited to have unit determinant, the group is SU(n).

Problem II.3.d–1: Why is the group of all $n \times n$ unitary matrices the same as the unitary group acting on complex vectors? Verify that it is. Check that these are actually groups, for both U(n) and SU(n).

II.3.e The Lorentz group

The (pseudo-)rotation group SO(3,1) is called the Lorentz group. It is the group of transformations, keeping a point fixed, of spacetime with three space and one time dimension (that is of relativity). It consists thus of all rotations around the three axes plus all boosts (changes of velocity) along these axes. It is a (pseudo-)rotation group because it acts in a space with metric of signature 3+1.

Problem II.3.e–1: The matrices [Ledermann (1953), p. 26]

$$A(v) = \beta^{-\frac{1}{2}} \begin{pmatrix} 1 & -v \\ -v/c^2 & 1 \end{pmatrix}; \tag{II.3.e-1}$$

$$\beta = 1 - v^2/c^2, \tag{II.3.e-2}$$

where c is a constant (called?), and

$$-c < v < c, \tag{II.3.e-3}$$

form a group. Show that this is true using the product rule

$$A(v) = A(v_1)A(v_2); \tag{II.3.e-4}$$

this is ordinary matrix multiplication, with

$$v = \frac{v_1 + v_2}{1 + v_1 v_2/c^2}. \tag{II.3.e-5}$$

Readers familiar with special relativity will recognize these relations (and so their relationship to the Lorentz group). This shows the generality of the concept of product. What would happen if v were outside the range indicated? Would these matrices still form a group? Could a set of matrices (like these) be found that forms a group?

II.3.f The translation group

Besides rotating and boosting coordinate systems, we can also translate them, that is move the origin. Translations form a group with four parameters (in 3+1-space) giving the four components of the displacement vector. The reader will not change if he moves to a different point in space, or a different point in time. Translations form an invariance group of space.

Problem II.3.f-1: Of course the last is different, since most people do change when growing old. What is a correct example?

II.3.g Time translations

Growing old does change things (the book, the reader and unfortunately the author), but not (much) over a very short time. However what really matters is that the laws of physics do not contain the time; presumably the laws of physics do not grow old (we ignore the expansion of the universe; but this is a change in a material object — the universe).

This discussion of growing old hints that a more precise definition of time translation is needed. We consider two identical observers who perform the same experiment on identical systems at different times (thus it does not matter whether the observer grows old as long as he can continue to do the experiment properly). The transformation from one observer to the other is a time translation. This transformation (unlike growing old) has an inverse. We can take the observations of an earlier observer and then time translate by t to find those of a later observer, or can consider those of the later observer and translate by $-t$ to find those of the earlier one.

Problem II.3.g-1: Define a space translation. What is its inverse? In a very strong wind will a space translation have an inverse?

II.3.h The Poincaré group

All that the reader can do without changing himself is turn around (about an axis that has to be specified by its three components), step unto a moving platform (with velocity specified by its three components), like a car or train, move (the displacement is specified by three components) and grow old (one component) — really only change the zero of his clock. These give ten transformations. The important point is that these change the reader in a way that with his eyes closed he would not notice (provided he did not pay attention while the transformations were being performed); the only change is in the relationship between the reader and other material objects. But he would notice if he were given an acceleration.

Thus on (homogeneous, isotropic) spacetime (that is on a coordinate system in space) we can perform ten transformations that leave the laws of physics invariant, the six of the Lorentz group plus four translations along any of the four axes. We can also perform other transformations, such as to accelerating frames. But these do not leave the laws of physics invariant — laws seen by an accelerated observer (seem to) differ from those seen by an inertial one.

This group of ten transformations is called the (proper) inhomogeneous Lorentz group or the (proper) Poincaré group [Mirman (1993, 1994, 1995b)]. Because it is the invariance group of space it is of the most fundamental importance [Mirman (1995a)]. (The complete Poincaré group includes reflections).

Homogeneous ("generated like") means of a similar kind, or of a single type. Space is homogeneous because all points are the same; they are all generated like each other. A homogeneous group has all operators the same (in some sense). Inhomogeneous (not homogeneous) means the operators differ; the inhomogeneous Lorentz group (the Poincaré group) is inhomogeneous because translations are included in the group, else it would be the homogeneous Lorentz group. The transformations of the Lorentz group are all of the same kind; they do not commute with each other. Translations do commute with each other (but not with all homogeneous transformations), so the inhomogeneous group contains transformations that are not all of a similar kind.

Problem II.3.h-1: Show that the rotation, Lorentz, translation and Poincaré groups are actually groups, and that the first three are subgroups of the last, and that the rotation group is a subgroup of the Lorentz group. What are the inverses for each of these transformations? Check experimentally, and algebraically (as far as possible), that the group axioms, including associativity, are obeyed. These are groups of mathematical operations, and groups of physical ones. What has to be checked to show that the mathematical operations form a group, and what to show that the physical (or geometrical) ones do?

II.3.i Inversions

There are transformations that may appear esoteric, although in studying the next set having a mirror should eliminate that feeling. Besides rotations and translations, another transformation of space is mirror reflection; actually reflection in a plane, for the operation is geometrical, and only secondarily physical — though for greater visualizability we often refer to mirrors, with planes understood. Some objects (can) remain invariant under reflection (spheres, obviously), some not (the words printed here, or hands).

Problem II.3.i-1: Compare apples and oranges.

Since reflections are in a plane there are three (and combinations), in the x, y, and z planes. Applying all together gives an inversion (through the origin), the coordinates of each point are replaced by their negatives; for an object, each (infinitesimal) piece at x, y, z is replaced by the piece whose coordinates are $-x, -y, -z$.

What about the laws of physics, are they invariant? What about an atom, is its mirror image the same, and if so what does that say about nature and the nature of the atom; and if it is not the same? Until 1957 it was believed that the laws of physics are inversion invariant. Most are, but not all. Both the invariance and the noninvariance have important effects. To use the proper terminology (which is defined when these transformations are discussed) parity (given by the operation labeled P) is conserved, except by weak interactions (parity, because the object and its reflection, being on par with each other, form a pair).

There is another kind of inversion, the interchange of matter and anti-matter. This is closely tied to the conservation (and nonconservation) of parity. The laws are mostly invariant under this inversion, charge conjugation, C, but not all.

The third kind of inversion is time reversal, T, an appealing idea to those of us who are growing old, but unfortunately that is not what is meant. It means that the (fundamental) laws of physics are invariant when the sign of the time is changed. Newton's second law involves the acceleration, the second derivative with respect to time. There is no change if we substitute $-t$ for t. Schrödinger's, and Dirac's, equations involve the first derivative of the time, but they too are invariant, though the meaning of time reversal is more subtle, so has to wait for a full discussion. Physical laws are invariant under time reversal, with a few (experimental) exceptions.

We can also apply all three inversions at once giving the inversion TCP (or CPT or PTC, or whatever order we like — the law is invariant under the group of permutations of these symbols). Fortunately all physical laws seem invariant under it. Fortunately, because there is no known way of obtaining a physical theory that is not invariant, since it is a consequence of Lorentz invariance [Mirman (1995b)]. If this invariance were experimentally violated, it would be highly unpleasant. So far, so good.

Problem II.3.i-2: There are objects that remain invariant under reflections in all planes through some point, and inversion (obviously not many), and ones that are unchanged by reflection in some planes, but changed by reflections in others. Take a few of the objects mentioned above. For each, find the planes, if any, for which reflection is a symmetry, stating the maximum number (not including combinations). Which objects are invariant under inversion through some origin? Is there any object that is unchanged by inversion no matter what the origin?

Problem II.3.i-3: Show (experimentally) that the product of two (different) inversions is a rotation. What rotation? How do they differ? Is this true of reflections?

Problem II.3.i-4: In a mirror right and left are interchanged, but not top and bottom (when you look in a mirror you do not see yourself standing on your head). Why? Is it possible to find a mirror that reverses top and bottom, but not left and right? How about one that does both? Can you find a mirror that reverses the direction of any (given) vector? Write its equation in terms of that of the vector. Take two vectors; is there any mirror that reverses both of them?

Problem II.3.i-5: It should be clear that the group of four inversions (even including time reversal) is a group. Show that the complete rotation, Lorentz and Poincaré groups, the groups including all inversions are groups. Which are subgroups of which?

II.3.j Infinite discrete groups

The number of elements can be infinite, but discrete; then the group is an infinite group [Baumslag and Chandler (1968), p. 51; Grossman and Magnus (1992), p. 15, 4 6, 71, 80; Landin (1989), p. 67; Ledermann (1953), p. 7; Stephenson (1986), p. 138]. To make life more interesting there are also mixed groups, combinations of these. The group of rotations plus reflections is a mixed group, having a continuous (rotation) subgroup plus a finite (reflection) subgroup. These are important and are discussed later.

Problem II.3.j-1: Show that the integers form a group, which is discrete and infinite, with the product being addition. What is the unit element? If the product is multiplication would the integers form a group? For these products, do the rationals form a group? The reals? What are the answers if only positive numbers were considered? Which of these groups are infinite and discrete? Vectors form an infinite discrete group (with what law of combination?). Is this result new, of was it stated in a previous sentence? For these groups, give the unit elements, and the inverse of each element. Prove that they are actually groups.

II.4 GROUPS OF MATHEMATICAL OBJECTS

While physical objects help in visualizing group operations, mathematical ones indicate the breadth of the theory; also in developing the theory of groups (as with other mathematical systems) it is useful to consider their effect on abstract entities. So here we turn to those groups, many the same as those given above though perhaps with different names and notations, whose transformations we interpret as acting on mathematical symbols.

II.4.a Abelian groups

The elements of a group need not commute; the order of terms in the product is important. The product of rotations about different axes (for the rotation group), depends on the order. We apply the transformation written on the right first, and then apply the transformation on the left to the result (sec. I.4.a, p. 14); $T(\theta)T(\theta')$ says apply $T(\theta')$ (to the statefunction) and to the result apply $T(\theta)$. For the rotation group, we rotate through θ' first, and then rotate the resultant object through θ — about, in general, a different axis.

If all group elements commute,

$$T(\theta)T(\theta') = T(\theta')T(\theta), \qquad \text{(II.4-1)}$$

the set of $T(\theta)$'s is called an Abelian group [Baumslag and Chandler (1968), p. 177]. SO(3) is non-Abelian.

An example of an Abelian group is the group of translations; these commute. This is a continuous group, the translations being functions of continuous parameters. The water group is also Abelian.

Problem II.4.a-1: Which groups discussed above are Abelian?

II.4.b Cyclic groups

A cyclic group has elements (e is the unit)

$$e, a, a^2, a^3, \ldots, a^n = e \qquad \text{(II.4.b-1)}$$

(the elements go in a cycle or circle) [Baumslag and Chandler (1968), p. 101; Grossman and Magnus (1992), p. 42, 56; Landin (1989), p. 71]. These groups are thus generated by a single element plus all its powers up to the $n-1$'th, with the n'th equaling the unit element (where "generated by" means that all elements can be written as products of a few, here one, the generators [Grossman and Magnus (1992), p. 41]). The length of the cycle is n. So for $a^2 = e$, the cycle length is 2. A cyclic group is determined by a single parameter, its cycle length.

Problem II.4.b-1: Show that a cyclic group is Abelian. Also, the cycle length is the order of the group (the number of elements). Prove that a cyclic group is actually a group. What is the inverse of an element?

Problem II.4.b-2: Prove that any element, a, of any finite group, plus all its powers forms a cyclic group. So for any finite group, a transformation applied enough times gives the original state ($a^k = e$, the identity, and k is called the order of element a).

Problem II.4.b-3: For the square group (sec. II.2.f, p. 37), show that each element gives a cyclic group (try it for each of the groups with this name). Do this both from the group table and by carrying out the transformations experimentally. Repeat this for the group of the double rectangle (sec. II.2.g, p. 39). What is the set of cycle lengths for these groups? Are they the same? What information can you get by comparing them? Can the two groups be isomorphic? Can you tell from the geometry, by looking at the objects, what the cycle lengths are, and that they are different?

Problem II.4.b-4: Is there a rule giving all cyclic groups of given order?

Problem II.4.b-5: Now that we have simple examples of Abelian and cyclic groups, the reader can show that not all Abelian groups are cyclic by writing the group table for several that are not (the smallest has order four). It would be more fun to find an algorithm or rule to find the group table for all Abelian groups of given order. If this can be done than the next step is to write a computer program based on it and actually compute group tables. Finding broad rules implied by the results would be interesting.

Problem II.4.b-6: That an element gives the cyclic group is true for finite and for compact continuous groups such as that of rotations (compact groups are groups for which this is true; a better definition is not needed here). It is not true for noncompact groups (for example the Lorentz group), for translations or generally for a continuous Abelian group. Give an example of a continuous Abelian group. Prove that every element of all rotation groups forms a cyclic group while no translation does. Physically why do we expect that not all elements of the Lorentz group form cyclic groups? Which do? Which do not?

II.4.c Dihedral groups

The simplest examples of dihedral groups [Armstrong (1988), p. 15; Baumslag and Chandler (1968), p. 75; Burn (1991), p. 30, 37, 67, 75; Grossman and Magnus (1992), p. 54, 69; Hall (1959), p. 19; Jones (1990), p. 7; Lomont (1961), p. 32, 78; Tung (1985), p. 13; Aivazis (1991), p. 3] are the group of the rectangle, D_2 (pb. II.2.f-3, p. 37), and the group of the triangle and the double-pyramidal group, D_3 (pb. II.2.e-2, p. 36). The D stands for dihedral ("two sides"); for the triangle there is an axis (the z axis) of three symmetry operations, and also a $\frac{\pi}{2}$ rotation interchanging its "two sides", that leave it invariant. It has trigonal ("three corners" or "three angles") symmetry. These can be generalized to D_n, for each (integer) n, explaining its subscript.

Problem II.4.c-1: Prove, experimentally, and mathematically, that D_n

is a group, and, for $n > 2$, is not Abelian. What are the group tables (for several values of n)?

Problem II.4.c-2: Show that the group of the square (sec. II.2.f, p. 37) (which?) is (isomorphic to) D_4.

Problem II.4.c-3: For D_2, following the symbols of pb. II.2.e-2, p. 36, and defining C_2' and σ, check that the group table is

	E	C_2	C_2'	σ
E	E	C_2	C_2'	σ
C_2	C_2	E	σ	C_2'
C_2'	C_2'	σ	E	C_2
σ	σ	C_2'	C_2	E

Table II.4.c-1: Group Table for D_2

Problem II.4.c-4: For D_3, the group table (using a reasonable interpretation of the subscripts, or none) should be

	E	C_3	C_3^2	σ_v	$\sigma_{v'}$	$\sigma_{v''}$
E	E	C_3	C_3^2	σ_v	$\sigma_{v'}$	$\sigma_{v''}$
C_3	C_3	C_3^2	E	$\sigma_{v''}$	σ_v	$\sigma_{v'}$
C_3^2	C_3^2	E	C_3	$\sigma_{v'}$	$\sigma_{v''}$	σ_v
σ_v	σ_v	$\sigma_{v'}$	$\sigma_{v''}$	E	C_3	C_3^2
$\sigma_{v'}$	$\sigma_{v'}$	$\sigma_{v''}$	σ_v	C_3^2	E	C_3
$\sigma_{v''}$	$\sigma_{v''}$	σ_v	$\sigma_{v'}$	C_3	C_3^2	E

Table II.4.c-2: Group Table for D_3

Problem II.4.c-5: Show that U, V, and of course E, subject to

$$U^2 = V^k = (UV)^2 = E, \tag{II.4.c-1}$$

form a group, dihedral group D_k, where $k \geq 3$ (why?) [Dixon (1973), p. 5]. Check that this gives the table for D_3 [Baumslag and Chandler (1968), p. 76].

Problem II.4.c-6: Prove that the set of operations on integers taking one to any other, plus that of replacing integers by their negatives, is a group. This is a dihedral group, but with an infinite number of (discrete) elements, the infinite dihedral group, D_∞ [Armstrong (1988), p. 23; Grossman and Magnus (1992), p. 71].

Problem II.4.c-7: Which dihedral groups are cyclic groups?

II.4.d Symmetric groups

The triangles, squares and so on that we have been discussing can be considered as rigid objects, perhaps toys with balls at the vertices and sticks connecting them. The transformations are then rigid-body rotations. They can also be considered as molecules for which the sticks do not exist. So besides viewing the transformations as on a rigid body, we can regard them as a shuffling of objects at the vertices, atoms, say. There are other objects besides atoms that we might shuffle, such as indices. Thus for electron "a" with angular momentum l_1 and electron "b" with angular momentum l_2 the statefunction is $\psi_1(a)\psi_2(b)$. But the electrons are identical so another one is $\psi_1(b)\psi_2(a)(= \psi_2(a)\psi_1(b))$. Thus we get one statefunction from another by shuffling, or more formally permuting, indices, either a, b or 1,2. These hint that the shuffling transformations, permutations, are important. And they show some, but only some, of the reasons.

We define, S_n, the symmetric group on n objects [Baumslag and Chandler (1968), p. 56; Grossman and Magnus (1992), p. 141; Maxfield and Maxfield (1992), p. 49], as the set of all permutations of these objects. (Permutation group would be better, but symmetric group is the established name; symmetric — "like measure" — a word which applies to many things, especially in group theory, suggests invariance, which need not hold, permutation — "move through" — is less general, and has less of a misleading connotation. Also a permutation is a transmutation — "move across" — but a specific type.)

We can consider the objects lined up and then a permutation rearranges the objects in the line:

$$P = \begin{pmatrix} 123\ldots n \\ 421\ldots 7 \end{pmatrix}, \qquad \text{(II.4.d-1)}$$

for example. But the top row is really unnecessary, so we write the permutations by giving in parentheses the new ordering of the symbols, it being understood that 1 (that is the object, atom, symbol, index, ..., labeled 1) is replaced by the first symbol in the parentheses, 2 by the second, and so on. For P, 1 is replaced by 4, 2 by 2, 3 by $1, \ldots, n$ by 7. Permutations may mix only subsets; so we write, say, (147)(25)(6), or usually (147)(25). This says that 1 is replaced by 4, 4 by 7 and 7 by 1. Also 2 and 5 are interchanged and 6 remains where it was (so need not be shown). In general $(ijk\ldots s)$ means replace i by j, j by $k, \ldots s$ by i. If there are l numerals in the parentheses this permutation is said to be of cycle length l.

Problem II.4.d-1: This notation gives a result which looks different from that at the beginning of the paragraph; in what way? Does it matter?

Problem II.4.d-2: A cycle is defined as a permutation that mixes a single set of numerals, (1379). The permutation (136)(24) is a product of two cycles. A general permutation then is written as a product of cycles with b_1 having length l_1, b_2 with length l_2, ..., b_k of length l_k (with cycles of length 1 generally suppressed), so for S_n,

$$b_1l_1 + b_2l_2 + \cdots = n. \qquad \text{(II.4.d-2)}$$

Why is the permutation written as a product? Verify that if instead of applying a single permutation we applied first the permutation given by the first cycle, then that given by the second, and so on, we would get the same final arrangement of objects as that given by the single permutation.

Problem II.4.d-3: Find (12)(23), and (23)(12), and notice that they are not the same. The permutations do not commute; symmetric groups are non-Abelian.

Problem II.4.d-4: Write as products of (disjoint) cycles, and give the order of each (operation P has order k if P^k = the identity):

$$\begin{pmatrix} 1234567 \\ 4213657 \end{pmatrix}, \begin{pmatrix} 123456789 \\ 243915678 \end{pmatrix}, \begin{pmatrix} abcdef \\ fbadce \end{pmatrix}, \begin{pmatrix} abcdefghi \\ geiahbdcf \end{pmatrix}. \qquad \text{(II.4.d-3)}$$

Problem II.4.d-5: Show that cycles that do not share a numeral, say (165) and (278), commute, while cycles that share one numeral, like (123) and (246), do not commute. Suppose they share more than one?

Problem II.4.d-6: A permutation of the form (ij), that is a cycle of length 2, is a transposition (it transposes the two symbols — it places their positions across each other). If i and j are neighboring, say 4 and 5, or just objects next to each other in the original line (two ways of saying the same thing), then (ij) is a neighboring transposition. For S_n, show that there are $n-1$ neighboring transpositions. How many transpositions are there all together?

Problem II.4.d-7: Prove that transpositions form a cycle (thus a cyclic group) of length 2, and a permutation with n numerals in one pair of parentheses forms a cycle of length n. What is the cycle length of one with l_1 numerals in the first pair of parentheses, l_2 in the second, ...?

Problem II.4.d-8: Prove that there are $n!$ distinct permutations of n objects, so that S_n has n! elements.

Problem II.4.d-9: Show that the group axioms are obeyed for symmetric groups.

Problem II.4.d-10: The square group (sec. II.2.f, p. 37) and the group of the double rectangle (sec. II.2.g, p. 39) are subgroups of symmetric groups (of course?). Do their cycle lengths give information about which symmetric-group elements the transformations of these groups correspond to?

Problem II.4.d–11: Instead of using the elements given, consider linear combinations, say $p_1 = (12) + (132), p_2 = (23) - (123) - (132)$, and so on. What does the "+" mean here? Find the group table. Which set of elements do you prefer? Why? Are the group axioms obeyed for these sums? Do they form symmetric groups?

II.4.d.i *The ordering convention*

A convention for ordering permutations in a product is needed. For $(abc\ldots)(xyz\ldots)T_1$, we take $(xyz\ldots)$ to act first, giving

$$T_2 = (xyz\ldots)T_1, \qquad \text{(II.4.d.i–1)}$$

and then $(abc\ldots)$ acts on this, so

$$T_3 = (abc\ldots)T_2. \qquad \text{(II.4.d.i–2)}$$

The opposite convention — in which $(abc\ldots)$ acts first — is also used, so care is needed in comparing different discussions.

Problem II.4.d.i–1: What property of symmetric groups necessitates an ordering convention?

II.4.d.ii *Alternating groups*

Of the subgroups of S_n there is one closely related to it, the alternating group A_n [Baumslag and Chandler (1968), p. 60; Grossman and Magnus (1992), p. 146; Hall (1959), p. 59; Hamermesh (1962), p. 14; Kurosh (1960a), p. 39; Landin (1989), p. 94; Ledermann (1953), p. 77; Maxfield and Maxfield (1992), p. 75]. Its order is half that of S_n, that is $n!/2$. It consists of all permutations on n objects that can be written as a product of an even number of transpositions. The elements of A_3 are

$$E = (1)(2)(3), (123), (132), \qquad \text{(II.4.d.ii–1)}$$

differing from S_3 in not including the transpositions.

Problem II.4.d.ii–1: Give the elements of A_2, A_3 and A_4, and check that their orders are correct. Is A_n really a group?

II.4.e The group of the roots of unity

There are n n'th roots of 1,

$$r_2 = exp(2\pi ik/n), \quad k = 0, 1, \ldots, ?. \qquad \text{(II.4.e–1)}$$

Problem II.4.e–1: Prove that these roots form a group [Baumslag and Chandler (1968), p. 53; Maxfield and Maxfield (1992), p. 189]. What is the product? What are the multiplication tables? Are any of these isomorphic to groups that we have discussed? Are these cyclic groups?

II.4.f The groups of numbers

Since numbers can be combined, additively and multiplicatively, they should form groups. Under what conditions do they do so?

Problem II.4.f-1: Which of the following sets of numbers form groups, under addition; under multiplication [Armstrong (1988), p. 11; Baumslag and Chandler (1968), p. 51; Falicov (1966), p. 9; Joshi (1982), p. 25; Kurosh (1960a), p. 37]? For those that do not, what goes wrong, which group axioms are not obeyed?

a) the integers, all, all positive, all nonzero?
b) the even integers, all, all positive, all nonzero?
c) the odd integers, all, all positive, all nonzero?
d) the rational numbers, all, all positive, all nonzero?
e) the prime numbers, all, all positive, all nonzero?
f) the non-integers, all, all positive, all nonzero? Any?
g) the real numbers, all, all positive, all nonzero?
h) the complex numbers, all, all nonzero?
i) the complex numbers of constant absolute value?
j) the complex numbers, written $Rexp(i\phi)$, for constant ϕ?
k) the numbers 1, -1, i, $-i$?
l) the numbers 1, -1, ai, -ai, where a is some number?

m) the integers, all, all positive, all nonzero, with the product multiplication (and similarly addition) modulo some integer [Grossman and Magnus (1992), p. 22]? What would happen if this were not restricted to integers? Modulo, from the Latin word for measure, means, in effect, that the measurement goes around in circles. Thus if we consider multiplication modulo n, and undoubtedly we want this to be an integer, then $ab = c$, if $c < n$, and $ab = c - kn$, where k is chosen so $c - kn < n$. So for $n = 5$, $2 \times 2 = 4$, but $2 \times 3 = 1$. Addition is similar. Clock arithmetic is a good example [Maxfield and Maxfield (1992), p. 72].

Problem II.4.f-2: Do any (all) of these form groups if the product is division? Why is there a difference, if there is, between multiplication and division?

Problem II.4.f-3: Can any of these sets that do not form groups be extended to groups (by adding elements in some "reasonable" way)? If they do not form groups, are there subsets that do?

Problem II.4.f-4: The set of numbers $(cos\theta + isin\theta)$ for all θ, might be expected to form a group (with what product rule?). However show that it (still) forms a group if θ is restricted to be rational.

II.4.g Matrix groups

A set of matrices that obeys the group axioms forms a group, called a matrix group. We can give each matrix a symbol and write a multiplication table for the symbols that is the same as the one for the matrices, so the set of these symbols forms an abstract group. The matrices are called a representation of the abstract group. The law of combination is matrix multiplication (so emphasizing that the word "multiplication" in the group axioms refers to different ways of combining objects). Since the representations of a group are the physically important objects these are discussed more fully below.

Numbers and matrices may not obviously be transformations. But of course they are, on mathematical objects. However, these mathematical objects can represent physical ones. So there is a relationship, and we will see that it is of fundamental importance, between these numbers (matrices are numbers) and physical transformations.

Problem II.4.g-1: Do the Pauli matrices, plus the identity,

$$E = \begin{pmatrix} 1 & 0 \\ 0 & 1 \end{pmatrix}, \sigma_z = \begin{pmatrix} 1 & 0 \\ 0 & -1 \end{pmatrix}, \sigma_+ = \begin{pmatrix} 0 & 1 \\ 0 & 0 \end{pmatrix}, \sigma_- = \begin{pmatrix} 0 & 0 \\ 1 & 0 \end{pmatrix}, \tag{II.4.g-1}$$

form a group? If not, is there a group that includes these as elements? Try this also using, instead of σ_+ and σ_-,

$$\sigma_x = \sigma_+ + \sigma_-, \quad \sigma_y = i(\sigma_+ - \sigma_-). \tag{II.4.g-2}$$

What would the situation be if all elements in the σ's were replaced by $i(= \sqrt{-1})$? How about if only those in σ_+ and σ_- were? Would it be possible to replace only some by i and get a group? Suppose they were replaced by arbitrary numbers?

Problem II.4.g-2: Which of the following form groups, using matrix multiplication as the product? The set of all

a) matrices?
b) nonsingular matrices?
c) singular matrices?
d) orthogonal matrices?
e) unitary matrices?
f) hermitian matrices?
g) symmetric matrices?
h) invertible matrices with integer elements?
h) invertible matrices with rational elements?
i) invertible matrices with positive elements?

j) diagonal matrices?

k) diagonal matrices with nonzero elements?

l) matrices all of whose diagonal elements are zero?

m) triangular matrices (those with all nonzero elements below, or above, the diagonal)? Suppose there are nonzero elements on the diagonal also?

n) matrices with constant determinant (say 1)?

o) matrices with constant trace?

p) any of the above with constant determinant?

q) any of the above with constant trace?

Problem II.4.g-3: Do any of these form groups using matrix addition as the combination rule?

Problem II.4.g-4: For any sets that do not form groups, are there subsets that do? Are there any that can be extended (simply) to groups?

Problem II.4.g-5: A correspondence can be set up between types of numbers and types of matrices. So, for example, hermitian matrices correspond to real numbers. For each type of matrix listed, is there a corresponding type of number, and conversely? For each type matrix that forms (does not form) a group, does the corresponding type of number form a group, and conversely? Why? Need their product rules be related? For each of these types of numbers and matrices, give the unit elements, and the inverse of each element. Prove that they are actually groups. Which are finite? Which are infinite and discrete? Which are continuous? Are there any significant differences between the answers for numbers and for matrices? For addition, and for multiplication, as the combination rule? Why?

Problem II.4.g-6: Show that the matrices (Ledermann (1953), p. 26)

$$\begin{pmatrix} 1 & 0 \\ 0 & 1 \end{pmatrix}, \begin{pmatrix} \omega & 0 \\ 0 & \omega^2 \end{pmatrix}, \begin{pmatrix} \omega^2 & 0 \\ 0 & \omega \end{pmatrix}, \begin{pmatrix} 0 & 1 \\ 1 & 0 \end{pmatrix}, \begin{pmatrix} 0 & \omega^2 \\ \omega & 0 \end{pmatrix}, \begin{pmatrix} 0 & \omega \\ \omega^2 & 0 \end{pmatrix}, \tag{II.4.g-3}$$

form a group; $\omega^3 = 1, \omega \neq 1$. Identify it. What is it if $\omega = 1$?

Problem II.4.g-7: Besides square matrices we can consider row and column matrices — vectors. Do vectors, in two, three, ..., dimensions form groups with the combination rule addition; the scalar (dot) product; the vector (cross) product; the tensor product? The tensor product of two vectors with components V_i and W_j, has components

$$T_{ij} = V_i W_j = W_j V_j. \tag{II.4.g-4}$$

What are the identities of such groups? If (any of) these do not form groups can they be extended (by adjoining elements) to groups?

II.4.h The group of a polynomial

Physical objects are described by (wave)functions of space coordinates so transformations can (really must) act on these. There are other types of transformations. To illustrate this and show that ways of getting groups are limited only by the reader's imagination, we take examples of groups that leave functions invariant [Joshi (1982), p. 28].

Problem II.4.h-1: Consider the functions

$$f = x_1x_2 + x_3 + x_4, \quad F = x_1x_2 + x_3x_4. \tag{II.4.h-1}$$

Find the sets of operators that leave f, and similarly F, invariant. Show that these form groups. Must they? These groups consist of subsets (why?) of operators of S_n (sec. II.4.d, p. 51) obeying the product rules given by the S_n group table (they are subgroups of S_n). Give the operators of S_n that make up the subgroups. Which subgroup is larger? Why? Does that contain the smaller one as a subgroup of itself? Why? How are the answers changed if the x's are noncommuting (matrices, say)?

Problem II.4.h-2: Are there any functions (say, polynomials) in the coordinates that are left invariant by all, or by some of, the transformations of the group of the square (sec. II.2.f, p. 37)? Do, must, these form a group (a subgroup)? What is its multiplication table? Is it a group that we have met so far? Repeat this for the pyramidal group (sec. II.2.e, p. 36).

Problem II.4.h-3: Since these objects can be taken as molecules or crystals, their statefunctions are of interest. One might wonder how the difference in group multiplication rules affects these functions. Compare (some) functions (particularly polynomials) in the coordinates (which have to be defined) left invariant by all (symmetry) transformations of the square, and of the double rectangle (sec. II.2.g, p. 39). Is there any difference or relationship between these two sets of functions? In studying statefunctions we also have to consider sets that are related by the group, not merely left unchanged.

Problem II.4.h-4: It is interesting to see how answers differ, if they do, for different dihedral groups (sec. II.4.c, p. 50), for these problems. Starting with the group D_2, for as many dihedral groups as possible, find if there are functions (particularly polynomials) of coordinates that are invariant under all transformations.

II.4.i Functions forming groups

The concept of a transformation is very general. We have an intuitive idea of what it means to rotate or interchange corners or indices on a function. And the idea of transforming using matrices should also be familiar. But to prevent any misconceptions that these are the only type, or the only ways of getting groups, we consider some functions [Joshi (1982), p. 28; Ledermann (1953), p. 14].

Problem II.4.i-1: Take

$$f_1(x) = x, \quad f_2 = 1 - x, \quad f_3 = \frac{x}{x-1},$$

$$f_4 = \frac{1}{x}, \quad f_5 = \frac{1}{1-x}, \quad f_6 = \frac{x-1}{x}. \tag{II.4.i-1}$$

The product is defined as the function of the function;

$$f_1 f_2 = f_1(f_2(x)). \tag{II.4.i-2}$$

So

$$f_3 f_4 = \frac{\frac{1}{x}}{\frac{1}{x} - 1}. \tag{II.4.i-3}$$

Work out all products, find the group table, and prove that it is a group. What is the unit element? The inverse of each element? It has six elements. Thus you would suspect it is isomorphic to what group? Prove that it is.

II.4.j Vector Spaces

For group theory the concept of a vector space is needed. We do not give a rigorous definition. Rather we just regard it as a set of functions (or more generally, symbols), $f_1, f_2, \ldots$, such that $\Sigma a_i f_i$ is defined and meaningful (and do not consider whether this last word is redundant) for all reasonable sets of numbers a_i. Undoubtedly someone can cause a lot of trouble by thinking of unreasonable sets, but that will definitely not occur in this book. Besides adding vectors we also have to take their scalar product; we do not define it except when we consider specific examples.

A maximal set of linearly independent vectors is a basis of the space (every vector can be written as a sum of these, and no vector of the set can be written as a sum of the others).

Problem II.4.j-1: Check that the following form vector spaces: displacements; velocities; forces; all continuous square-integrable solutions of an n-th order ordinary linear homogeneous differential equation; the set of all real (and likewise complex) square matrices of each dimension.

Problem II.4.j-2: Prove that a group (of physical transformations) gives a vector space, where the f's are the group operators and it is an interesting exercise to consider what is meant by addition of group operators. This problem can be done by considering the groups discussed and defining addition for each. Do abstract groups give vector spaces? Must they?

Problem II.4.j-3: Vector spaces are important in group theory because groups act on vectors. But there is another reason. Show that a vector space is a group. What is the law of combination? The identity?

Problem II.4.j-4: It might be wondered what type of space is not a vector space. Is the space we live in a vector space? Why? Would there be a difference if the space were a (hyper)sphere? Suppose space were extended to include the wind velocity at each point? How enjoyable would it be to live in a vector space?

II.5 THE GROUPS OF PEOPLE AND OTHER FAMILIAR OBJECTS

Not only is symmetry all around us, it is in us. The human body, its organs, and many other familiar objects, have symmetry. We can enlarge the concept of a symmetry group, one leaving the object unchanged, to a revision group, one transforming the object into a reasonably similar object (unfortunately this term is never used). So there are many everyday things whose groups are interesting.

In the following find the groups — the sets of transformations (which transformations?), their combination rules, and their group tables, and determine whether these are symmetry groups or revision groups, and find all groups (discussed) to which they are isomorphic.

Problem II.5-1: What is the group of a bicycle wheel? What is the group of a bicycle?

Problem II.5-2: What are the groups of an orange, an apple, a banana, a pear, a starfruit?

Problem II.5-3: Does an egg have a symmetry group? Does a chicken? Is there a relationship?

Problem II.5-4: Can a screw have a group?

Problem II.5-5: A hard rubber ball, as far as we can tell visually, a perfect sphere, has symmetry group O(3), the rotation group. Show that this is the group of the sphere. Other balls have less obvious groups. Does a baseball have a symmetry group? A bat? Are there groups of symmetries, even approximate ones, of a basketball? A football?

Problem II.5-6: The reader can pick his favorite game (even chess and checkers) and find the groups of the objects involved.

Problem II.5-7: What is the invariance group of a chess board? Suppose it is infinite? Does a fold in the middle make a difference?

Problem II.5-8: Digital clocks, and other devices, write numbers as sets of lines, thereby introducing symmetries that may not exist when they are written in more curvilinear form. Look at such a clock and determine the symmetry group of 1, 2, 3, 5, 8. Do any other integers have symmetry groups? What is the symmetry group of 12:51? of 12.51? of 12;51? Does it matter whether it is AM or PM?

Problem II.5-9: Is there a difference between the symmetry groups of an engagement and a wedding ring? What is its romantic significance?

Problem II.5-10: What is the group of an empty filing cabinet?

Problem II.5-11: Find the group of this book.

Problem II.5-12: Are there parts of a bicycle that have (finite) cyclic groups? What flowers have cyclic groups? Do any flowers have noncyclic groups?

Problem II.5-13: What is the group of an automobile? How does it differ from that of a railroad car? Does it differ from that of an airplane? Are there any functional reasons for any differences?

Problem II.5-14: Do the symmetry groups of a spoon, a fork and a knife differ? Why?

Problem II.5-15: What letters have symmetry groups [Lyndon (1989), p. 2]? What numbers (of any number of digits)?

Problem II.5-16: Are there words with noncommutative symmetry groups? Does this have any effect on their meaning?

Problem II.5-17: What is the group of a pair of hands? Can a single hand have a group? What is the group of a face (both with and without a part in the hair)?

Problem II.5-18: Is there a group of the human body? What effect does the heart being on the left have on the group of the body? Might the group of the brain be of interest? The retina of the eye consists of the cells in back of the eye which convert light to nerve impulses. What is its group? How might the group be related to its ability to understand the world it is looking at? Could it function with a larger, or smaller, group? Would its functioning be easier or more difficult? If you were a physician would you be interested in knowing the groups of the body, the brain, the eye, other organs? How might these help? Could they have physiological significance? Diagnostic significance? Consider some organ whose group you have determined. Is it possible to redesign it to give a different group; would it likely perform its function more, or less, efficiently?

Problem II.5-19: The groups of these objects consist of transformations, but that does not mean that these can be carried out easily or even physically. There is (often) no problem in carrying out the symmetry transformations on a wheel. In other cases it may be necessary to take the object apart, cell-by-cell, and put it back together again, cell-by-cell, with different connections between the parts. For all the groups considered above discuss whether it is possible to carry out these transformations, at increasing levels of sophistication in the types of transformations that are possible. Show how they can be performed.

Problem II.5-20: Which of the objects considered before, like the foods, give examples of broken symmetry?

Problem II.5-21: Transformations are often called mappings. Do the mappings (why plural?) of the Earth to a globe form a group? Which? How about the mappings to a flat piece of paper? There are various types of these mappings (different types of projections). Does the answer depend on the projection? Do these sets of transformations have identities? Products? Inverses? Are they unique?

Problem II.5-22: Which of the symmetry transformations on objects discussed can be performed using mirrors? Where would the mirrors have to be placed? Would the system of mirrors have a symmetry group? Does it have any relationship to the group of the object?

Problem II.5-23: While we have been considering groups of transformations leaving objects unchanged there can be other transformations of interest. Describe the group of transformations produced by a curved mirror (or lens). Do these transformations actually form a group (in what sense?)? How about sets of these? Might there be materials from which these can be constructed so the transformations do not form a group? Is it possible to use them (which?) to obtain a set of transformations that do not include the identity? Which cannot be extended to include the identity? Can the associativity condition be violated?

Problem II.5-24: Can you construct a set of mirrors (or lenses or both) which produce a noncommutative group of transformations?

Problem II.5-25: Reflection in a plane mirror leaves an object almost invariant (and many objects are invariant). A curved mirror or lens changes the object but it still looks much the same. A photograph looks like the (two-dimensional) object. One taken through a curved mirror looks much, but not completely, like the object. Suppose the photograph were a rubber sheet that was stretched. This would look less and less like the object as the sheet was stretched more and more, yet the resemblance would not be completely eliminated. It would still be possible to distinguish (very) different objects from their images. If the sheet were torn, it might no longer be possible to recognize the object, especially if there were many pieces. Further, while the transformations produced by the mirrors and the stretching have inverses, that of tearing does not (although jigsaw-puzzle fans may disagree). This emphasizes that while transformations that leave a system invariant tell much about it, and are thus of interest, there may be others that leave only certain aspects of the system unchanged, and may also be interesting. Finding the properties of the system that are invariant under these various transformations, and the groups that they form, can undoubtedly be stimulating problems. We cannot discuss these here, but clearly our present discussion is only the very start of the study of invariance, transformations, and group theory. Nevertheless the reader might wish to explore this further, say by considering various objects and studying the set of transformations on them produced by actions (like those) mentioned. Do these form groups; under what conditions; for which objects; what do they tell us about the objects?; about the actions?; about the laws of nature?; about what actions are possible? Do the answers to these questions depend on whether the sets of transformations form groups or not? Why?

Problem II.5-26: Invariance ("not varying"), which usually means that an object looks the same after a transformation as before, can have other meanings. For example we might consider computer programs that function the same if DO-loops or subroutines are permuted. In redesigning a program to make it run efficiently on a parallel computer, we would be interested in such invariance groups. Write computer programs that are (functionally) invariant under all groups (of small order). Would knowing the groups of these programs help in making them more efficient; in efficiently converting them to run on a parallel processor? How? Might some of these groups be noncommutative? Can any interesting results or rules be found from these?

Problem II.5-27: Are there groups of transformations that physically change an automobile but leave it functionally invariant?; Which? Why? Under what conditions? Must these be commutative?

Problem II.5-28: These examples should give some idea of the meaning of transformations, symmetry and invariance. Undoubtedly there are many other meanings. Think of all possible situations, of objects and of concepts, that can be transformed, and for which we can talk about invariance and symmetry. To what extent such situations are physical, or can be studied mathematically, are different questions. But these concepts are more wide ranging and subtle then they may seem. An open mind and a good imagination would be proper in the study of (not only) group theory.

II.6 THE PHYSICS OF GROUP TABLES

A group consists of elements — transformations — and a rule giving their products. This rule has physical significance, there are reasons why the products give the transformations they do; these are determined (partly) by the nature of the object. It is interesting to see how.

After reading the following chapters the reader may want to come back and try these problems again, and see whether the answers change.

Problem II.6-1: The smallest order giving more than one group is four. There are two groups, both of four transformations, but different products. One was given in pb. II.4.c-3, p. 51. Construct two different types of bananas having these two order-four groups as their groups. How can you tell by looking at them that their groups are different and what these are? What is the significance of the difference in the product rules? Which banana would you prefer to eat? Are there fruits or vegetables whose groups are isomorphic to either of these two? Can you tell what their groups are by looking at them? Would you expect any, few, or many, answers to these questions? Are there any organs that have either of these groups as their symmetry groups (at least approximately)? Why physiologically do they have these groups? Does the symmetry help them to function? How? Suppose the organ had the other group, of the same order, would that affect its functioning? How? Would it make it more or less efficient?

Problem II.6-2: Repeat this problem for other (small) orders for which there is more than one group.

Problem II.6-3: For order six there is a group, S_3, which is noncommutative (sec. II.4.d, p. 51). Find objects whose (at least approximate) symmetry, or revision, group is isomorphic to this. In what way does the noncommutivity of the products show itself in these objects?

Problem II.6-4: Why is the group of the square noncommutative? Is this related to the properties of the square? Why? Can you tell by looking at the square and the pyramid given above that they have different groups? Can you tell by looking at the square, double rectangle, and cube, which have different groups, and which have the same?

Problem II.6-5: Consider a filing cabinet, partially filled and partially painted. Can you fill and paint it so its transformations form a group isomorphic to S_3? What is the significance of the noncommutivity? Can you tell what the products are, and that they are noncommutative, by looking at the cabinet? Can you repaint the cabinet, and rearrange the contents, so that it still has a transformation group of order six, but one not isomorphic to S_3?

Problem II.6-6: Try these questions for groups of higher order.

Problem II.6-7: Repeat these for cyclic groups, starting with the smallest.

Problem II.6-8: Are there any organs of plants or animals that have noncommutative symmetry or transformation groups? What is the physiological significance of the noncommutivity? How would they function with different groups (especially of the same order)? Design an animal with a noncommutative group. How likely is it that such an animal would find life enjoyable? Design a plant with a noncommutative group. How difficult would life be for it? Design a flower with a noncommutative group. What can you say about its esthetics? Would you want to use it for wallpaper? Would you be able to?

Problem II.6-9: Canyou design an automobile that has a cyclic group? Can you design one with the other order-four group? How about an automobile that has a noncommutative group? Are you willing to drive, or ride in, any of these? What happens to the groups of the automobiles if passengers are in them?

Problem II.6-10: Is it possible to design a lens, or a combination of lenses, with finite cyclic groups? How about other symmetry or transformation groups? Are noncommutative groups possible? What type of images would these form? Do the images have symmetry groups? Which? Why? How are they related to the groups of the lens? Try this in particular with the order-four and order-six groups, and comment on the differences resulting from the different group tables.

Problem II.6-11: The reader should study his environment and notice the groups of the objects and systems within it, especially noncommutative ones, and should ponder their significance and try to determine why these objects have the groups they do, and how it relates to their purpose and their history.

Problem II.6-12: Undoubtedly trying to find answers to many of these problems involves searching through various possibilities and being clever enough to figure out what the groups are (many symmetries may be subtle), and attempting to understand how the groups limit or imply the shapes of objects and how there can be real, functioning objects with these shapes. Computers probably can help in these searches and constructions. Write a (or many, if necessary) computer program(s) to search for objects having the various (low-order) symmetry groups. Write one to construct an object, of two, three, four or more dimensions, with each (small order) symmetry group. In writing the program does the commutivity of the group matter? Would there be any complications due to there, often, being more than one group of each order?

Problem II.6-13: Is it possible to write a computer program that, given a group (and other conditions), can design a system capable of carrying out useful, say mechanical or physiological, functions?

II.6.a How the dimension of space affects the group

The groups of symmetry of objects, and their consequences, are determined in part by the properties of space. One important property is its dimension. Of course there are differences between, say, the group of the square and the group of its generalization to three dimensions, the cube. However while the square is a two-dimensional figure it exists in three dimensions. This raises another question about the effect of the dimension.

Problem II.6.a-1: Consider a square in two-dimensional space — the only transformations allowed now are in the plane; what is its group? How is it related to the group of the square that contains transformations in three dimensions? Repeat this for other plane figures, triangles, pentagons, and so on, with various restrictions on them. Does this suggest any general rules?

Problem II.6.a-2: Is there a general relationship between the groups of plane figures with transformations limited to the plane, and the groups for transformations in three dimensions? How about if the plane figures are in a space of n dimensions? What is the relationship between the group of a k-dimensional figure in k-dimensional space, and that for the same figure in n-dimensional space?

II.6.b The connection between groups and objects

These questions emphasize the close connection between the groups of transformations allowed (whatever this means) by a system and the properties of the system, and shows how we can get information about the system from its group. Of course we cannot get complete information. Many very different systems have the same group. Their properties, although not completely determined by their symmetry, in many cases are strongly limited by it. Often there are good reasons why objects have the symmetry, and the groups, that they do. Understanding these, and the relationship between the properties of an object and its group, is a major step toward understanding the object. But then what? How can we use this understanding and information? It turns out that this is useful mainly in quantum mechanics. Perhaps there are reasons for this. Or could it be that the classical value of groups has been overlooked? The reader may wish to think about this. (Sudarshan and Mukunda (1983) give a general discussion of groups in classical physics.)

Chapter III

Groups as Mathematical Objects

III.1 EXAMPLES AND RELATIONS

Up to this point we have been discussing what groups are, and how they are related to geometrical and physical objects. Now we have to consider them as mathematical entities, studying their structures and relationships. It is useful to start with examples, where these now are not objects, but relationships that give groups, like group multiplication tables. We also introduce various concepts related to group theory, plus definitions and discussions of terms and ideas from the theory itself which we can build on to understand it, and to develop tools for applications.

After these examples, and investigations suggested by them, we consider how groups are related, how they differ, why many are the same, though perhaps dressed differently, and how they are combined.

We emphasize problems; these are designed to raise questions that we wish, and need, to answer, and points to be discussed. It is worthwhile for the reader to try them; they (and problems in general) are good preparation for studying material that follows. Many readers may find them difficult at this stage, though hopefully less so later. But it is still valuable to go through them; incompletely worked problems can be educational, and can prepare readers for subsequent discussions. There are problems that everyone will find, at best, difficult, even after studying the entire book with great care. If they are solved, their solutions should definitely be published. We do not indicate which these are; if someone does not know that a problem is impossible he may actually solve it.

III.2 THE LEAST COMPLICATED GROUPS

Groups are uncomplicated (simple means something else, as we will see) because they have few elements, or because the elements are related in some way that we can easily deal with, or state. We start by listing a few of these (ways and groups).

The simplest group is that of a single element, the identity. That this is a group should be noted, but there is nothing further to say.

III.2.a Cyclic groups

The simplest nontrivial group has two elements, E and I; as one is the identity, E, it is Abelian. It is a cyclic group (sec. II.4.b, p. 49). A cyclic group is one whose elements form a cycle (from the Greek word for circle). The order-n cyclic group C_n has elements

$$E, G, G^2, \ldots, G^{n-1}, G^n = E. \tag{III.2.a-1}$$

It is of course Abelian and, for each n, unique. (An example is the set of rotations — around a circle — through angle $2\pi(m/n), m = 0, \ldots, n-1$). The group table is

	E	G	G^2	$\ldots$	G^{n-1}
E	E	G	G^2	$\ldots$	G^{n-1}
G	G	G^2	$\ldots$	$\ldots$	E
G^2	G^2		$\ldots$		G
$\ldots$			$\ldots$		
G^k	G^k	G^{k+1}	$\ldots$	$\ldots$	G^{k-1}
$\ldots$			$\ldots$		
G^{n-1}	G^{n-1}	E	G	$\ldots$	G^{n-2} .

Table III.2.a-1: Group Table for C_n

Each element A of a finite group is itself a cyclic group (so is a subgroup) — the number of group elements is finite so for some k,

$$A^k = E. \tag{III.2.a-2}$$

Then k is called the order of element A, and thus it is the order of the subgroup formed by A (pb. II.4.b-2, p. 49). These cyclic subgroups can limit the group. If all elements are of order 2, the group is Abelian, for, with A, B any two elements,

$$C = AB, \; C^2 = ABAB = E. \tag{III.2.a-3}$$

So

$$AB = B^{-1}A^{-1} = BA. \tag{III.2.a-4}$$

Problem III.2.a-1: Check the table. Verify that it is a group table — the axioms are obeyed — first for small orders, then in general.

Problem III.2.a-2: For every order there is, up to isomorphism, only one cyclic group, that of a single generator and its powers. Prove this. Give examples of cyclic groups.

Problem III.2.a-3: It should be clear that, for any group, an element and its inverse have the same order, also TV and VT have the same order [Hall (1959), p. 24].

Problem III.2.a-4: If a group has only one element that is of order 2, T, then T commutes with all group elements [Hall (1959), p. 24].

Problem III.2.a-5: Two commuting elements have orders p and q,

$$T^p = V^q = E, \text{ and } TV = VT; \tag{III.2.a-5}$$

then show that

$$(TV)^l = E, \tag{III.2.a-6}$$

where l is the least common multiple of p and q [Hall (1959), p. 24]. Is the converse true? This is not true if T and V do not commute; find counter-examples.

Problem III.2.a-6: Describe all groups whose order is a prime number. Show that there is only one for each prime. In considering groups we have thus finished with orders 1, 2, 3, 5, 7, 11, 13, 17, 19, ...

III.2.b Abelian groups

Abelian groups [Baumslag and Chandler (1968), p. 177; Grossman and Magnus (1992), p. 29, 74; Hall (1959), p. 35; Scott (1987), p. 89] are ones in which all elements commute (sec. II.4.a, p. 49). We give these in general, though leaving the explanation to the references [Baumslag and Chandler (1968), p. 196; Kurosh (1960a), p. 137; Ledermann (1953), p. 140; Ledermann (1987), p. 54]. Each element of any group forms a cyclic group, so Abelian group A, being Abelian, can be written as a direct product of cyclic (finite) groups,

$$A = C_1^{\alpha}C_2^{\beta}\ldots; \tag{III.2.b-1}$$

C_1^{α} is the cyclic group of order α, C_2^{β} is of order β, and so on. An Abelian group is thus specified by the set $\alpha, \beta, \ldots$, with different sets giving different Abelian groups. (A direct product means that each group element is a product of the elements of the subgroups, and the subgroups commute.)

What can be said about these sets, what restrictions (if any) are on them? The order g of A, like every integer, can be written as

$$g = p^a q^b r^c \ldots; \tag{III.2.b-2}$$

p, q, and r are (distinct) primes. Likewise the order of cyclic group C_1^{α} is

$$\alpha = w^k, \tag{III.2.b-3}$$

where w is prime. Further for order g of Abelian group A,

$$g = \alpha\beta \ldots; \tag{III.2.b-4}$$

the product of the orders of the cyclic groups making up A. Combining these,

$$g = w^k v^{k'} \ldots, \tag{III.2.b-5}$$

so that the order of A is the product of the powers of the primes appearing in the orders of its constituent cyclic groups. This can now be extended to give that every finite Abelian group is the direct product of cyclic groups whose orders are primes [Ledermann (1953), p. 142]. Thus not only can the order of the group be written as the product of powers of primes, but the group itself can also.

Problem III.2.b-1: Check this with the Abelian groups given and for each (finite?) Abelian group as it is encountered below.

Problem III.2.b-2: The restrictions on the orders can often give much information. So prove that if the order of an Abelian group is not divisible by a square, the group is cyclic [Ledermann (1953), p. 156]. Show that there are cyclic groups whose order is not so divisible. Check the orders of the groups as they are met to verify that there are no counterexamples.

Problem III.2.b-3: Prove that an Abelian group of order g has a subgroup of each order dividing g [Ledermann (1953), p. 156]. Find them for the various Abelian groups.

Problem III.2.b-4: Prove that a group of order pq, with p and q both primes, cannot have two distinct subgroups of order q if $p < q$ [Hall (1959), p. 24]. Can this occur if $p = q$? If $p > q$? Give examples.

Problem III.2.b-5: It is worthwhile mentioning notation. A group has a combination rule, usually called a product. But it need not be multiplication. For Abelian groups it is often called addition, since we think of addition, as we do not necessarily of multiplication, as commutative. Thus instead of writing a "product" as ab, we write it as $a + b$; the identity is written as 0, and can be called zero. The order of element a is the smallest value of m such that $ma = 0$ (suggesting that Abelian groups are useful in modular arithmetic; pb. II.4.f-1, p. 55). It is worth verifying that the group axioms are satisfied using this notation (and perhaps check whether these axioms might have hidden assumptions that the notation might make more, or less, noticeable). We can thus also say that every finite Abelian group is the direct sum of cyclic groups whose orders are primes. To treat all groups the same, so make the discussion more compact, we do not distinguish groups by the notation. It should however be kept in mind.

III.2.c Groups of order 4

The smallest order requiring a nontrivial analysis is 4 [Joshi (1982), p. 23; Ledermann (1953), p. 49]. We label the elements E, A, B, C. If A is of order 4, it gives a cyclic group,

$$B = A^2, \quad C = A^3, \quad E = A^4. \tag{III.2.c-1}$$

Are there any other groups? The order of the group, 4, is an integral multiple of that of every subgroup, so A cannot have order 3. If all elements have order 1 the group is the trivial (and always excluded) product of four identities. Thus take the elements of order 2;

$$A^2 = B^2 = C^2 = E. \tag{III.2.c-2}$$

Now AB is one of these elements; it cannot be A or B for that would give the other the identity. So

$$AB = C. \tag{III.2.c-3}$$

The group is Abelian. The group table is (pb. II.4.c-3, p. 51)

	E	A	B	C
E	E	A	B	C
A	A	E	C	B
B	B	C	E	A
C	C	B	A	E .

Table III.2.c-1: Group Table for the Four-Group

This is sometimes called (for obvious reasons) the four-group [Grossman and Magnus (1992), p. 70].

Problem III.2.c-1: Why is the order of a group an integral multiple of that of every subgroup (pb. III.3-15, p. 84)?

Problem III.2.c-2: Check that the group table is correct, is a group table, follows from the conditions, and that the group is Abelian. Find the subgroups. How are they related? Does the group table have a symmetry group? Why? This shows these two are the only order-4 groups (up to isomorphism), both Abelian, one cyclic (always), the other, of course, not.

Problem III.2.c-3: Repeat this for order 3 [Tung (1985), p. 25; Aivazis (1991), p. 2].

III.2.d Groups of order 6

The same argument can be applied to order 6 (and so on, but it soon becomes impractical). The operators are E, A, B, C, D, F. For the cyclic group the latter five are of order 6, and equal powers of the others. The other possible orders of the elements are 2 and 3. If all have order 2, the group is Abelian. However then the subgroup E, A, B, $AB = BA$, is of order 4. But 6/4 is not an integer. So we take A of order 3, and elements C, D, F of order 2, giving

$$E, A, C, D, F, B = A^2,\ AB = A^3 = E,\ C^2 = E,\ D^2 = E,\ F^2 = E. \quad \text{(III.2.d-1)}$$

We need $AC, CA, AD, DA, CD, DC, \ldots$. We can set

$$D = AC. \quad \text{(III.2.d-2)}$$

Does

$$D = CA, \text{ or } F = CA? \quad \text{(III.2.d-3)}$$

For the former,

$$D^2 = CACA = E = C^2A^2 = A^2, \quad \text{(III.2.d-4)}$$

which is incorrect. Hence A does not commute with C, and similarly with D or F, giving a non-Abelian group. This determines the group.

Thus for order 6, there are two groups, the (Abelian) cyclic one, and the other which is noncyclic, and is (isomorphic to) symmetric group S_3 (sec. II.4.d, p. 51); this is of order 6, and non-Abelian, and there is only one non-Abelian order-6 group.

Problem III.2.d-1: What is the order of B? Why can other elements not have order 3?

Problem III.2.d-2: Verify that these relations do give the table, and conversely, and that it is a group table (the axioms are satisfied). Is it Abelian? Can this be easily seen from the table? This table is not symmetric about the diagonal. Why? Does it have any symmetries? What is its symmetry group? Interpret it. Identify the symbols here with the S_3 transformations and check that S_3 does satisfy this group table. The table is

	E	A	B	C	D	F
E	E	A	B	C	D	F
A	A	B	E	D	F	C
B	B	E	A	F	C	D
C	C	F	D	E	B	A
D	D	C	F	A	E	B
F	F	D	C	B	A	E .

Table III.2.d-1: Group Table for S_3

Problem III.2.d-3: What would go wrong if we took the three-element subgroup, with elements all of order 2, plus element $C(= AB)$ and tried to construct a group from them? What would the group table look like? Would it be obvious that this is not a group?

Problem III.2.d-4: Why do we not consider groups of order 5?

Problem III.2.d-5: The table

	E	T	T^2	I	IT	IT^2
E	E	T	T^2	I	IT	IT^2
T	T	T^2	E	IT	IT^2	I
T^2	T^2	E	T	IT^2	I	IT
I	I	IT	IT^2	E	T	T^2
IT	IT	IT^2	I	T	T^2	E
IT^2	IT^2	I	IT	T^2	E	T ,

Table III.2.d-2: Group Table for What Group?

is that of a group of order 6 which is Abelian and the direct product (sec. III.2.b, p. 69) of groups of orders 2 and 3. However there is no other order-six group. What is wrong? Can you get interesting general information or rules from this?

III.2.e Groups of order 8

For order 8 [Ledermann (1953), p. 51; Lomont (1961), p. 34] there are five groups of which three are Abelian, the (Abelian) cyclic group plus two others.

Problem III.2.e-1: Write the group table for the cyclic group of order 8 [Ledermann (1953), p. 54].

Problem III.2.e-2: There is an Abelian group [Ledermann (1953), p. 54] that is the direct product of a cyclic group of order 4 and one of order 2. Its group table is

	E	T	T^2	T^3	I	IT	IT^2	IT^3
E	E	T	T^2	T^3	I	IT	IT^2	IT^3
T	T	T^2	T^3	E	IT	IT^2	IT^3	I
T^2	T^2	T^3	E	T	IT^2	IT^3	I	IT
T^3	T^3	E	T	T^2	IT^3	I	IT	IT^2
I	I	IT	IT^2	IT^3	E	T	T^2	T^3
IT	IT	IT^2	IT^3	I	T	T^2	T^3	E
IT^2	IT^2	IT^3	I	IT	T^2	T^3	E	T
IT^3	IT^3	I	IT	IT^2	T^3	E	T	T^2 .

Table III.2.e-1: Group Table for an Abelian Order-Eight Group

This can be obtained by adding to rotations through angle $2\pi(m/n)$, $m = 0, \ldots, n-1$, for example, a reflection or an inversion, where here n is 4, and there are similar groups for any n. Explain why this shows, without direct checking, that it is a group table. What group is this? What are the orders of the elements? Show that the table gives

$$IT = TI. \tag{III.2.e-1}$$

Does this table have any symmetry group? Why? Give its group table.

Problem III.2.e-3: The third Abelian group [Ledermann (1953), p. 51, 55] is the direct product of three cyclic groups all of order 2,

$$A^2 = B^2 = C^2 = E, \tag{III.2.e-2}$$

with all elements commuting. Work out the group table. Verify that it is a group and is not isomorphic to either of the above.

Problem III.2.e-4: One non-Abelian group of order 8 is the quaternion group [Ledermann (1953), p. 52, 55]. For its group table we give just the subscripts (so $A_5 \Rightarrow 5$, etc.). Does it matter whether only subscripts are given? Verify that the table is correct, that each element appears once, and only once in each row and column, that it obeys the axioms, and that the multiplication rules give the table. To verify the table use quaternions, E, i, j, k (generalizations of $\sqrt{-1}$), with multiplication rules,

$$i^2 = j^2 = k^2 = -1, \quad ij = -ji, \quad ik = -ki, \quad jk = -kj. \tag{III.2.e-3}$$

Identify

$$E \Rightarrow 1, \quad A_1 \Rightarrow -1, \quad A_2 \Rightarrow i, \quad A_3 \Rightarrow -i,$$

$$A_4 \Rightarrow j, \quad A_5 \Rightarrow -j, \quad A_6 \Rightarrow k, \quad A_7 \Rightarrow -k. \tag{III.2.e-4}$$

What is ij? Is this group Abelian? This is also (isomorphic to) a symmetry group of a geometric figure. Which? Use the transformations of that to check the table. Why is the group table not symmetric about the diagonal? Does this group table have any (obvious) symmetry group? What is that group? Would you expect it from the quaternion interpretation? From the geometrical one? Could it have symmetry beyond what might be given by these interpretations? The table is [Lomont (1961), p. 18]

	E	1	2	3	4	5	6	7
E	E	1	2	3	4	5	6	7
1	1	E	3	2	5	4	7	6
2	2	3	1	E	6	7	5	4
3	3	2	E	1	7	6	4	5
4	4	5	7	6	1	E	2	3
5	5	4	6	7	E	1	3	2
6	6	7	4	5	3	2	1	E
7	7	6	5	4	2	3	E	1 .

Table III.2.e-2: Group Table for the Quaternion Group

Problem III.2.e-5: The generalization of the quaternion algebra is the octonion algebra, with eight elements,

$$q_1 = E, \ q_i, \ i = 1, \ldots, 7; \ q_i^2 = -1, \ i \neq 1. \qquad \text{(III.2.e-5)}$$

These anticommute, for $i \neq j$,

$$q_i q_j = -q_j q_i, \qquad \text{(III.2.e-6)}$$

$$q_i q_j = \varepsilon_{ijk} q_k, \qquad \text{(III.2.e-7)}$$

where ε_{ijk} is the completely antisymmetric symbol,

$$\varepsilon_{ijk} = 1, \ \varepsilon_{jik} = -1, \ \varepsilon_{iik} = 0; \qquad \text{(III.2.e-8)}$$

it is 0 if two indices are equal, 1 if the indices are in numerical order (123, 124, 235, etc.) and -1 if they are not (213, 421, etc.). The exact convention giving q_k for each i, j is not important here. Show that the product of these symbols is not associative. Hence these do not form a group. Can you tell by (quickly) looking at the multiplication table that this is not a group? Try to identify these symbols with geometric transformations. Is it possible? What would the effect of applying these to an object be? Is it possible to think of a geometry (or a set of physical laws) for which transformations would be (like) these? How would it differ from our geometry, and physics?

Problem III.2.e-6: There is another non-Abelian group of order eight [Falicov (1966), p. 61], given by the table below. What geometrical transformations do these symbols stand for? For which figure is this the symmetry group? Verify that the table is correct, it obeys the group axioms, and check the multiplication rules experimentally. The elements can also be written as products of two, making it similar to the table in pb. III.2.e-2, p. 73. Rewrite that, and check that even though the top rows are the same, the tables (so the groups) are different. This group has a name, which was introduced before [Ledermann (1953), p. 55]. What is it? Check that this is not isomorphic to the quaternion group. Is there a symmetry group of the group table? Why should that be the group? The table of this non-Abelian group is

	e	a_x	a_y	d_1	d_2	b	c_1	c_2
e	e	a_x	a_y	d_1	d_2	b	c_1	c_2
a_x	a_x	e	b	c_2	c_1	a_y	d_2	d_1
a_y	a_y	b	e	c_1	c_2	a_x	d_1	d_2
d_1	d_1	c_1	c_2	e	b	d_2	a_x	a_y
d_2	d_2	c_2	c_1	b	e	d_1	a_y	a_x
b	b	a_y	a_x	d_2	d_1	e	c_2	c_1
c_1	c_1	d_1	d_2	a_y	a_x	c_2	b	e
c_2	c_2	d_2	d_1	a_x	a_y	c_1	e	b

Table III.2.e-3: Group Table for an Order-8 Group

Problem III.2.e-7: Prove that these are the only groups of order 8.

III.2.f The tetrahedral group

A tetrahedron ("four sides") is a pyramid with a triangular base and three triangular sides, these four triangles congruent (though in different planes). It has four vertices and is invariant under their permutation [Armstrong (1988), p. 1, 38]. The tetrahedral group is the subset of these permutations that can be carried out physically, by rotations (without tearing the figure apart and putting it back together again). The symmetric group S_4, the set of permutations of four objects, has 24 elements. Only half of these can be carried out on the solid tetrahedron, so the tetrahedral group has twelve elements [Grossman and Magnus (1992), p. 115; Falicov (1966), p. 14, 66].

Problem III.2.f-1: Verify that this is a group table (using just subscripts) [Janssen (1973), p. 79; Lomont (1961), p. 30]. Should this be too much work, write a computer program to do so. Apply it to other group tables. Identify the subscripts with operators and check that this is the correct group of symmetries of the tetrahedron. Is it Abelian? Does this agree with what you would expect from looking at the tetrahedron (found by drawing, or by constructing with, say, sticks)? Is there a symmetry group of this table? Why should that be the group? The table is

	E	1	2	3	4	5	6	7	8	9	10	11
E	E	1	2	3	4	5	6	7	8	9	10	11
1	1	E	3	2	5	4	7	6	9	8	11	10
2	2	4	6	8	9	7	E	10	11	1	5	3
3	3	5	7	9	8	6	1	11	10	E	4	2
4	4	2	8	6	7	9	10	E	1	11	3	5
5	5	3	9	7	6	8	11	1	E	10	2	4
6	6	9	E	11	1	10	2	5	3	4	7	8
7	7	8	1	10	E	11	3	4	2	5	6	9
8	8	7	10	1	11	E	4	3	5	2	9	6
9	9	6	11	E	10	1	5	2	4	3	8	7
10	10	11	4	5	2	3	8	9	6	7	E	1
11	11	10	5	4	3	2	9	8	7	6	1	E .

Table III.2.f-1: Group Table for the Tetrahedral Group

Problem III.2.f-2: Give the vertices coordinates, and find the set of their permutations (write these as both permutations and as transformations of the coordinates; $x_1 \Rightarrow ax_1 + bx_2 + \ldots + h$, and so on, with the coefficients to be determined). Which can be implemented as rotations? How would the others be? Can you see the answers from the transformations of the coordinates? Build a tetrahedron using sticks, say, and check this, verifying that there are twelve transformations that are possible without dismantling the object, and twelve others that require disassembly. Find the group table and verify that it is that of a subgroup of S_4. This group is the alternating group A_4, an order-12 subgroup of S_4 (sec. II.4.d.ii, p. 54).

Problem III.2.f-3: The tetrahedral group has order 12. For this order, there are five (nonisomorphic) groups, of which two are Abelian, the cyclic group (sec. III.2.a, p. 68) and a direct product (sec. III.2.b, p. 69) of order-6 and order-2 groups. Prove that there are only two order-12 groups that contain an element of order 6 [Ledermann (1953), p. 61]. These are given by

$$A^6 = B^2 = (AB)^2 = E,$$

$$A^6 = E, \quad B^2 = (AB)^2 = A^3; \qquad \text{(III.2.f-1)}$$

E is the identity. Which is the tetrahedral group? For it, identify the symbols with those in the table, and with the geometrical transformations. Find the multiplication table for the other. Is there a geometrical figure for which it is a symmetry group? If so, draw it.

Problem III.2.f-4: Does the direct product of S_3 and an order two group give a group (nonisomorphic to the other groups of order 12)?

Problem III.2.f-5: Why is there no new Abelian group that is a direct product of groups of order 3 and order 4 (say the cyclic groups of these orders)?

III.2.g The cubic and icosahedral groups

The regular octahedron ("eight faces") is a solid with eight sides, each an equilateral triangle; it is two tetrahedra with bases coincident. The regular icosahedron ("twenty faces") is a solid with twenty sides, again each an equilateral triangle. Their symmetry groups [Janssen (1973), p. 79; Ledermann (1953), p. 92] are the octahedral (or hexahedral) group, and the icosahedral (also known as the dodecahedral) group. The regular hexahedron is, clearly, a solid with six faces — a cube. The regular dodecahedron ("twelve faces") is a solid with twelve, each an equilateral pentagon.

Problem III.2.g-1: Construct these figures [Armstrong (1988), p. 38] and find their symmetry groups. Verify that they have the symmetry stated and shown:

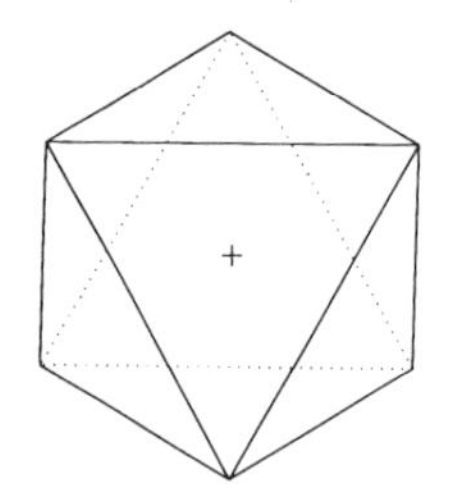

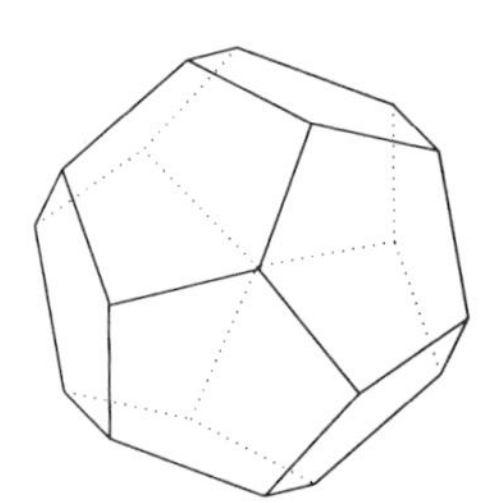

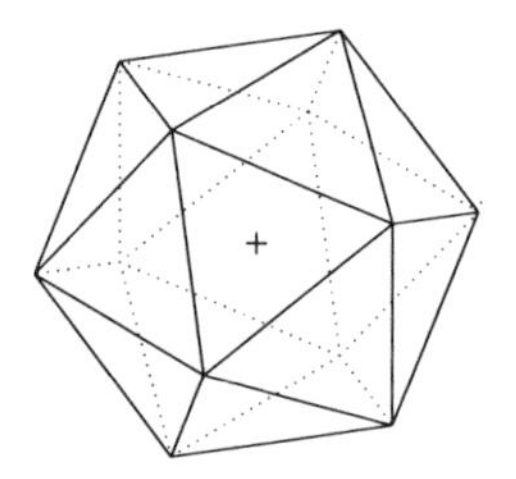

Figure III.2.g-1: THE OCTAHEDRON, DODECAHEDRON

AND ICOSAHEDRON

Also give coordinates for their vertices, and the groups that permute them. Draw lines connecting the centers of the faces of the octahedron; these should form a cube. Also the lines connecting the centers of the sides of the cube form an octahedron. Thus the symmetry groups of these two figures are the same. Similarly connecting the centers of the faces of the icosahedron (dodecahedron) gives a dodecahedron (icosahedron), so their symmetry groups are the same. This group is of order sixty. It is also isomorphic to alternating group A_5 (sec. II.4.d.ii, p. 54) [Ledermann (1953), p. 94, 123; Ledermann (1987), p. 75]. From the group tables found by analyzing the symmetry of the figures (or more directly, if tables can be avoided) show that the groups of the figures are the same (or should it be isomorphic?) and are isomorphic to A_5.

III.2.h Dihedral groups

Dihedral groups (sec. II.4.c, p. 50) are found by adjoining to cyclic groups (sec. III.2.a, p. 68) an element, I, of cycle length 2;

$$I^2 = E. \qquad \text{(III.2.h-1)}$$

Dihedral group $\mathsf{D_n}$ has 2n elements, T, of cycle length n, and its powers, and IT^k, $k = 0, \ldots,$ n-1. Elements T and I do not commute, and are related by

$$ITI = T^{-1}. \qquad \text{(III.2.h-2)}$$

Dihedral groups, other than $\mathsf{D_2}$, are non-Abelian; if $IT = TI$ then

$$ITI = TII = T = T^{-1}, \qquad \text{(III.2.h-3)}$$

so T would be of order two. The group table for $\mathsf{D_2}$ is

	E	T	I	IT
E	E	T	I	IT
T	T	E	IT	I
I	I	IT	E	T
IT	IT	I	T	E

Table III.2.h-1: Group Table for $\mathsf{D_2}$

This is another way of writing the order-four group (sec. III.2.c, p. 71).

These are symmetry groups of figures having an n-fold axis of symmetry (that is, are invariant under rotations about the axis of an integer multiple of $2\pi/n$), and another axis at right angles to it, with two-fold symmetry. For $\mathsf{D_3}$ consider an equilateral triangle. A rotation around a perpendicular axis through its center by $T = 2\pi/3$ leaves it invariant, as does T^2 and T^3. If we take it as a (two-faced) solid figure, π-rotations about axes in its plane (perpendicular to the first) through its vertices interchange the two faces, and also leaves it invariant. Interchanging the faces, rotating by $2\pi/3$, then interchanging the faces again is equivalent to a rotation of $2(2\pi/3) = -2\pi/3$, so

$$T^2 = T^{-1}, \qquad \text{(III.2.h-4)}$$

the inverse of a $2\pi/3$ rotation. Thus the symmetry group is $\mathsf{D_3}$.

Problem III.2.h-1: Verify this experimentally. Also verify that IT and TI give different positions of the figure. Thus this group is not Abelian. Since it is non-Abelian of order six it must be isomorphic to symmetric group S_3 (sec. II.4.d, p. 51). Find the group multiplication table experimentally and check that it is. Also, the group multiplication table can be obtained from eq. III.2.h-2, p. 79. Check that it is the same as that found experimentally.

Problem III.2.h-2: Show that the group table for D_4, using only subscripts is [Baumslag and Chandler (1968), p. 77]

	1	2	3	4	5	6	7	8
1	1	2	3	4	5	6	7	8
2	2	3	4	1	8	5	6	7
3	3	4	1	2	7	8	5	6
4	4	1	2	3	6	7	8	5
5	5	6	7	8	1	2	3	4
6	6	7	8	5	4	1	2	3
7	7	8	5	6	3	4	1	2
8	8	5	6	7	2	3	4	1

Table III.2.h-2: Group Table for D_4

Problem III.2.h-3: Check that these are group tables, and of dihedral groups; in particular check D_4. These transformations can be taken as a $\pi/2$ rotation about an axis, rotations about a perpendicular axis, and products of these. Which subscripts stand for which operators? Define symbols for them and rewrite the table using the symbols. Are there any ambiguities? Do these suggest anything about a symmetry group of the table? What is it? Find geometrical figures with these symmetry groups. What are the geometrical implications of ambiguities in the identification of subscripts with operators?

Problem III.2.h-4: Construct group tables for D_5 and D_6. Are these Abelian? Think of figures that have these symmetries and find the group tables experimentally and using the defining relations (eq. III.2.h-2, p. 79). Answer the questions in the previous problem for these.

Problem III.2.h-5: Is D_6 the same as (isomorphic to) any of the above order-twelve groups (sec. III.2.f, p. 76)?

III.2.i Groups and their tables

A group is a set with structure, and this structure (for finite groups) can be expressed in its group table, the list of products. The group axioms limit the possible multiplication rules. These limitations may seem obvious but we try to phrase them as to make them less so — what is clear is not always correct, and being too obvious can inhibit what is most needed, thought. If the reader thinks (about them) he will become involved enough to see whether knowing how products are limited by the axioms helps him understand groups (and even their physics).

Problem III.2.i-1: Show that each group element appears, and only once, in every row, and similarly column, of the group table (pb. II.2.f-10, p. 39). How does this follow from the axioms? Need this be true for a semigroup? A groupoid?

Problem III.2.i-2: Is there any kind of group whose table is diagonal?

Problem III.2.i-3: Can you tell by looking at the table that a group is Abelian? That it is non-Abelian? That it is cyclic? Just by looking (quickly) at the table can you see that a group is a symmetric group (sec. II.4.d, p. 51)? How about dihedral?

Problem III.2.i-4: Why is this not a group table [Ledermann (1953), p. 14]:

	E	A	B	C	D
E	E	A	B	C	D
A	A	E	D	B	C
B	B	C	E	D	A
C	C	D	A	E	B
D	D	B	C	A	E ?

Table III.2.i-1: Group Table for a Nongroup

Problem III.2.i-5: Must (can) there be symmetry in a group table, say reflection in a line — does a group table have its own group? Is it related to the group? Are there restrictions on what a group of a group table can be? Are they related to the type of group given by the table? If one is Abelian can, must, the other be; non-Abelian; cyclic; symmetric; general?

Problem III.2.i-6: Does the symmetry of the group table depend on the ordering of the group elements? What would happen if the rows and columns were ordered differently? How would the answers to the problems of this section be changed if a different ordering were picked for the group elements? Suppose the columns were labeled by the group elements, the rows by their inverses. What would the table look like? Would there be any advantages, or disadvantages, to this? Try it with the tables given in this chapter.

III.3 GENERATORS AND PRESENTATIONS

It is clear that while groups can be given by their tables as done above, this is not a good method in general. For a group of high order, it takes too much effort, and too much paper, to write the table explicitly. And it would be so large that getting information would be difficult. Is there any other method for giving the group, a simpler, and shorter, way of presenting the information in the table?

For a cyclic group (sec. III.2.a, p. 68) a table is clearly unnecessary. The group can be given by a single number, the cycle length. Knowing that, we know everything about the group. And as seen above, there are other groups that can be presented by a few equations (eq. III.2.c-1, p. 71; eq. III.2.e-3, p. 74; eq. III.2.f-1, p. 77), these being equivalent to the group table.

Moreover, we do not need symbols for all group elements. For a cyclic group only one (besides the identity) is needed. Every other is just a power of it. Similarly, as we saw, for many groups, at least, only a few symbols are needed; the others can be written as products and powers of these. Thus for any group we introduce a set of symbols, the generators of the group, in terms of which all group elements can be written. The minimum number of generators needed is the rank of the group, and the smallest set of these the basis of the group. Also there is a set of relations among the generators (say their cycle lengths), giving elements as products of generators (the order of terms in the product is usually relevant). This set of generators and relations is called a presentation of the group [Armstrong (1988), p. 166; Baumslag and Chandler (1968), p. 253; Coxeter and Moser (1980); Grossman and Magnus (1992), p. 56; Magnus, Karrass and Solitar (1966); Scott (1987), p. 187]. Given a presentation, usually there are several, we can write all group elements in terms of the generators, and the group table then follows. A presentation is more, usually much more, compact than the table.

Problem III.3.-1: What is a presentation — the generators and relations — of a cyclic group? A dihedral group (sec. III.2.h, p. 79) [Baumslag and Chandler (1968), p. 257]? What are the ranks? Are there other presentations besides those of the basis?

Problem III.3.-2: Find a presentation, and a basis, for an arbitrary Abelian group.

Problem III.3.-3: Check that for symmetric groups, generators are the neighboring transpositions (pb. II.4.d-6, p. 53) [Coxeter and Moser (1980), p. 63, 137]. Also as is the case for S_3, there is a presentation with two generators (pb. III.2.h-1, p. 80), a transposition and, for S_n, an (any?) element of cycle length n [Baumslag and Chandler (1968),

p. 256]. Presentations are not unique. Find others [Dixon (1973), p. 13, 84; Lyapin, Aizenshtat, and Lesokhin (1972), p. 64], and one(s) for the alternating group [Dixon (1973), p. 13, 84; Lyapin, Aizenshtat, and Lesokhin (1972), p. 65]. Why are they different?

Problem III.3.-4: Find [Jansen and Boon (1967), p. 72] the multiplication table of the group with generators E, A and B,

$$A^3 = E, \ \ C^2 = E, \ \text{and} \ \ CA = A^{-1}C. \tag{III.3-1}$$

What group is it (isomorphic to)?

Problem III.3.-5: Consider a group given by two elements [Dixon (1973), p. 4, 75] x, y, subject to the conditions,

$$x^3 = y^2 = (xy)^2 = 1; \tag{III.3-2}$$

these supplying an example of relations on generators. Find its order, rank, group table and all subgroups. Do the same if the conditions are

$$x^3 = y^2 = (xy)^3 = 1. \tag{III.3-3}$$

Can you give a simple explanation of the results?

Problem III.3.-6: Show that the matrices

$$a = \begin{pmatrix} 0 & 1 \\ -1 & 0 \end{pmatrix}, \ \ b = \begin{pmatrix} 0 & i \\ i & 0 \end{pmatrix} \tag{III.3-4}$$

generate the quaternion group (pb. III.2.e-4, p. 74) [Dixon (1973), p. 4]. Show also that this group is generated by x, y, z, u with

$$x^2 = y^2 = z^2 = u, \ \ u^2 = 1, \ \ xy = z, \ \ yz = x, \ \ zx = y. \tag{III.3-5}$$

How are these related to the matrices? Is this a basis of the group? These look similar to Pauli matrices (pb. II.4.g-1, p. 56). Do those generate a group? If so, which? Is there a difference? Why?

Problem III.3.-7: For the quaternion group [Grossman and Magnus (1992), p. 137] is the set of generators given (anywhere) the minimum (the basis)? If not, what is a minimal one? What is the rank?

Problem III.3.-8: Consider [Dixon (1973), p. 5, 75] generators with the conditions

$$a^2 = b^k = (ab)^2 = 1. \tag{III.3-6}$$

Show that these generate the dihedral groups D_k, for $k \geq 3$ (sec. III.2.h, p. 79). What happens if $k = 2$? Another set of generators is

$$a = \begin{pmatrix} 0 & 1 \\ 1 & 0 \end{pmatrix}, \ \ b = \begin{pmatrix} \zeta & 0 \\ 0 & \zeta^{-1} \end{pmatrix} \tag{III.3-7}$$

where $\zeta = exp(2\pi i/k)$. Is this true for $k = 2$? Show from this that the group has order $2k$.

Problem III.3.-9: Find the group given by the generators [Dixon (1973), p. 4]

$$a = \begin{pmatrix} 0 & 1 \\ -1 & 0 \end{pmatrix}, \quad b = \begin{pmatrix} 0 & 1 \\ 1 & 0 \end{pmatrix}; \tag{III.3-8}$$

check that it has order 8 and is non-Abelian, but not isomorphic to the quaternion group. Is there a simple way of explaining why they are not isomorphic? Show that it is isomorphic to $\mathsf{D_n}$ (for what n?). Obtain the group of symmetries of its group table (from a presentation).

Problem III.3.-10: For the previously-discussed groups of geometrical figures, find geometrically the (minimum) set of generators and relations, giving presentations of them. Check that the correct group multiplication tables are obtained. What are their ranks? Are there any geometrical reasons for the ranks? For the bases? Find the group-table symmetry groups from a presentation.

Problem III.3.-11: Repeat this for the groups of different number systems (sec. II.4.f, p. 55).

Problem III.3.-12: Repeat also for any group(s) found from the Pauli matrices (pb. II.4.g-1, p. 56).

Problem III.3.-13: Can a presentation be given for the matrix groups considered (sec. II.4.g, p. 56)? What are their ranks? Is there a relationship between the ranks of different matrix groups? Some of these groups are continuous rather than finite. Does this matter?

Problem III.3.-14: Show that for an Abelian group with a (minimal?) basis of n elements, each subgroup has a (minimal?) basis of $\leq n$ elements [Dixon (1973), p. 21, 95]. This may seem reasonably trivial for a cyclic group, but is it so for an arbitrary Abelian group?

Problem III.3.-15: From the group axioms it follows that group G is generated from the cyclic subgroup of any element (including the identity?) by applying all the group operators to the elements of the subgroup. The order of G is then an integral multiple of the order of the cyclic subgroup — the order of the element. And in general (sec. IV.6.b.i, p. 131), the order of a group is an integral multiple of the order of every subgroup (Lagrange's theorem) [Baumslag and Chandler (1968), p. 109; Grossman and Magnus (1992), p. 82; Hall (1959), p. 10; Hamermesh (1962), p. 20; Landin (1989), p. 110; Scott (1987), p. 20]. Verify these statements.

III.4 RELATIONS AMONG SETS SUCH AS GROUPS — MORPHISMS

Morphism (there seems to have been no such word without a prefix) refers to shape. The word appears in various contexts: amorphous solids are popular in physics, as is morphology in biology, metamorphic rocks in geology, and amorphous ideas in too many fields. These words refer to shapes; in mathematics a morphism connotes a relationship, that is a mapping (transformation), between "shapes" or "forms" [Maxfield and Maxfield (1992), p. 78].

Groups are related if they have similar "forms"; this can happen in various ways. Mathematicians have developed many types of morphisms: isomorphism, homomorphism, automorphism, endomorphism, meromorphism, diffeomorphism, ... — the list seems endless. It seems that if the time were taken to learn all the different kinds of morphisms there would be none left for the rest of the subject. Therefore we just give brief definitions of those needed here. Should one be accidentally omitted its definition can easily be found by digging through the literature.

III.4.a Equivalence relations

Morphisms imply relations, and relations themselves can be categorized. One type is distinctively important, an equivalence relation. A relationship is an equivalence relation, $a \sim b$, if it is

a) reflexive (as in a mirror) $a \sim a$; a is equivalent to itself,

b) symmetric $a \sim b \Rightarrow b \sim a$, and

c) transitive (cf. translation) $a \sim b$ and $b \sim c \Rightarrow a \sim c$. The relationship is transitive because the equivalence of a to b and b to c translates ("carries across") to $a \sim c$. Thus a and c are equivalent.

Problem III.4.a-1: An equivalence relation is given by a set of transformations taking the objects into each other. Explain why these form a group. Do all groups give equivalence relations? Are the number of axioms the same? If there are only three objects, what group is it? Prove. For this case is associativity relevant?

III.4.b Isomorphisms

Two groups that are identical except for different symbols, are isomorphic (sec. II.1, p. 30); each element of one corresponds to exactly one element of the other and their multiplication rules are the same — the products of corresponding elements correspond. So if

$$A \Rightarrow \alpha, \quad B \Rightarrow \beta \text{ and } C \Rightarrow \gamma \qquad \text{(III.4.b-1)}$$

and

$$AB = C, \text{ then } \alpha\beta = \gamma, \qquad \text{(III.4.b-2)}$$

for all elements. An isomorphism ("equal form or shape") is a mapping of the elements of one group onto those of another — a one-to-one correspondence between the elements (and the products) of the groups.

All two-element groups are identical, as abstract objects — they have the same elements (the symbols are irrelevant) and the same group table. However the groups can be that of symbols, of the identity and a transposition (S_2), the identity plus a reflection, the identity plus a rotation of π, the group (1,-1), the identity matrix and a matrix whose square is the identity, and so on. In this sense they are different, they are identical only in their multiplication tables — they are all isomorphic (or to use a term defined below, they are different realizations of the same abstract group). Whether groups are the same, or merely isomorphic, can depend on the subject under study; in (abstract) group theory there is no (real) difference between the two-element groups, one with a reflection, the other with an inversion, in geometry, and physics, there is. Not only can we consider such a relationship, an isomorphism, between two groups, we can, and shall, consider an isomorphism between a group and itself — an automorphism.

It might seem that isomorphism is just another of the many possible relations between sets. However, its significance goes far deeper. There are many types of operators that form groups: tables, differential operators, matrices, ..., operations (including permutations, both on lists and physically) on symbols, on meaningless symbols, meaningful symbols, indices, functions, atoms, people, ..., giving an immense array of mathematical structures and physical applications. To develop and use these would be impossible, except for one fortunate fact. All these groups, though they may have vastly different expression and usage, all fall into a few classes, the members of which are all isomorphic — abstractly (essentially) the same. It is the concept, and occurrence, of isomorphism which allows mathematics and physics.

Problem III.4.b-1: Can a group table be considered an operator? Why can differential operators form groups?

Problem III.4.b-2: Show that all order-2, and likewise all order-3, groups are isomorphic [Baumslag and Chandler (1968), p. 95; Landin (1989), p. 119].

Problem III.4.b-3: Is an isomorphism an equivalence relation [Baumslag and Chandler (1968), p. 97]? What are the objects?

Problem III.4.b-4: Give examples of larger groups, like ones we have studied, that are different, but isomorphic. Are there other examples of geometrical symmetry groups, isomorphic, but different enough so their symbols represent unlike transformations?

Problem III.4.b-5: Readers familiar with logarithms can find an isomorphism between the group of positive numbers (combination being multiplication) and the group of real numbers (combination being addition) [Baumslag and Chandler (1968), p. 119; Grossman and Magnus (1992), p. 103; Landin (1989), p. 123]. Is there is an operator that carries out this mapping. Can it be an element of the group? Note that here we take the two groups isomorphic even though they have different products. However the lists of products (which cannot be written down completely) are the same.

Problem III.4.b-6: Can symmetric groups (sec. II.4.d, p. 51) S_n and $S_m, n \neq m$, be isomorphic [Baumslag and Chandler (1968), p. 97; Landin (1989), p. 124; Scott (1987), p. 27]? How about different cyclic groups (sec. III.2.a, p. 68)? Under what conditions are two Abelian groups isomorphic [Scott (1987), p. 29]? Are nonisomorphic Abelian groups of the same order possible? If so, give examples.

Problem III.4.b-7: Show that there are an infinite number of nonisomorphic groups [Baumslag and Chandler (1968), p. 97].

Problem III.4.b-8: Let p and q be different primes. There is only one group with order p, and one with order q. Why? Find the number of nonisomorphic groups for orders $p^2, p^3, p^4, p^5, p^2q^3, qp^2, qp^3, qp^4, qp^5$ [Dixon (1973), p. 20, 94]. Try it with $p, q = 2, 3, 5$.

Problem III.4.b-9: Show that [Dixon (1973), p. 20, 94] the nonisomorphic groups of order 12 (sec. III.2.f, p. 76) are two Abelian groups (describe them), the alternating group A_4 (sec. II.4.d.ii, p. 54), the dihedral group (sec. III.2.h, p. 79) D_6, and a group given by the relations

$$y^6 = 1, \ \ x^{-1}yx = y^{-1}, \ \ x^2 = y^3. \tag{III.4.b-3}$$

What are their group tables?

Problem III.4.b-10: For two groups there can be different isomorphisms from one onto another. That is, different elements of one can be correlated with (mapped to) each of the elements of the other. Thus, there are a set of isomorphisms. Dihedral group D_3 is isomorphic to S_3. Is there more than one way of assigning to each D_3 element an S_3 element? Write down all such sets of assignments. Do these form a group? Does it mean anything to take the product of two sets? What? (That is, an isomorphism is a transformation between two objects, here D_3 and S_3, so we should be able to apply rules for handling transformations to these.) Does the set of isomorphisms between two groups ever form a group?

Problem III.4.b-11: In our daily lives, to what extent are the days isomorphic? How necessary is this for life to be possible?

III.4.c Homomorphisms

A homomorphism ("like form" or "like shape") is a mapping (here from one group to another) preserving the multiplication rules. It is not required to be one-to-one. To each element of one group there may correspond more than one of the other; several elements of one can map into a single element of the other. An isomorphism (not merely "like", but "equal", shape) — which is one-to-one — is a special case. If,

$$A_1 \Rightarrow \alpha,\ B_1 \Rightarrow \beta,\ C_1 \Rightarrow \gamma,\ \text{and}\ A_2 \Rightarrow \alpha,\ B_2 \Rightarrow \beta,\ C_2 \Rightarrow \gamma, \qquad \text{(III.4.c-1)}$$

and

$$\alpha\beta = \gamma, \qquad \text{(III.4.c-2)}$$

the preservation of the product requires that

$$A_1B_1 = C_1 \text{ and also } A_2B_2 = C_2. \qquad \text{(III.4.c-3)}$$

A homomorphism is a mapping of the elements of one group *into* the elements of another. If we put something into something else (like a container) we imply that it is smaller so that it can fit in. When we put something on (here onto) there is more of a suggestion of an exact fit. Strictly, "into" means that all elements of set L into which set S is poured need not have elements from S hitting them, while "onto" means that every element of set L is hit by at least one, but maybe more, from set S. An isomorphism is one-to-one and onto, so an isomorphism works both ways, while a homomorphism can be many-to-one and into, so need go in only one direction. (If not all elements of G_2 have an element of G_1 mapped to them, then G_1 is homomorphic to a subgroup of G_2).

All groups are homomorphic to the group of a single element (the identity), so all are (also) homomorphic to this subgroup. Thus that groups are homomorphic may tell us nothing about their relationship to each other. For isomorphism the situation is quite different.

Problem III.4.c-1: Is onto an equivalence relation (sec. III.4.a, p. 85)? Is into? Is a homomorphism an equivalence relation?

Problem III.4.c-2: If one group is mapped into another — several elements of the first are mapped into the second — does the container analogy fit? For a homomorphism must every element of one group map into at least one of the second? Might different elements of a group have different numbers of elements of the other group mapped into them, or conversely? Why? Explain why the identity must be mapped to the identity, but other elements can also be. Does this differ for an isomorphism? Also the homomorphic image of the inverse is the inverse of the image [Grossman and Magnus (1992), p. 101]. Why? The homomorphic image (only one?) of an element (only one?) is the element of the other group to which it is mapped.

Problem III.4.c-3: Can we consider isomorphisms as transformations leading to groups of isomorphisms? How about homomorphisms? Does the set of homomorphisms (operators) of one group to another itself form a group? Always? Ever? If there is a homomorphism of one group to a second, must (can) there be a homomorphism of the second to the first? If not what type of object is a set of homomorphisms?

Problem III.4.c-4: Could a group be homomorphic to a subgroup? Must it be? A subgroup is proper if it is neither the identity nor the entire group. Can a group be isomorphic to a proper subgroup [Grossman and Magnus (1992), p. 105]? Homomorphic? Give (counter-)examples.

Problem III.4.c-5: A set of relationships between the elements of one group and that of another is an operator — at least in the sense that we can make a table and regard it so. However given two groups, it may be that there are different such tables — different homomorphisms of one to the other, so different operators. Give examples, that is define operators or tables mapping the elements of one to those of the other, for which there is more than one homomorphism. Under what conditions can such a set of operators form a group?

Problem III.4.c-6: If group G is homomorphic to group F, prove that a subgroup of G is homomorphic to one of F. Could the latter homomorphism be an isomorphism? Determine (if possible) what the subgroup of F is. Are there operators that produce these homomorphisms (that is which act, perhaps by multiplication or a similarity transformation, on the elements of one group to give those of the other)? What are their properties?

Problem III.4.c-7: Can the groups of order 4 (sec. III.2.c, p. 71) be homomorphic? If so, is there an operator (from the groups) producing the homomorphism?

Problem III.4.c-8: Is S_3 homomorphic to S_2? S_4? How about S_4 and S_3? Show that S_n is homomorphic to S_2 (and S_3, for any n?). Can two symmetric groups (of different order) be homomorphic? Are there rules telling which S_n is homomorphic to S_m, for any n and m? Are there operators that produce these homomorphisms? If so, what are their properties?

Problem III.4.c-9: Find the (a?) homomorphism from the order-4 cyclic group (sec. III.2.a, p. 68) to the order-2 group (for each element of the latter give the list of elements of the former that are mapped to it).

Problem III.4.c-10: Can an Abelian group be homomorphic to a non-Abelian one? Interchange Abelian and non-Abelian in the last sentence. Can an operator carrying out such a homomorphism be given, representable by a single symbol, perhaps identical to one from these groups?

Problem III.4.c-11: Are there pairs of cyclic groups that are not homomorphic? Give a rule stating when two cyclic groups are homomorphic [Baumslag and Chandler (1968), p. 106, 107, 118], and perhaps also operators (from where)?

Problem III.4.c-12: Show that every dihedral group (sec. III.2.h, p. 79) is homomorphic to the order-2 group [Hall (1959), p. 34].

Problem III.4.c-13: For a cube [Hall (1959), p. 34] draw three lines connecting the midpoints of opposite faces. Find the group (of symmetries of the cube) that permutes these lines. Show that it is a group of order six (which?, is it Abelian?), and that it is a homomorphic image of the group of the cube (sec. III.2.g, p. 78) — the group of the cube is homomorphic to it. This sentence goes in one direction only, why?

Problem III.4.c-14: Can two groups not be homomorphic? Under what conditions?

Problem III.4.c-15: Are there procedures to determine when one group is homomorphic to another? Isomorphic [Kurosh (1960a), p. 76]? Can these be stated as a (computer-implementable) algorithm?

Problem III.4.c-16: A homomorphism of G into K takes subset F of G into subset L of K; if F is a subgroup of G, must L be a subgroup of K? If L is a subgroup of K, must F be a subgroup of G? Can a homomorphism of G into K result in an isomorphism of subsets F onto L? Under what conditions?

Problem III.4.c-17: Are all infinite (discrete) groups (sec. II.3.j, p. 48) isomorphic to those of the integers (the reals)? Homomorphic?

Problem III.4.c-18: Is it possible for two groups of the same (different) order to be homomorphic but not isomorphic?

Problem III.4.c-19: Can rules, or algorithms, be given that tell if two groups are homomorphic but not isomorphic? Is the order relevant? Can such algorithms be computer implementable?

III.4.d Endomorphisms

Since endo means inner, an endomorphism is a homomorphism of a group into (a subgroup of) itself [Landin (1989), p. 138; Lyapin, Aizenshtat, and Lesokhin (1972), p. 65; Scott (1987), p. 24].

Problem III.4.d-1: Show that raising every element of an Abelian group to the power n gives an endomorphism [Scott (1987), p. 29]. Is it an isomorphism?

Problem III.4.d-2: For the water group (sec. II.2.d, p. 35), the pyramidal group (sec. II.2.e, p. 36), the groups of the square (sec. II.2.f, p. 37), and the double rectangle (sec. II.2.g, p. 39), and for S_2, S_3, S_4 (sec. II.4.d, p. 51), construct all endomorphisms of each into their subgroups. Construct means give a set of relations like "elements A, B, C are mapped into (correspond to) element α ". Also give operators, if any exist, and tell where they come from.

Problem III.4.d-3: Is it possible to give a general rule for constructing, perhaps with operators, all endomorphisms for any S_n?

Problem III.4.d-4: Under what conditions is it possible to have an endomorphism of a group into a subgroup?

Problem III.4.d-5: Can there be different endomorphisms of a group into the same subgroup? What defines the difference if such exists?

III.4.e Automorphisms

Auto means self (as in automobile) so an automorphism is a self-isomorphism, a mapping of a group onto itself, a shuffling of symbols [Armstrong (1988), p. 131; Baumslag and Chandler (1968), p. 84, 87; Burn (1991), p. 161, 166; Dixon (1973), p. 19; Hall (1959), p. 84; Kurosh (1960a), p. 85; Landin (1989), p. 138; Lyapin, Aizenshtat, and Lesokhin (1972), p. 99; Scott (1987)]. It may thus not seem like we are doing anything with an automorphism; however it is not trivial. The automorphisms of a group are limited and provide much information about it. After all we cannot shuffle symbols arbitrarily and still have the same, or even *a*, group. An automorphism is a special type of endomorphism since it is a special type of homomorphism, an isomorphism.

Problem III.4.e-1: Can symbols be shuffled arbitrarily and still have a group, though perhaps a different one? Why? Try it.

Problem III.4.e-2: For S_3, show that permuting the symbols acted on, say, $1 \Rightarrow 2 \Rightarrow 3$, gives an automorphism. This permutes the transpositions $(12) \Rightarrow (23)$, and so on. Thus a permutation of the transpositions gives the same group, so is an automorphism. Why? What happens to the two cycles of length 3? Clearly the interchange of a transposition and the identity is not an automorphism. Check that neither is the interchange of a transposition and a cycle of length 3. No automorphism can interchange cycles of different length. Thus some, but not all, permutations of the group operations preserve the multiplication rules; automorphisms are limited, and closely tied to the structure of the group. List the corresponding elements given by this automorphism of S_3.

Problem III.4.e-3: For S_n, prove that an automorphism does not change the number or size of the cycles making up a permutation.

Problem III.4.e-4: Show that the subset of elements, the identity, (12), (34) and (12)(34), forms a subgroup of S_4. Find its automorphisms and show that they form a group. Are there operators producing these automorphisms? Are they group elements?

Problem III.4.e-5: Find all automorphisms (including operators carrying them out, if they exist) of cyclic groups [Dixon (1973), p. 21, 95; Lyapin, Aizenshtat, and Lesokhin (1972), p. 100, 219; Scott (1987), p. 117]. Do the same for dihedral groups (sec. III.2.h, p. 79) [Hall (1959), p. 90].

Problem III.4.e-6: The complex numbers form a group. Why? With what combination law? Check that taking the complex conjugate of each element gives an automorphism [Scott (1987), p. 29]. What is a geometrical interpretation?

Problem III.4.e-7: If a product of a pair of homomorphisms gives the identity mapping (transformation) of the first group onto itself, show that both homomorphisms are isomorphisms [Baumslag and Chandler (1968), p. 96]. Would this be true if the product were an automorphism?

III.4.e.i *Inner and outer automorphisms*

Since a morphism is a transformation there is an operator to carry it out (not necessarily a mathematical operator; see the above operations changing English to Greek). This leads to two types of automorphisms, inner and outer. Obviously, these are carried out by operators in (belonging to), and outside of (not belonging to), the group.

Problem III.4.e.i-1: Find the S_3 operator carrying out the automorphism of pb. III.4.e-2, p. 91, and verify that it belongs to S_3, thereby showing that it is inner. Show that all S_3 automorphisms are inner. Repeat for S_4 [Dixon (1973), p. 21, 96]. A symmetric group permutes symbols. But a permutation of the symbols permutes the group operators, though this is not the reason that there are inner automorphisms for these groups.

Problem III.4.e.i-2: Find the inner automorphisms of the quaternion group (pb. III.2.e-4, p. 74) [Lyapin, Aizenshtat, and Lesokhin (1972), p. 101, 220]. Explain why the structure of the group gives automorphisms, and why inner automorphisms.

Problem III.4.e.i-3: Consider the group with identity E and commuting generators, R, I, both of second order. Prove that any permutation of elements R, I and RI, is an automorphism, and that it cannot be produced by an operator of this group, hence is outer. This shows the need, and rationale, for outer automorphisms. Interpreting the operations as a reflection and inversion, their interchange cannot alter the group structure, since their group properties do not distinguish them, giving thus an automorphism. However it is clear, not only from the group but from the geometrical interpretation, and from their effect on coordinates, that no product of a reflection and inversion can convert a reflection to an inversion, thus the automorphism is outer.

Problem III.4.e.i-4: One interesting way of constructing an isomorphism (here an automorphism) is to map each element onto its inverse (all elements are replaced by their inverses). Show that this is an isomorphism for, and only for, Abelian groups [Armstrong (1988), p. 131; Burn (1991), p. 161, 166; Lyapin, Aizenshtat, and Lesokhin (1972), p. 100]. For an arbitrary Abelian group, is there an operator (of the group?) taking each element to its inverse? Test this using cyclic groups (sec. III.2.a, p. 68), the four-group (sec. III.2.c, p. 71), S_2 and S_3. What is the relationship between S_3 and the group of inverses? Do their group tables differ? Can an operator be found that carries out this mapping? In general? For S_2? S_3? S_n? Is it an element of the (a) group? These give examples of automorphisms that must be outer. Explain why they cannot be inner.

Problem III.4.e.i–5: For alternating group A_4 (sec. II.4.d.ii, p. 54), show that replacing each permutation p, by $(12)p(12)$, is an automorphism [Burn (1991), p. 161, 166]. Since transpositions are not in the group, it is outer.

Problem III.4.e.i–6: Symmetric groups (sec. II.4.d, p. 51) have only inner automorphisms [Kurosh (1960a), p. 92], except for S_6, which also has an outer automorphism. It interchanges the two classes of elements of order 3 [Hall (1959), p. 90]. Find it.

Problem III.4.e.i–7: Consider the group [Hall (1959), p. 90] with generators a, b, c and defining relations

$$a^8 = b^8 = c^4 = 1, \quad b^{-1}ab = a^5, \quad c^{-1}bc = a^6 b. \qquad \text{(III.4.e.i–1)}$$

Show that this gives a group of order 256 with elements of the form $a^i b^j c^k$, and that it has an outer automorphism given by

$$a \Rightarrow a^5, \quad b \Rightarrow b, \quad c \Rightarrow c. \qquad \text{(III.4.e.i–2)}$$

Notice what happens to the defining relations under this transformation. Also check that there is no group operator that carries out this transformation, so it must be outer.

Problem III.4.e.i–8: If a subset of operators of a group produce automorphisms must all the operators?

Problem III.4.e.i–9: For which groups are the inner automorphisms limited to the identity [Lyapin, Aizenshtat, and Lesokhin (1972), p. 101, 220]?

III.4.e.ii *The automorphism group of a group*

Group theory is (mostly) about transformations, and among these are ones on the groups themselves. Automorphisms are transformations, and, since they transform a set to itself, products can be defined. Thus, the automorphisms of groups form groups [Armstrong (1988), p.

131; Baumslag and Chandler (1968), p. 83, 87; Dixon (1973), p. 21, 95; Hall (1959), p. 84; Lyapin, Aizenshtat, and Lesokhin (1972), p. 65, 99]. Might the automorphism group of a group provide information about it?

Problem III.4.e.ii-1: Must (can?) the set of automorphisms of a group form a group? Why? How is it related to the group? Are isomorphisms different? Must (can?) the inner automorphisms form a group [Kurosh (1960a), p. 88]? The outer ones? Only both together? Explain why the automorphism group of group G is the same as the group of symmetries of the group table of G. Need it be a subgroup of G? If the automorphisms form a group must the order of this group be finite (for a finite group)? Must its order be larger than, equal to, less than, that of the group, or does it depend on the group? What about the group of inner automorphisms? Given either a group or its set of automorphisms, can the other be found (group tables, and if they exist operators — from the other group) [Kurosh (1960a), p. 88]? How much can be determined about a group from its automorphism group? Is there a (computer-implementable) algorithm to do so?

Problem III.4.e.ii-2: Prove that the automorphism group of a finite group is finite [Baumslag and Chandler (1968), p. 93; Kurosh (1960a), p. 88].

Problem III.4.e.ii-3: An automorphism is an endomorphism. Can endomorphisms that are not automorphisms form groups?

Problem III.4.e.ii-4: If a group is Abelian we might expect its group of automorphisms to be also. However, show that for every (finite) noncyclic, Abelian group (sec. III.2.b, p. 69), the automorphism group is non-Abelian [Dixon (1973), p. 21, 96; Hall (1959), p. 90]. What about a cyclic group? Check this for the four-group (sec. III.2.c, p. 71). This shows why an automorphism group of an Abelian group is non-Abelian. The group elements are symbols, subject to restrictions, but with an Abelian group these are not severe. Thus the group is invariant under permutations, and for the four-group under S_3. But symmetric groups are non-Abelian.

Problem III.4.e.ii-5: Show that for the cyclic group of order n, the automorphisms consist of $T \Rightarrow T^k$, for all elements T, where n and k are relatively prime [Lyapin, Aizenshtat, and Lesokhin (1972), p. 100]. Determine the automorphism group.

Problem III.4.e.ii-6: Prove that the automorphism group of a non-Abelian group is not cyclic [Lyapin, Aizenshtat, and Lesokhin (1972), p. 102]. Is the converse true? Also the automorphism group of a non-Abelian group has order greater than one (that is, every non-Abelian group has a nontrivial automorphism group) [Baumslag and Chandler (1968), p. 93].

Problem III.4.e.ii-7: Find all automorphisms of S_2, A_2, S_3, A_3, S_4, A_4 [Baumslag and Chandler (1968), p. 84, 85; Dixon (1973), p. 21, 96]. Can the automorphisms of S_n be determined [Hall (1959), p. 90; Kurosh (1960a), p. 92; Lyapin, Aizenshtat, and Lesokhin (1972), p. 101, 220; Mirman (1991e), Appendix C]? Do these automorphisms form groups? How are they related to the groups whose automorphisms they are? Are there operators that carry out the mappings? Do they belong to a group [Scott (1987), p. 309]? Which? What do the these answers show, what information do they give, about (the) groups? Why do they show that knowing automorphisms is not trivial, that we cannot shuffle symbols arbitrarily?

Problem III.4.e.ii-8: Prove that the dihedral group of order 8 is isomorphic to its group of automorphisms [Hall (1959), p. 90]. Is this true for any other order? Why is 8 special ?

Problem III.4.e.ii-9: For the quaternion group (pb. III.2.e-4, p. 74, pb. III.4.e.i-2, p. 92), find all automorphisms [Lyapin, Aizenshtat, and Lesokhin (1972), p. 100, 101, 219, 220]. Can the automorphism group of the quaternions be obtained from a presentation?

Problem III.4.e.ii-10: Do the groups of pbs. III.3-5-8, p. 83 have automorphism groups? Can they be obtained from the (or any) presentations?

Problem III.4.e.ii-11: Find the automorphisms of the water (sec. II.2.d, p. 35) and pyramidal (sec. II.2.e, p. 36) groups. Can these be produced by operators? By group elements? Do they form groups? How are these related to the groups whose automorphisms they are? What information do they give about them; about the geometrical figures whose symmetry groups they are? If we take an automorphism group of a geometric symmetry group (for example, sec. II.2.d, p. 35–sec. II.2.h, p. 40), and regard it, if possible, as itself such a group, that is we find a figure whose symmetry group the automorphism group is, do we get relationships between the figures? Can we get information about one figure from the other? Give explicit answers for these two cases.

Problem III.4.e.ii-12: What are the automorphisms of the groups of the square (sec. II.2.f, p. 37) [Hall (1959), p. 90] and the double rectangle (sec. II.2.g, p. 39)? How do they differ? What does this (difference) say about the geometry of the figures? About their group tables? Are there operators? Discuss the physics (geometry?) of all similarities and differences. Answer the questions of the last problem.

Problem III.4.e.ii-13: Are there examples of nonisomorphic groups with isomorphic automorphism groups [Dixon (1973), p. 21, 96; Kurosh (1960a), p. 89]? If the nonisomorphic groups are taken as symmetry groups of figures, what do the isomorphisms of their automorphism groups say about the relationship of the figures? Are there nonisomorphic automorphism groups whose own automorphism groups are isomorphic? What does this say about the relationship of the groups for which these are automorphism groups?

Problem III.4.e.ii-14: Can any groups not be automorphism groups [Kurosh (1960a), p. 89]?

Problem III.4.e.ii-15: Given a (minimal?) basis B of group G. An automorphism of the group produces (?) an automorphism of the basis, these denoted by G^a and B^a. It should be clear that B^a generates G^a; also B^a generates G [Dixon (1973), p. 21, 95]. Is the last clause redundant?

Problem III.4.e.ii-16: If an automorphism does not leave any element of the group fixed (except the identity), then the group consists of elements A^{-1}A', where A runs over all group elements, and A' is the automorphic image of A [Dixon (1973), p. 21, 97]. What is the automorphic image?

Problem III.4.e.ii-17: Show that for finite group G, if its automorphism group has an element of order 2 that leaves only the identity invariant, G is Abelian, and of odd order [Dixon (1973), p. 21, 97].

Problem III.4.e.ii-18: Find a (general) algorithm giving the automorphism group for any group; and also one using its (minimal) presentation.

III.4.e.iii *Automorphisms and isomorphisms*

Is there really any difference between isomorphism and automorphism — does it matter whether we change the symbols or only shuffle them? If we regard the group as a set of abstract symbols, there is not much difference. But there is an important difference between rotations of a particle's spin and rotations of its isospin. The transformation to a different set of axes is an automorphism of the SU(2) group describing spin. But the relationship between this group and the SU(2) group describing isospin is an isomorphism.

Even regarding the groups as symbols there is a difference, also quite important, between a set of letters and a set of matrices. An abstract group, a collection of letters or other symbols, can be isomorphic to a group of matrices. This will bring us to a concept at the core of this work, representations.

Problem III.4.e.iii-1: Knowing the automorphisms of a group can you find all isomorphisms between it and another group? Can the reverse be done?

Problem III.4.e.iii-2: The combination of morphisms suggests the great generality of the concept of a product. A group can consist of many types of things, with many ways of combining them. However one might suspect that there are relations between these different types of objects, and different types of products. Thus the morphisms (mappings) act on groups, but they do so by acting on their elements, setting up correspondences between different elements, or elements of different sets. The group elements can be placed in row or column matrices — that is what we do in a group table. This reflects the ability to write transformations given by the morphisms as matrices, and that combining morphisms means combining matrices, presumably by matrix multiplication. Review the sets of transformations considered above. Can these be written as matrices? Verify that if this is done the rule of combination is matrix multiplication — the morphism that is the product of two, corresponds (in what way?) to the matrix that is the product (using matrix multiplication) of the two matrices, corresponding to the morphisms in the product. Are there restrictions on these matrices? Are they square? Nonsquare? Singular? Nonsingular? Unique? Nonunique? In what way(s)? Under what conditions? How? Why? Do the answers depend on the type of morphism? Do all morphisms of one type give the same answers? How about other restrictions, if any? Are these morphism dependent? Element dependent? Thus we (might) have, for each group of morphisms a set of matrices that represents it — that corresponds to it. But these groups of morphisms are just finite groups, with a particular interpretation, and application. There is nothing to distinguish them, in abstract, from groups given other interpretations, or none at all. This implies that for all (finite) groups we can (hopefully) find a set of matrices that represents it, that corresponds to it. This is an interesting point to which we shall have to return and discuss in great depth.

III.4.f Enantiomorphisms

A right hand reflected in a mirror is different, being then a left hand. Thus it is with many objects, including molecules. If an object is different from its mirror image, then the two forms, the object and the image, are enantiomorphic (having "opposite form" or "opposite shape", from the Greek words *anti* — against — and *enantios* — opposite); they are enantiomorphs (or enantiomers). Though certainly not antagonistic, not being at opposite corners (of any polygon) since they are only reflected in a plane, they do have shapes, and these are opposed to each other in a mirror. Each of the two enantiomorphic forms are isomers ("equal parts").

Many objects, macroscopic and microscopic, are related by a reflection. The different enantiomorphic objects (forms would be redundant), distinguishable, though not really different, often occur with vastly different probabilities, especially in biological systems. An interesting point, with no certain explanation.

Problem III.4.f–1: Find (draw) objects that are members of enantiomorphic pairs [Hargittai and Hargittai (1987), p. 57, 63; Martin (1987), p. 192], and determine when (always?) it is possible to find operators that convert one member to the other (say by acting on coordinates of points on the objects). If there is also a symmetry group, can that operator be related to the group? How do the answers to these questions depend on where the mirror is?

Problem III.4.f–2: Consider the objects that have the (low order) symmetry groups discussed above. Do any of these exist in enantiomorphic pairs? Is it possible for objects to have (any of) these groups as (broken?) symmetries while being members of such pairs? How do the answers depend on where the mirror is placed? Do the answers say anything about the symmetry groups?

III.4.g Isometries

An isometry ("equal measure") is a transformation that keeps distance invariant [Armstrong (1988), p. 136; Baumslag and Chandler (1968), p. 64; Burn (1991); Lyapin, Aizenshtat, and Lesokhin (1972), p. 70; Martin (1987)]. To have an isometry we must have a distance defined. But in many cases, certainly in physics, there is a natural definition. Thus a square is often a physical object and there is an obvious (and unavoidable) definition of the distance between points on it. Or if we regard it as a geometrical figure, then there is an implication of Euclidean (or at least some other such) geometry. An isometry of the square leaves the distance between points unchanged.

Most of the groups we have considered are isometries, as we have interpreted them (as rigid body transformations, or reflections and inversions). But there can also be groups isomorphic to these with different physical meanings; these may not be isometries. Transformations by a curved mirror are not, nor are those obtained by blowing up a balloon.

Another example is a molecule in the form of an isosceles triangle, with the atom at a vertex different than the two at the base. If we permute the atoms, so the one at the vertex becomes the same as one at the base, but different from the other, than the distance between the atoms changes. Geometrically this transformation leaves the figure unchanged, so is an isometry. Physically, using a different definition of distance, it is not. Of course this does not leave the molecule invariant, which is not unrelated to it not being an isometry.

Problem III.4.g–1: Verify that reflections and inversions are isometries. Find other examples that are, and are not.

Problem III.4.g-2: Can the groups we have considered be interpreted so that they are not isometries?

Problem III.4.g-3: Are all sets of isometries groups? Are there conditions under which a set of isometries is (is not) a group?

III.5 COMBINATIONS OF SETS AND GROUPS

A group consists of a set of transformations plus rules for combining them. From an assortment of sets we can construct other sets, so from an assortment of groups, other groups. This we have been doing all along so the reader should be familiar with the concepts. Yet it is useful to give the definitions formally, which is the excuse for including this section.

A group is a set [Baumslag and Chandler (1968), p. 1; Landin (1989), p. 1], with conditions. We consider various subsets. These can be combined in diverse ways (to give the group), with two general types. The sum, or union, of (sub)sets is the (sub)set containing all the elements appearing in any of them; the intersection is the set containing (all) elements appearing in both. The product of (sub)sets is the (sub)set containing all products of all elements in the (sub)sets.

These definitions have an interesting consequence. Let the sets be $S_1, S_2, \ldots$, with elements $a_1(i), a_2(i), \ldots$, where i labels the elements of the set, and its range is need not be the same for all the sets. Then the sum of the sets is the one containing all elements

$$a_1(i), a_2(j), \ldots, \text{ all } i, j. \tag{III.5-1}$$

The product is the set containing

$$a_1(i)a_1(j), a_1(i)a_2(k), \ldots, \text{ all } i, j, k. \tag{III.5-2}$$

Given these subsets we can form various sums,

$$A_{12} = S_1 + S_2, \quad A_{13} = S_1 + S_3, \quad A_{123} = S_1 + S_2 + S_3, \quad \ldots, \tag{III.5-3}$$

and similarly for products,

$$P_{12} = S_1 S_2, \tag{III.5-4}$$

and so on. Having done this, that is having made such a list of all sums or of all products, we can forget that these symbols represent sets, and just regard them as abstract symbols (or transformations, or whatever). Thus, we have a set of symbols with a product rule. If this rule obeys the group axioms, then this set of symbols — this set of sets — itself forms a group. This provides an interesting hint, which we develop below.

Problem III.5-1: Prove that the product of subsets of a group is associative. Of course the sum is.

Problem III.5-2: Why is the intersection of subgroups a subgroup [Baumslag and Chandler (1968), p. 55]?

Problem III.5-3: Show that if an element of a group commutes with all elements of a subset, so does the inverse of that element.

Problem III.5-4: Under what conditions does a set of subsets of a group itself form a group, for the combination rule the set product, and also the set sum?

III.5.a Sums of sets

For a system of two atoms there is a set of transformations consisting of the rotations that act on one atom (leaving the other unchanged) plus a similar set acting on the second atom. The set consisting of both of these rotations is an example of a set that is a sum of sets — it contains all operators appearing in any of the sets, and no other.

This is a special case of a sum of sets, a direct sum. For a direct sum, no operator in one set is in any other, and the operators of the different sets commute. The number of operators in the sum set is the sum of the numbers of operators in each of the sets. Rotations of different atoms commute, and a rotation of one does not affect others.

Problem III.5.a-1: Examples of sums that are not direct sums are provided by the symmetric groups (sec. II.4.d, p. 51). The sum of S_{n-1} and the S_n operators not in it is S_n, but not all elements of one set commute with all of the other. Check this, for some symmetric groups. That set consisting of the permutations of 1, 2 and of those of 3, 4 is a direct sum; it is a subgroup of S_4. (There is a subtle point, about the identity, which is ignored here.)

Problem III.5.a-2: Must the sum of two groups be a group? If the sum of sets is a group must all the sets in the sum be? Must some of them be? Does it matter if the sum is a direct sum? Give examples.

III.5.b Products of sets and groups

From a collection of sets we can form another; that consisting of all products of a member of one set with any other element — the product of the sets, the set of all elements formed by taking all products of elements in any of the sets. If the sets are groups, then each has an identity, so the product includes all the sets (here groups), plus all products of all elements of each with all of all others.

Consider an electron moving around a nucleus; it has both orbital angular momentum and spin. There are two sets of operators acting on it: those that rotate its orbital angular momentum, and those that rotate its spin. The group acting on the electron is the product of these two; a statefunction of the electron is the product of two terms, one giving the spatial dependence (so orbital angular momentum) and the other the dependence on spin direction. A rotation operator acting on the statefunction is therefore also a product of two operators, each acting on one of the terms in the product statefunction. The set of rotation operators, the operators of the rotation group acting on the electron, is the product of the sets of operators acting on the orbital, and on the spin angular momentum. Here the three groups are isomorphic to each other, but they need not be.

Problem III.5.b-1: How is the group of rotations of observers related to these three groups?

Problem III.5.b-2: What is the product formed from S_2 acting on 1, 2, and S_2 acting on 3, 4? On 1, 2 and on 2, 3? How about S_2 and S_3 acting on 1, 2 and 3, 4, 6? Are these product groups isomorphic to any terms in the product? Homomorphic?

Problem III.5.b-3: Is the product of two groups always a group? Can a group be formed from a product of sets none of which is a group? Some of which are groups?

III.5.c Direct products

A special product of groups is the direct product [Baumslag and Chandler (1968), p. 143; Grossman and Magnus (1992), p. 72; Hamermesh (1962), p. 30]. A group is a direct product of its subgroups if all the subgroups commute (every element of one commutes with every element of every other one), and if each element of the group can be expressed in only one way (up to ordering) as a product of elements of the subgroups. Essentially, a direct product is a set of operators, each a product of operators such that the transformations in the product do not affect each other.

So a nucleon can be transformed in two distinct ways; its spin can be rotated and its isospin can be rotated (a neutron can go into a sum of a neutron plus a proton, and similarly for a proton). We can perform rotations and isospin transformations independently; the group acting on the nucleon is a direct product of these two groups. To be adventurous, we can consider these two as subgroups of a larger group, like SU(6), by introducing other transformations. Ignoring such transformations, the group acting on the nucleon is a direct product, though this product group is actually a subgroup of a larger one.

The set of transformations of the Poincaré group (sec. II.3.h, p. 45) is a product of the sets of transformations of the Lorentz group and the translation group, but it is not a direct product since the transformations of these subgroups do not commute.

A semi-direct product of two groups is one in which the two sets share no element (except the identity), and the product of an element of group G_1 with one of G_2 is always a member of G_2. In this sense G_2 is invariant. Inhomogeneous groups, like the Poincaré group, are semi-direct products. There are also examples of finite groups, such as crystallographic groups.

Problem III.5.c-1: For the group of the square (sec. II.2.f, p. 37) are there two subgroups whose product gives the group? Is this a direct product? Semi-direct? Is there any way of writing the group as a direct product of subgroups? Try this for the other (small) groups considered, especially the geometrical ones (is this a limitation?). Is there a (geometrical) interpretation of the answers?

Problem III.5.c-2: Show that the direct product of a finite number of finite cyclic groups (those with only a finite number of elements) is cyclic, if the orders of the groups in the product are relatively prime in pairs (the order of any one divided by the order of any other is not an integer) [Dixon (1973), p. 20]. Would the group be cyclic if the orders were not all relatively prime? Are there counter-examples?

Problem III.5.c-3: Show that an Abelian group whose basis has n elements, can be written as a direct product of m cyclic subgroups, where $m \leq n$ [Dixon (1973), p. 20]. Give examples.

Problem III.5.c-4: A (finite) Abelian group G can be written as the direct product of m cyclic subgroups. Show that with m the smallest possible number of generators of G (the ? of G), the order of cyclic subgroup C_i divides (the quotient is an integer) that of C_{i+1}, for all $i < m$ [Dixon (1973), p. 20, 95]. How are the orders of C_i and C_j related if $j \neq i + 1$?

Problem III.5.c-5: It should (?) be clear [Dixon (1973), p. 21, 95] that the group of all rational numbers, using addition as the product (of course?) cannot be written as the direct sum of trivial subgroups. How about the direct product? Would the result be different if other types of numbers were used, all numbers, all reals, all integers, all complex, the others listed in sec. II.4.f, p. 55? Would it matter if the product were multiplication (if possible?)?

Problem III.5.c-6: How is the order of a group related to the orders of the groups whose direct product it is? Would the answer be different if the product were not a direct product? What would it be?

Problem III.5.c-7: Does the direct product of groups have to be a group? Are the groups in the direct product always (ever) subgroups of the complete group? Can a group be formed from a direct product of sets none of which is a group? Some of which are groups? Answer these for a semi-direct product. A general one.

III.6 DEVELOPING ALGORITHMS TO FIND GROUPS

One way to study mathematics is to develop algorithms and programs to prove theorems, provide examples and produce illustrations and diagrams. These require thinking about the systems, stating their properties in different language than that usual in mathematics, and in a different way, and they require the problem to be analyzed differently, often more precisely and with more specificity. Reducing theorems to algorithms requires making the abstract concrete, the general specific and exact, and forces thinking, and understanding. It requires theorems to be rewritten, which helps understanding, as does putting them in different forms, completely different ways of stating an approach to a problem. Here then are some suggestions, but these approaches should also be used for the other systems and problems discussed, and for those the reader is imaginative enough to think of himself (or herself, or itself — if the reader can program a computer to have the imagination to think of useful examples).

Problem III.6-1: The order (number of transformations) of a group, g, does not determine the group; there can be more than one nonisomorphic group. A table of the number of (fairly small) groups, plus references, is given by Lomont (1961), p. 34, and different groups, for low order, are listed by Joshi (1982), p. 22. However, there are limitations. There are only a finite number of distinct groups of any order — there are only a finite number of ways to arrange objects in a table, and most of these do not give groups anyway. These groups determine the possible number and types of objects invariant under a given number of transformations. In sec. II.2.i, p. 40, we mentioned two different such objects. Give other examples of such groups, and such objects, for orders larger than have been discussed previously. There is one case in which g completely specifies a group. If g is a prime number there can only be a single (up to isomorphism) group. Why? What type of group is it? Write down the group table. Can this be written so it applies to any prime g? For every order there exists at least one Abelian group. Trivially, for $g = 1, 2, 3$ there can be only one group. Give their group tables. Prove that for $g = 4$ there are two nonisomorphic groups, both Abelian. How do they differ? What are their tables? Show that there is no way of reordering, or renaming, the symbols to make them isomorphic (this, of course, is the definition of nonisomorphic). Show that for $g = 6$ there are (only) two nonisomorphic groups. What are they? State their tables. For $g = 8$ there are five nonisomorphic groups of which three are Abelian. Give them.

Problem III.6–2: Could you tell by (quickly) looking at the tables of two groups if they are isomorphic? Would the symmetry of the tables be relevant? What would happen if the elements of one were reordered but not those of the other? Is there any way of reordering that would make an isomorphism, or a nonisomorphism, easy or difficult to see?

Problem III.6–3: Are there rules or algorithms to tell whether two (finite) groups are isomorphic? Can these be implemented as a computer program? Do they lead to theorems?

Problem III.6–4: Derive a formula (or algorithm) for the number of nonisomorphic groups as a function of the order. How many are non-Abelian? A computer program to calculate the number (and even better, the group tables) given g would be quite nice. Rules following from these would be even nicer.

Problem III.6–5: Write an algorithm and a computer program to take a list of words, and find those with non-Abelian symmetry groups, finding the names of the groups (insert a word and let the program identify the group, and so the name). Also have it find the generators, work out from the properties of the word the group table, show the action of the generators (as a movie, perhaps coloring different letters or parts of letters, and showing how they change under the action of the generators). Also have it show the difference between the action of the generators if parts of the word have different colors, and if all are the same color, thus showing the symmetry and how it is broken by the coloring. This can also be done for Abelian groups. Are there any noticeable differences between Abelian and non-Abelian cases? Can you watch the effect of the generators, or of the movie, and tell whether the group is Abelian or non-Abelian? Can you watch and tell what the group is? What its group table is? The number of its generators?

Problem III.6–6: Repeat this for the other objects discussed.

Problem III.6–7: For each (small) finite group find objects with symmetry under it. Repeat the previous problems. How can the properties of the group be used to find such objects?

Problem III.6–8: Write a program to draw objects on a screen, read the object from the screen, then find the group as above. Do this also for various molecules and crystals. What is the difference in the algorithms and programs for two-dimensional and for three-dimensional objects? How can the program tell that an object is three dimensional and its symmetry, from this two-dimensional projection?

Problem III.6–9: What are the best ways of generating two-dimensional projections of a three-dimensional object to show its symmetry; to allow the group to be determined? Are there algorithms to find the three-dimensional symmetry from two-dimensional projections? If so, use them (in programs) on the objects discussed.

Problem III.6–10: Use the group properties to find objects with symmetry described by the group, for each (small) finite group. Have the program draw a line, then show the action of the group generators on it, until a closed set is obtained. How is the resultant shape related to the shape of the object with the given symmetry? Will this method work for all finite groups? Starting with other shapes? For all finite groups?

Problem III.6–11: Repeat all this for whatever continuous groups can give objects with symmetry (like the rotation group).

Problem III.6–12: Take a symmetric object. Have the algorithm start with some shape and then draw the object. Show the action of the group generators. Do the same for some given object and also for a given group and try to find the object.

Chapter IV

Groups, Combinations, Subsets

IV.1 USES OF SUBSETS

Transformations on physical objects, on geometrical figures, are of different types, and are related in different ways. By studying a group of transformations of an object as a whole, we gain useful information about it. But we might expect further insight from analyzing various subsets of transformations, and by finding the types of subsets that a group can be divided into, and the ways it can be divided. Group operators can be split into subsets using various (useful) criteria; these provide tools for understanding not only the physics (and geometry), but for studying the structure of the group itself. Here we consider what information can be obtained from different ways of partitioning groups, and how to extract it. And we see how this can provide information, not only about particular groups, but about the structure of groups in general, how it limits, and generates, groups. Conversely, groups and their subsets can be combined; studying this is interesting, and necessary. This we start here also.

Problem IV.1-1: Consider the geometrical groups (sec. II.2.e, p. 36 — sec. II.2.h, p. 40). Based on the geometry, physics or chemistry (taking the objects as molecules or crystals) what subsets of transformations appear interesting or seem likely to provide useful information?

Problem IV.1-2: How about symmetric groups (sec. II.4.d, p. 51)?

IV.2 OPERATIONS ON OPERATORS: SIMILARITY TRANSFORMATIONS

Group elements perform operations (for example on vectors). However they can also be acted on themselves, by operators not in the group, and also by other group elements (else products could not be defined). We can use these to break a group into subsets; one type of operation is notably important.

A similarity transformation of T with operator P is given by

$$T' = PTP^{-1}. \tag{IV.2-1}$$

P can, but need not, be a member of the group. To transform *a group* by a similarity transformation, all group elements are acted on, using a single operator P — the same for all elements.

Problem IV.2-1: For example, the rotation-group operators are the rotations around three axes. However these axes are arbitrary. To go to a new coordinate system the operators are transformed. Explain why this is done by a similarity transformation.

Problem IV.2-2: Rotations produce interesting transformations of coordinates, but not the only ones. There is also inversion through the origin of the axes (which reverses the sign of all coordinates), and reflections in mirrors (reversing the sign of one). Hold three fingers of your right hand mutually perpendicular (thus forming a set of orthogonal axes) and reflect the hand in a mirror (the reflection is a left hand) to see the reflected set of axes. Show that the left hand cannot be obtained from the right hand by any rotation. Inversions and reflections are not members of SO(3) — to make sure this is understood we put an S in the group's name; without it they would be operators of the group. They give similarity transformations of the group operators using operators outside the group. How is the inversion (through the origin) related to the reflections in mirrors perpendicular to each of the axes (say, as your hand defines them)? Try all three reflections; describe the results. How would you carry out experimentally the product of all three? To go from rotations on a set of axes to those on its mirror image, we perform a similarity transformation on the rotation operators using a reflection operator. The rotation as seen in the mirror is found from the unreflected one by a similarity transformation. What is the rotation given by $IR(\theta)I^{-1}$, where I is an inversion and $R(\theta)$ a rotation through angle θ about, say, z? What are the rotations obtained using the three reflections in the mirrors perpendicular to the axes? What effect do these have on axes? Can these be answered for reflections in arbitrary mirrors?

Problem IV.2-3: For S_n, under similarity transformations by permutations (of S_n), transpositions go into different operators; show that these are also transpositions. Also any permutation goes into another with the same cycle structure (like one of three numbers, (123)(4), or two sets of two, (12)(34) — these are described by the number of numerals in each parentheses); a similarity transformation changes the symbols, but not the number of them, in each pair of parentheses.

Problem IV.2-4: Problems in previous chapters asked whether there is an operator that produces the relevant transformation. It is worth going back to these and see if there is an operator for a similarity transformation that gives the result. Under what conditions (all?, none?) are there operators giving a transformation by multiplication (perhaps matrix multiplication), and also ones that give it by a similarity transformation? Are these operators related? Always? Never? For what cases? How? Why? Under what conditions is there only one way of transforming?

Problem IV.2-5: Show that the necessary condition for an inner automorphism is that the transformation be a similarity transformation. Check it for S_3 and S_4 (and other S_n groups). Are all similarity transformations automorphisms (sec. III.4.e, p. 91)? Would it matter if they are produced by a group operator? Can an outer automorphism not be a similarity transformation?

Problem IV.2-6: One automorphism of the group of complex numbers is complex conjugation. It is outer. Why? Find the operator that produces it by a similarity transformation. It might help to write complex numbers as 2×2 matrices, as every complex number can be. Why?

IV.3 SUBGROUPS

A subset of the transformations of a group may also form a group; this then is a subgroup of the group. There are many examples above, although often not explicit. Subgroups are important because they supply information about the groups, so about the objects they describe, and because they relate, and so limit, groups. And they provide tools to analyze and exploit the theory. The subject of subgroups is closely related to those of group structure and typology.

There are two obvious subgroups, the identity and the group itself, neither interesting; these are improper (or trivial) subgroups. All others are proper subgroups.

Problem IV.3-1: After (and during) reading this chapter (and book) it is worthwhile returning here to see whether, and to what extent, these statements have been justified, and examples provided.

Problem IV.3-2: Given a subset of the group elements, with a product rule the same as that of the group; what two conditions must be placed on the set for it to be a subgroup [Grossman and Magnus (1992), p. 77; Hamermesh (1962), p. 15; Jansen and Boon (1967), p. 13; Kurosh (1960a), p. 42; Landin (1989), p. 87; Maxfield and Maxfield (1992), p. 59; Scott (1987), p. 16]? These are needed to satisfy two of the four group axioms. Which? Why are the other two then satisfied? Show that for finite groups, only one of these conditions is necessary; the other follows from it. However take the positive integers, an infinite set, with addition as the product (combination rule). These satisfy one, but not the other, so do not form a group, thus this set is not a subgroup of all integers, which do form a group with this product. Are the even integers a subgroup? The odd ones? Combine the two conditions into one [Baumslag and Chandler (1968), p. 54].

Problem IV.3-3: Show that the rational numbers form a group using addition, but not using multiplication, as a product. What is the identity? The positive rationals are a subset. They form a group with multiplication as the product, but not with addition (why?) . Thus the positive rationals are not a subgroup of the rationals. While they form a subset which is a group, the rule of combination is different. A subset is a subgroup only if the same product rule is used.

Problem IV.3-4: It is reasonable that groups should have subgroups and also reasonable that they should give information about the group and be useful. But does a group always have a (proper) subgroup [Baumslag and Chandler (1968), p. 106; Grossman and Magnus (1992), p. 81]? Characterize those that do not.

IV.3.a Conjugate subgroups

Subgroups have properties that are interesting because they aid in obtaining and relating information. And subgroups often are related; such relations can themselves provide important information. A similarity transformation of subset S of a group, using group element t, gives subset S';

$$S' = tSt^{-1} \qquad \text{(IV.3.a-1)}$$

means that the similarity transformation acts on all elements of set S to get the elements of S' — sets S and S' are then conjugates. Likewise on subgroup H of group G a similarity transformation with element t of G (not in H),

$$H' = tHt^{-1}, \qquad \text{(IV.3.a-2)}$$

gives set H' — the set tht^{-1}, for all h in H. These two subgroups are conjugate (under t).

Problem IV.3.a-1: Prove that S and S' have the same number of elements [Baumslag and Chandler (1968), p. 132]. Also the conjugate of a subgroup is a subgroup, and they are isomorphic. This inner automorphism of a group gives isomorphisms of its subgroups. How do the subgroups differ? Try examples. Most of the information about groups, and the information they provide, is invariant under these transformations (why?) and this includes the subgroup structure. Why is t not in H? What is the conjugate of the subgroup consisting of the identity? Check that the sum of the conjugate subgroups of an arbitrary subgroup (the set of all elements in at least one of them) need not be the entire group. Are there any (nontrivial) cases in which it is? Two sets are equal if they contain the same elements (though these may be ordered differently). Show that the transforms of H, using all elements of G, can all be the same subgroup H. The set of all conjugates of a subgroup need not exhaust the group.

Problem IV.3.a-2: For the (low-order) symmetric groups [Armstrong (1988), p. 74], the geometrical groups, the rotation group*s* (sec. II.3.a, p. 41), find the subgroups, and all conjugates of each. Explain, mathematically, geometrically, and physically (taking physical models) which sets of subgroups are invariant under similarity transformations, and why, and relationships between transformed and untransformed subgroups. Find which (if any) do, and do not, give the entire group. Are there reasons? When are conjugates of different subgroups equal?

Problem IV.3.a-3: Is there a rule that tells whether a set of conjugates gives the entire group? Is this ever possible?

IV.3.b Intersections and unions

Subgroups can be found in various ways. One is by combining subgroups; here two ways of doing so are considered to see whether they give other subgroups.

Problem IV.3.b-1: The intersection of two sets is the set all of whose elements belong to both sets (so it is a subset of both). The intersection of two roads contains both roads. Prove that the intersection of two subgroups is a subgroup [Baumslag and Chandler (1968), p. 55; Kurosh (1960a), p. 46]. Is this true for any number of subgroups? Can distinct subgroups (differing in at least some elements), have a (nontrivial) intersection? If so, give examples, and interpret them (physically). If one of two sets and their intersection are both subgroups, must the second set be a subgroup?

Problem IV.3.b-2: The union of two sets is the set that contains all the elements of both (so the two sets are both subsets of their union). Does the union of distinct subgroups have to be a (proper) subgroup? Find an example of a union of two subgroups that is not a subgroup [Baumslag and Chandler (1968), p. 56, 91]. The intersection of two subgroups is a subgroup, why is the union not (necessarily)?

Problem IV.3.b-3: Let $G_1, G_2, G_3, \ldots,$ be subgroups of group G, such that G_1 is contained in (is a subgroup of) G_2, which itself is contained in G_3, and so on. Then show that the union of all these subgroups is a subgroup. What is the union?

IV.3.c Cyclic groups and cyclic subgroups

The simplest groups are the cyclic groups. For these we can tell most about their subgroups, and something about the groups of which they are subgroups.

Problem IV.3.c-1: It is obvious, but is it trivial, that all subgroups of cyclic groups are cyclic [Jansen and Boon (1967), p. 71]? Are the orders of the group and the (possible) orders of its subgroups related?

Problem IV.3.c-2: What is the relationship between the orders of A and A^{-1}? Show that AB and BA have the same order, whether they commute or not [Grossman and Magnus (1992), p. 88]. How is it related to the orders of A and B?

Problem IV.3.c-3: If every element of a group (except the identity of course) is of order 2, the group is Abelian. Why?

Problem IV.3.c-4: For a finite group, the set of powers of every element, a cyclic group, is a subgroup. From this check that if the product of every pair of elements of a subset is in that subset, then it is a subgroup — this is the only group axiom that need hold for a subset to be a subgroup, with the other three then following (from where?) . Give examples to show that for infinite groups, both discrete and continuous, not all elements give cyclic groups. Can any? If a subset has in it all products of all its elements, for the subset to be a subgroup, which other group axiom(s), that follow for finite groups, need also be verified for infinite ones? Show that a subset is a subgroup if for every pair of elements, T, T', in it, $T'T^{-1}$ is also in it [Cornwell (1984), p. 24]. Why does this fail if instead $T'T$ is in it, for all T, T'? Would the latter condition be enough for a finite group? If elements S and T of a group both belong to a subset, is the condition that TS^{-1} belongs to the subset necessary for the subset be a subgroup, sufficient or both? Why does the inverse of S, rather than S, appear?

Problem IV.3.c-5: The necessary and sufficient condition for a (finite) group to have a proper subgroup, is that it is not a cyclic group whose order is a prime number. Is this true or false? Having established this the next thing is obvious: find all proper subgroups of a cyclic group [Baumslag and Chandler (1968), p. 105, 126; Kurosh (1960a), p. 47].

IV.3.d Subgroups of symmetric groups

Symmetric groups are fundamental enough to provide information about many areas of mathematics, making their subgroups especially interesting. Beyond providing useful examples, these give information about, not merely symmetric groups, but all finite groups. (One major result, not here relevant, is that there is no general formula for the roots of an algebraic equation of degree five or greater [Edwards (1984); Ledermann (1953), p. 123; Maxfield and Maxfield (1992), chap. 8-10].)

IV.3.d.i *Groups having, and being, symmetric subgroups*

Permutations can be divided into two kinds, those expressible as a product of an odd number of transpositions (odd parity), like (12), and ones with an even number (even parity) like (12)(34), or (123) = (12)(23). The identity is even.

Problem IV.3.d.i-1: Permutations can be written in various ways; show that (324) = (432) = (24)(43). The concept of parity (as in "pair"; it can be even or odd) would not be useful if it depended on how a permutation is expressed. Prove that the parity of a permutation is the same no matter how it is written as a product of transpositions, and then no matter how it is written [Baumslag and Chandler (1968), p. 60; Grossman and Magnus (1992), p. 145; Landin (1989), p. 94; Scott (1987), p. 267; Wigner (1959), p. 126]. Given the parity of elements in a product, what is the parity of the product?

Problem IV.3.d.i-2: For S_n, there is a subgroup, the alternating group A_n (sec. II.4.d.ii, p. 54), consisting of the permutations with even parity, like (124), (136)(245), or (1)(2)(3)(4)(5) — the identity. Show that it is a group, and that A_n is the largest (proper) subgroup of S_n. The odd elements do not form a subgroup. Why? Would they do so if the identity were adjoined? Show that transpositions do not form a group. What is the order of A_n? Verify that A_3 is a subgroup of S_3 and that A_3 and A_4 are subgroups of S_4. Of course A_m is a subgroup of A_n, as is S_m of S_n, $m < n$. Are there subgroups of S_m that are not subgroups of A_n? Are there S_n subgroups that are disjoint, except for the identity, from A_n (they share no elements except the identity)? For any subgroups of these two sentences, what are their intersections with A_n? Their unions? How many alternating groups are subgroups of S_n?

Problem IV.3.d.i-3: Find all subgroups of S_4 [Lyapin, Aizenshtat, and Lesokhin (1972), p. 231].

Problem IV.3.d.i-4: What are the cyclic subgroups of S_n? The Abelian subgroups? Are any of the geometrical groups (sec. II.2.e, p. 36 — sec. II.2.h, p. 40) subgroups of symmetric groups? What is the smallest each is a subgroup of? Does this provide any information about the geometry, or the physics, of the figure?

Problem IV.3.d.i-5: The symmetric group can itself be a subgroup, perhaps usefully so. For each group considered, particularly the geometrical ones, find the largest symmetric group that is a subgroup of it. What is the geometrical significance of this subgroup? The physical significance? Compare the answer, and the significance, to that of the previous problem.

Problem IV.3.d.i-6: Repeat the last two problems for the alternating groups. Compare the answers.

IV.3.d.ii *All finite groups are subgroups of symmetric groups*

A group operator acting on another from the group (the product of the two) gives a group operator, by definition of a group. Thus acting on all group operators it gives back all the operators — it simply permutes the group elements. But an operator that permutes a set of objects is an element of a symmetric group. Thus the operators of a finite group are elements of symmetric groups. This gives

THEOREM (Cayley's theorem): Every group of order n is (isomorphic to) a subgroup of the symmetric group on n objects, S_n [Baumslag and Chandler (1968), p. 214; Hall (1959), p. 9; Landin (1989), p. 126; Maxfield and Maxfield (1992), p. 58; Scott (1987), p. 48]).

PROOF: The theorem is obvious but we prove it anyway (at least to indicate that no matter how obvious something is, it is possible to give a proof that is not obvious). First we give a "formal" proof, and then to show that formal proofs do not always explain everything we give an informal one. In the group table the elements are listed on top, and each row contains the same set, but in different order. Thus each group operator acting on the set of all group operators gives the same set, but permuted. So each operator of the group is a permutation, and thus the group is a subgroup of a symmetric group. •

Problem IV.3.d.ii-1: To prove it what has to be shown? Given an operator T of the group, take its product with every other group element. The set of elements obtained consists of every element of the group, and each appears once. In particular, the product of T with distinct elements is distinct,

$$TU = TV \Rightarrow U = V. \qquad \text{(IV.3.d.ii-1)}$$

Prove these, and also that the identity element of a group corresponds to the identity element of the symmetric group of which it is a subgroup. T therefore acts as a permutation on the group elements. Two elements T_1 and T_2 thus act as permutations, called P_1 and P_2, as does their product, $T = T_1T_2$, which gives permutation P. Prove that

$$T = T_1T_2 \Rightarrow P = P_1P_2, \qquad \text{(IV.3.d.ii-2)}$$

by explicitly written the group symbols, seeing the effect of the T's on them, and finding the permutations [Hamermesh (1962), p. 16]. This shows that the group of symbols T, the group (of n elements) that we are considering, is (isomorphic to) a subgroup of S_n. Is there anything in this problem that is not in the previous paragraph, or is this just a different (better, worse?) way of stating the same thing? How do these compare with the proofs in the references?

IV.3.d.iii *The group of the square as a subgroup of a symmetric group*

To illustrate subgroups and Cayley's Theorem we take the group of the square (sec. II.2.f, p. 37). Since every element of this group appears exactly once in each row of the group table, all rows are permutations of the top one — of all elements listed in a predefined order. So each element permutes the elements in the top row; it is an element of the symmetric group. However this group is not a symmetric group because that contains all permutations of the elements; here are only some. The group D_4 has eight operators so is a subgroup of S_8 (sec. II.2.f, p. 38).

Problem IV.3.d.iii-1: List some transformations in S_8 that are not in D_4. Is D_{4h} a subgroup of S_8?

Problem IV.3.d.iii-2: The transformations relabel the (four) corners so the group of square is also a subgroup of (the 4! = 24 element group) S_4. The S_4 transformations outside the group of the square are those that cannot be carried out physically; interchanging the top corners but not the bottom cannot be done without cutting. Find physically all transformations in S_4 that are not in the square group [Maxfield and Maxfield (1992), p. 59, 200]. Check this using the group table.

Problem IV.3.d.iii-3: What are the transformations in S_8 that are not in S_4? They rename the elements of the square group; these are ordered (arbitrarily) at the top of the group table and the S_8 transformations reorder them. Show that operators of S_8 change the ordering of the group elements and that some of these reorderings are equivalent to relabeling corners (the elements that do this are in S_4). Give a rule for them. Check that there are elements in S_8 that can be carried out physically (some by rigid body transformations, others not) and also elements that have no physical significance.

Problem IV.3.d.iii-4: The generators not in S_4 relabel the elements of the square group (which is the significance of S_8) but leave the corners fixed. Why? The product rule is determined by, and defines, the group, but not the ordering of elements. In general an extra freedom implies a symmetry, so a larger group. Does this suggest that for the group of the square the group table has a group of symmetries? If so, what? How is it related to the group of the square? What is the smallest symmetric group of which it is a subgroup? Can the group of the square be a subgroup of a smaller (symmetric) group than S_8 or S_4? What? Why? How do these answers relate to those for the questions about symmetry of the group table in sec. III.2.i, p. 80, and the automorphism group of the group of the square (pb. III.4.e.ii-12, p. 95)?

Problem IV.3.d.iii-5: Repeat the questions for D_{4h}. Compare answers.

IV.3.d.iv *Implications of Cayley's Theorem*

Since S_n is a finite set of symbols, it has only a finite number of subsets, thus S_n has only a finite number of subgroups. So for each finite integer, there are only a finite number of (distinct — nonisomorphic) groups. In principle then, to find all finite groups we find all subgroups of S_n, for each n — we find each set of elements that forms a distinct subgroup. The product rule is known — it is inherited from S_n; if $P = P_1P_2$ in S_n, it is so in every subgroup containing these three generators.

Problem IV.3.d.iv-1: Can a subgroup contain only two of these?

Problem IV.3.d.iv-2: Find all subgroups of S_n, $n = 1, \ldots, 6$. In each case, find all nonisomorphic groups. Give the set of S_n elements forming them and their group tables. Can there be subgroups distinct (containing different sets of elements), but isomorphic? It would be a nice contribution to find an algorithm for doing this for arbitrary n, and then (if necessary) implementing it as a computer program.

Problem IV.3.d-3: Using Cayley's Theorem, prove that any element of any finite group, plus all its powers, forms a cyclic group.

Problem IV.3.d.iv-4: Find the inclusion rules (the rules telling which generator of a subgroup corresponds to a generator of the group) for the group of the double rectangle (sec. II.2.g, p. 39) in the smallest S_n of which it is a subgroup. Are there ambiguities? Why?

Problem IV.3.d.iv-5: The square group is also a subgroup of the rotation group. Of which rotation group? Give a rule for stating those elements of SO(2) and SO(3) that are in the square group. How are these related to the elements of the (which?) symmetric group that are also in the group of the square? Does this imply that symmetric groups are subgroups of the rotation group?

Problem IV.3.d.iv-6: Do a similar analysis for the pyramidal group (sec. II.2.e, p. 36). Is the pyramidal group a subgroup of the group of the square? If so, what are the corresponding elements? Repeat this for the water group (sec. II.2.d, p. 35). Is it a subgroup of the pyramidal group? Of what smallest symmetric group is it a subgroup? Answer these questions for the group of the double rectangle.

Problem IV.3.d.iv-7: Prove experimentally that the symmetry group of the peeled tangerine is a subgroup of the rotation group (which, SO(2) or SO(3)?) . Is it a subgroup of a symmetric group?

Problem IV.3.d.iv-8: These illustrate symmetry groups of geometric figures that are subgroups of symmetric groups. There are other groups of permutations that are symmetric-group subgroups. Give examples.

Problem IV.3.d.iv-9: Find the inclusion rules for a cyclic group in the smallest S_n of which it is a subgroup. Are there ambiguities? Repeat for a general Abelian group. For an order-k Abelian group, is the smallest S_n of which it is a subgroup determined by k, or does it depend on the Abelian group? Why? Do this explicitly for all (small) values of k. Find a (computer-implementable) algorithm to determine the smallest symmetric group each Abelian group is a subgroup of, and inclusion rules.

Problem IV.3.d.iv-10: Repeat for an arbitrary subgroup.

IV.4 CENTRALIZERS AND NORMALIZERS

The group elements that commute with an element, or set, give information about it [Baumslag and Chandler (1968), p. 112; Hall (1959), p. 14; Jansen and Boon (1967), p. 15; Joshi (1982), p. 27; Kurosh (1960a), p. 79, 84; Scott (1987)]. Is an element, or set, special if the (complete) set of other group members that commute with it is a nontrivial subgroup? What can be learned about a group from the set of its elements that commute with all others? Can, must, this be a subgroup? Is there anything interesting about it? Here we introduce concepts whose full significance we explore further below. We present these as problems so the reader can explore them first. They should be reviewed after reading this chapter (and book).

The subgroup that commutes with all members of a set (for a group there is always at least one such) is the centralizer of the set; set C is the centralizer of set S, if for all $c_i \varepsilon C$, and all $s_j \varepsilon S$,

$$c_i s_j = s_j c_i. \quad \text{(IV.4-1)}$$

The set of group elements commuting with all of a subgroup is the centralizer of the subgroup. The centralizer of the entire group is also called the center of the group; it is the set of all group elements that commute with every group element. If two elements commute, one gives a similarity transformation of the other into itself

$$ta = at \Longleftrightarrow a = tat^{-1}. \quad \text{(IV.4-2)}$$

A set, all of whose elements commute with all those of another set, is invariant under similarity transformations by members of that set.

The center of a group is invariant under the group, and the group is invariant under its center.

The subgroup consisting of elements all commuting with a particular element is called the normalizer of that element; N is the normalizer of t if for all $n_i \varepsilon N$,

$$n_i t = t n_i. \tag{IV.4-3}$$

A set of group elements, N, that commute with a subset S,

$$sN = Ns', \quad s, s' \varepsilon S, \tag{IV.4-4}$$

is the normalizer of set S. Note the difference between the centralizer and normalizer of a set. The elements of the centralizer commute with all elements of the set; those of the normalizer give a similarity transformation,

$$s' = nsn^{-1}, \quad s, s' \varepsilon S, n \varepsilon N, \tag{IV.4-5}$$

taking members of set S into other members, but not necessarily into themselves.

Problem IV.4-1: Clearly the centralizer is a subset of the normalizer. Prove also that the centralizer, and all normalizers, for any element, and any subset of a group, are subgroups of the group, and that the centralizer is a subgroup of the normalizer.

Problem IV.4-2: What is the normalizer of the identity?

Problem IV.4-3: If a group element is in a subgroup, must its normalizer be in the same subgroup? What would the normalizer be if only the subgroup were considered; would it be the same (sub)group obtained by considering the whole group?

Problem IV.4-4: How are the centers and the set of normalizers of a group and those of (all) its subgroups related? Is one set obtainable from the other? Is there an algorithm that does so?

Problem IV.4-5: Is it possible for an element to have more than one normalizer (where one is not a subgroup of the other)? If so, can the normalizers not commute with each other; are they disjoint (except for the identity)?

Problem IV.4-6: Is there a relationship between the normalizers of an element and of its inverse? Can they ever be the same group? What about the normalizer of a set and that of the set of inverses?

Problem IV.4-7: Given the normalizers of two elements, can the normalizer of their product be found? If two elements commute must their normalizers? Give examples.

Problem IV.4-8: Consider the set of elements consisting of all normalizers of all elements of a group. How is this set related to the center? Are there groups for which this set is identical to the group? What groups?

Problem IV.4-9: How are the normalizers of the elements of a cyclic subgroup related?

Problem IV.4-10: For the water group (sec. II.2.d, p. 35) and the groups of geometrical figures (sec. II.2.e, p. 36 — sec. II.2.h, p. 40), or molecules or crystals if we wish, find the centers and all normalizers, and the centralizers of all important subsets, including all subgroups. Interpret them geometrically and physically; is there special significance for, interesting information provided about, the system by the center of each group? Explain the importance of the normalizer of each element, and of the centralizers of each subset. Could you tell by looking at the object what the center of the group is, or the normalizer of an element? What is the difference, and the physical (or geometrical) significance, between the sets of normalizers for the group of the square and for the group of the double rectangle? For all groups of order eight (sec. III.2.e, p. 73), how are these answers related? Discuss the significance of this and of the similarities and differences.

Problem IV.4-11: Do symmetric groups have (nontrivial) centers? Do any of their elements, and subsets, have normalizers? Which? What? Try this for S_2, S_3, S_4, ..., and then in general. Answer the relevant questions of the previous problem for these.

Problem IV.4-12: Repeat for the dihedral groups (sec. III.2.h, p. 79).

Problem IV.4-13: What is the relationship between the center of a group and that of a subgroup? When must they be the same?

Problem IV.4-14: Must a centralizer, or normalizer, be Abelian? Can either be? Give examples. How about the center of a group? If a set of elements forms an Abelian subgroup of a group, are their normalizers related? How? Give examples.

Problem IV.4-15: Are there groups whose centers consist only of the identity? Characterize them. Is it possible for a group to have no elements whose normalizers are larger than the identity? Can (any) such groups be characterized? Is there a rule for distinguishing elements with normalizer the identity, from those with larger normalizers?

Problem IV.4-16: Given a group with a nontrivial center (one larger than the identity). Is it possible to construct from it a group whose center is the identity?

Problem IV.4-17: If two groups are homomorphic, are their centers and normalizers related? Would knowing the homomorphism (the list of corresponding operators) allow the development of an algorithm to get one set from the other? Can (must) a group be homomorphic to its center? How about isomorphic? Are there similar relations for the normalizers?

Problem IV.4-18: It would be interesting to determine whether any groups have the centralizers of all subgroups (or subsets, if there is a distinction) identical to their normalizers; if so develop an algorithm to find them (a deliberately ambiguous word), and computer implement it, if necessary.

Problem IV.4-19: Is there a prescription for the type of group that can be a center? How about a normalizer? How are these answers related to the type of group of which they are subgroups? Is there a relationship between the type of group and the type of its center, or of its normalizers? Might there be relationships between the structure, or symmetry, of the group table and those of its centers or normalizers? Must (can?) every group have a (nontrivial) center? How about a normalizer for every element? Any element? Are there rules or algorithms telling which groups, or elements, can have centers or normalizers? Can there be rules or (computer-implementable) algorithms to find the center of any group, or the normalizer of any element?

IV.5 CLASSES

Intuitively we expect a difference between the identity and other operators, and between rotations and inversions, or for the symmetric group, between transpositions and other permutations. There appears to be relevant dissimilarities between (12)(3)(4), (1234) and (12)(34). How can we define such differences, and why are they important? Are there similar ones for other groups (perhaps not as visually obvious)? Do these tell us anything valuable about a group or its applications — about the physics or geometry of a system?

To answer such questions we study the behavior of group elements under similarity transformations. This leads to the division of group elements into sets, called classes [Armstrong (1988), p. 61, 73; Burn (1991), p. 155; Duffey (1992), p. 45; Hall (1959), p. 13; Hamermesh (1962), p. 23; Jansen and Boon (1967), p. 24; Joshi (1982), p. 11; Kurosh (1960a), p. 79; Ledermann (1953), p. 96; Lomont (1961), p. 22; Lyapin, Aizenshtat, and Lesokhin (1972), p. 83; Scott (1987), p. 52; Wigner (1959), p. 65]. A class of a group is a subset of its operators that is invariant — all members going into another member of the subset — under all similarity transformations using all operators of the group. So

$$s' = tst^{-1}; \qquad \text{(IV.5-1)}$$

s and s' are in the same class. Using every t in the group acting on any s in a class gives all the members of the class. The group operators divide naturally into classes. The classes (except that of the unit element) do not form groups — they lack a unit element (at least).

Elements are in different classes because they are different, in identifiable ways, and these have significance not only for the group, but for what it describes, such as the geometry and the physics [Falicov (1966), p. 13; Tinkham (1964), p. 13, 16]. Elements in the same class are distinguished, but not truly different. The identity is in a class by itself. Rotations and reflections are in different classes — and they are different. And as we see by considering the geometrical groups, rotations, and similarly reflections, in the same class, are essentially the same. Transpositions are not really different, but differ from other permutations. They are in one class, alone. For S_n, all elements with the same cycle structure form a class, elements with different cycle structures are in different classes [Armstrong (1988), p. 74]; (12)(34), (13)(24), (15)(34) are in the same class, (1234) is in a different one.

Problem IV.5-1: Why is the identity always in a class by itself? Clearly the necessary and sufficient condition for an element to be in a class by itself is that it commutes with all group elements. Only what kind of group contains such a (nontrivial) element? What is the name of the set of elements of a group, each of which is in a class by itself? What does this imply about Abelian groups?

Problem IV.5-2: If

$$a' = tat^{-1}, \tag{IV.5-2}$$

a' and a are said to be conjugates (under t). Show that conjugation is an equivalence relation (sec. III.4.a, p. 85). All members of a class are equivalent to each other under conjugation. So classes are often called conjugate classes.

Problem IV.5-3: Two subsets of a finite group that are conjugate (each element of one is conjugate, under the same similarity transformation, to one of the other) have the same number of elements. And the conjugate of a subgroup is a subgroup. Check these. How are the properties of such subgroups related?

Problem IV.5-4: Every element of a group is a member of some class of that group [Cornwell (1984), p. 26]. No element can be a member of two classes. The group then is a sum of disjoint sets, each a class. For any class, except that of the identity, the set of elements not in it, is a subgroup. The number of elements in a class is a divisor of the order of the group [Hall (1959), p. 14].

Problem IV.5-5: How are the classes of a group and a subgroup related? If two elements are in the same class of a group can, must, they be in the same class of a subgroup? Is the converse true?

Problem IV.5-6: All elements of a class have the same order — for a finite group. Check that

$$u^k = E \iff (tut^{-1})^k = E. \tag{IV.5-3}$$

However if two elements have the same order they need not belong to the same class. Give an example.

Problem IV.5-7: Given groups of the same order, find a simple way of showing that they are nonisomorphic. Can this method be used to show that groups are isomorphic? Give examples.

Problem IV.5-8: For S_n, prove that all elements with the same cycle structure form a class, elements with different cycle structures are in different classes. That is, a class is determined by the set of numbers of symbols in each pair of parentheses (eq. II.4.d-2, p. 53); a similarity transformation changes the symbols in the parentheses, but not their number.

Problem IV.5-9: Find the classes for S_2, S_3, S_4, S_5, and S_n [Baumslag and Chandler (1968), p. 171; Lyapin, Aizenshtat, and Lesokhin (1972), p. 85, 218; Scott (1987), p. 298]. Repeat for the alternating groups (sec. II.4.d.ii, p. 54). How are the classes related? Give a (recursion) formula for the number of classes as a function of n [Jansen and Boon (1967), p. 72; Weyl (1931), p. 328]. What is the number of elements in each class?

Problem IV.5-10: Find the classes of the dihedral groups [Armstrong (1988), p. 73; Burn (1991), p. 156, 163; Dixon (1973), p. 5, 75], and give a formula for their number.

Problem IV.5-11: Find those of the quaternion group (pb. III.2.e-4, p. 74) [Lyapin, Aizenshtat, and Lesokhin (1972), p. 84, 217].

Problem IV.5-12: Give examples to show that elements of different classes do not commute in general. These are then examples of sums that are not direct sums: the sum of sets consisting of all elements in every class of a group, which is the group itself, is generally not a direct sum. Can we consider the set whose elements are the classes? What type of set is it — a group?

Problem IV.5-13: For a proper subgroup H of group G, consider the set (of subgroups?) found by taking all conjugates of H using all (only?) transformations in G [Hall (1959), p. 24]. Show that this set does not contain all operators in G. Explain.

Problem IV.5-14: Prove that there is only a single finite group with just two classes; it has order 2 [Hall (1959), p. 24; Lyapin, Aizenshtat, and Lesokhin (1972), p. 85, 217]. What are the classes?

Problem IV.5-15: Show that the set of inverse transformations of all elements of a single class themselves form a single class (they are all in one class and there is no other element in this class). Why do the inverses of elements in one class all have to be in a single class? Must (can) the classes of transformations and their inverses be different? A class that is its own inverse (the inverses of all elements of the class are in the class) is called ambivalent (as well it might be) [Jansen and Boon

(1967), p. 31].

Problem IV.5-16: Are there groups all of whose classes are ambivalent? Are there any for which no (nontrivial?) classes are? For which at least one is? Answer first for cyclic and Abelian groups, for dihedral groups, the small symmetric groups, then for all of these, for the geometric groups (sec. II.2.e, p. 36 — sec. II.2.h, p. 40) and then (if possible) in general. Develop a (computer-implementable) algorithm to do it, and to find all ambivalent classes for any group.

Problem IV.5-17: A group of odd order (with an odd number of elements) has only one (nontrivial?) ambivalent class (why?) ; find it. A group of even order contains at least one nontrivial ambivalent class [Jansen and Boon (1967), p. 32].

Problem IV.5-18: Give examples of two groups with the same order, and the same number of classes, that are not isomorphic [Joshi (1982), p. 26]. Suppose (in addition) that the orders of the elements in the classes correspond, would the groups be isomorphic? Find (simple) conditions, if they exist, whose satisfaction is necessary and sufficient for such groups to be isomorphic.

Problem IV.5-19: If a group of order $4n$, for any n, contains an element of order $2n$, it can have a maximum of two elements in each class [Jansen and Boon (1967), p. 74]. Can it have fewer? Prove and explain this. Give examples. Can this be generalized to other orders? Why?

Problem IV.5-20: How are the normalizers of elements of a single class related? Can a subgroup (not the center) be the normalizer of two different elements? Must these be related? Can (must) they be in different classes?

IV.5.a Why classes exist and are important

There are different types of transformations, rotations, reflections, inversions, say, for that is a property of geometry and (so) physics. Groups, to describe these, must have structures that reflect these differences. Thus, the elements must divide into classes. This is an essential, but not the only, reason classes are important. Key properties of matrices are invariant under similarity transformations. The trace of a matrix is one. (The traces of representation matrices, which we shall come to, are the characters of the representation, whose importance we consider later, and these are class functions — every operator in the class has the same character, for a representation). Thus, as we shall see in depth, elements of a class are largely the same, distinct though not really different; those of different classes are really different. Classes are needed because of, and given by, the requirement that the group structure reflect all these differences.

Problem IV.5.a-1: For the group of the square (sec. II.2.f, p. 37) find the classes, interpret them physically [Falicov (1966), p. 62], giving rules to do this, mathematical (does the structure of the group table distinguish classes?), and physical; explain the physical (geometrical?) distinction between elements of different classes. What is the meaning of similarity transformations between members of a class? Demonstrate that operators from different classes have different physical meanings, and that all of a class are physically equivalent. Show (including experimentally) that no similarity transformations, using group elements, on a rotation about an axis through the centers of the sides, give a rotation about a diagonal. Why, geometrically, is this true? Could this be done using operators from outside the group?

Problem IV.5.a-2: Repeat this for the group of the double rectangle (sec. II.2.g, p. 39). How do the class structures differ for these groups? What is the significance of this? Could you tell that there is a difference by looking at the figures? How? What information would this give you? How about looking at the group tables? Can the classes be found by these methods?

Problem IV.5.a-3: We shall not ask the reader to find the classes of the pyramidal group (sec. II.2.e, p. 36) since unfortunately these are given by the notation. Why? But it is worthwhile verifying that these do give classes, and that these are all the classes. The questions of the previous problems can usefully be pondered for this one.

IV.5.b How geometry gives the classes of the square group

Why are all operators of a class physically equivalent, and why do they differ physically from those of other classes? Why is there a distinction between elements of a group, and how is it related to similarity transformations? To illustrate the answers, consider the group of the square. The classes contain the identity, rotations about an axis through the center — and considering non-rigid body transformations — reflections in lines perpendicular to two sides, and reflections in lines through two corners. It is the symmetry of the square that breaks these elements into sets. Reflections about a line through the sides are distinguished from each other, but there is no difference between them. Rotating the square to interchange the pairs of sides, which leaves the figure invariant, interchanges these reflections. And so it is with the elements of the other classes. But there is no (symmetry?) operator that transforms a reflection in a line through the sides into one in a line through the corners, or into a rotation about an axis perpendicular to the square.

In general because of the symmetry of a figure, there are sets, each of transformations that are mutually distinguishable but not different. These are taken into each other by a similarity transformation, so these elements form a class. There are sets of transformations different from these, but not from the others in the same set; they have different effects on the figure and physically are carried out differently. They form different classes.

Problem IV.5.b-1: Why are classes given by similarity transformations, not by group products; why is an operation that carries a figure into itself, given by such a transformation, rather than by multiplication by a group element?

Problem IV.5.b-2: Repeat the analysis for the water group (sec. II.2.d, p. 35) and the geometrical groups (sec. II.2.e, p. 36 — sec. II.2.h, p. 40). Discuss the similarities and differences.

IV.5.c Classes are given by similarity transformations

From the geometry and physics then, there are different types of transformations and the group structure must, and (fortunately) does, reflect this. Why are transformations of a single type connected by a similarity transformation (and not, say, by multiplication of one operator by another)? Take a reflection in a line through a diagonal of a square; there are two. A transformation of the group either leaves the pair of vertices unchanged or changes them to the other pair. Applying a transformation and then the reflection, gives a reflection either in the same line, or the other (perpendicular) line (giving two changes of the square, that due to the transformation and that due to the reflection). Then applying the inverse of the transformation to the first brings the square back to its original orientation (undoing the first transformation) except for the change produced by the reflection. Thus the whole effect of these three operators is to produce a single change of the square — due to a reflection of the same type, but not necessarily the same one. This type of symmetry operator on the square cannot produce a transformation of a different type (say a rotation).

Problem IV.5.c-1: This would not be true if the transformation of the reflection was not a similarity transformation. Verify that the product of the reflection with some other transformation is not a reflection of the same type, and that this one is.

Problem IV.5.c-2: The argument for all types of transformation is identical, as should be proven. It is obvious for one, the identity, that the only way of getting one of the same type — for it, the same operation — is by a similarity transformation. This is sufficient to require all these (which?) be similarity transformations. Why?

Problem IV.5.c-3: Similarly for the symmetric group: essentially all transpositions are the same — we can change (12) to (47) by relabeling numerals, but this is different from a permutation that acts on three numerals. What is the effect of a similarity transformation on a permutation? Numerals are relabeled, the transposition acts by interchanging a pair of numerals, and then the relabeling is undone. The effect is to interchange a pair of numerals — a transposition. This transposition is different from the one that the similarity transformation acted on, but it is still a transposition. Verify this and repeat the argument for an arbitrary permutation. Check that the product of a permutation and a transposition is not a transposition, and similarly for an arbitrary permutation. For S_n, prove that the parity of every permutation is preserved under a similarity transformation (but need not be preserved under multiplication by a group operator), so is a class function. Compare these results to those of pb. IV.5-8, p. 121.

Problem IV.5.c-4: When a symmetry is broken, say by making two atoms at the vertices of a square different, what happens to the classes? Do classes break up into subclasses? Do different classes become the same? Why? Explain physically what happens.

IV.5.d Classes of rotation groups

For the rotation group, each class consists of the rotations through the same angle about every axis. The classes are labeled by the angle, the elements in each by the axis. Thus the set of rotations through $\pi/9$ about the z axis, the x axis, the axis in the xy plane making an angle of $\pi/4$ with the x axis and so on, are all in a single class. The number of elements in each class, and the number of classes, are both infinite. We have not yet treated this group so we just give a heuristic explanation.

Problem IV.5.d-1: Rotate a linear object (say a ruler) through angle θ about an edge. Then rotate it about some other line through $\pi/8$ then reverse the first transformation: rotate it through $-\theta$ about the edge. The ruler will be rotated from its original position by angle $\pi/8$, but around a different line then that of the original $\pi/8$ rotation. This similarity transformation produces a rotation of the same angle about a different line, so all these rotations are in the same class. Such a similarity transformation cannot produce a rotation through a different angle, so these lie in different classes. Next hold two fingers on each hand apart, giving two angles, and place one finger next to (but not parallel to) one of the other hand. The rotation from the outside finger of one hand to the outside one of the other then is the product of the rotations for the two pairs. Notice that the product is a rotation through an angle different from either, and about a different axis. Similarity transformations, but not products, preserve angles, so are the proper way of transforming to a different coordinate system. Also the identity is the same for all axes, and is invariant (only) under a similarity transformation. This relates rotation-group classes and transformations between coordinate systems: physically all coordinate systems are the same — and the same rotation, as seen in different systems, is given by operators of the same class.

Problem IV.5.d-2: While all rotations — of the rotation group — of the same angle are in the same class, this need not be true for its finite subgroups. Explain. Give examples [Cornwell (1984), p. 27].

IV.5.e Classes of groups and subgroups

Many groups have (proper) subgroups, and all (finite?) groups are. How are the classes of a group and those of its subgroups related, and what does this tell us?

Problem IV.5.e-1: Clearly each subgroup class is a subset of a class of the group — all operators of the subgroup class are in a single class of the group (there may be elements in the class of the group not in the subgroup). Are all subgroup operators in a single class of the group in the same subgroup class? Why? Give examples.

Problem IV.5.e-2: The group of a square (sec. II.2.f, p. 37) is a subgroup of a symmetric group (which itself is a subgroup of larger symmetric groups). Relate the classes of the smallest symmetric group to those of this subgroup. Compare these answers to those of the previous problem. For this subgroup, show that all members of a class have the same set of cycles (the cycle lengths are the same). Repeat for the pyramidal group (sec. II.2.e, p. 36) and for the group of the double rectangle (sec. II.2.g, p. 39). The groups of the square and the double rectangle are both subgroups of S_8. Are they both subgroups of S_4; if not, what is the smallest S_n for which they are? What (classes of) operators of S_8 become (classes of) operators of these groups? How do (the classes of) these operators differ; is there a rule giving them?

Problem IV.5.e-3: If two elements of S_n are in different classes in one subgroup, are they in different classes in all other subgroups?

Problem IV.5.e-4: If two elements permute different numbers of indices they are in different classes; give examples (for S_4 or larger) to show that if two elements permute the same number of indices they are not necessarily in the same class.

IV.5.f Class multiplication

Classes of objects are themselves objects, and can be multiplied

[Cornwell (1984), p. 304; Falicov (1966), p. 18; Jansen and Boon (1967), p. 32, 74; Joshi (1982), p. 13; Lomont (1961), p. 23; Tinkham (1964), p. 15]. What is their product? To multiply two classes, the (group) product is taken of every element of one, with every element of the other. This gives a set of elements; these belong, not to one, but to several classes, and each may appear several times — a group element can be given by more than one product. Thus, for the product of classes ξ_i and ξ_j,

$$\xi_i\xi_j = \Sigma c_{ij}^k \xi_k, \tag{IV.5.f-1}$$

where the sum is over all classes of the group, and c_{ij}^k is the number of appearances of class ξ_k — each element of ξ_k appears c_{ij}^k times in the set $\xi_i\xi_j$; the c_{ij}^k are the class constants, or the class multiplication coefficients. With N_i, N_j, the number of elements of ξ_i, ξ_j, the product has N_iN_j members (not necessarily distinct).

Problem IV.5.f-1: Show that if an element of class ξ appears, then all elements of ξ appear, and the same number of times; the sum is over complete classes, only. So c_{ij}^k is the same for all elements of ξ_k. Also, class multiplication is commutative. How does this follow from the definition of a class? To show that c_{ij}^k can be 0, take the class of the identity.

Problem IV.5.f-2: Can, must, classes form a group, using this product? Explain. Is there an identity?

Problem IV.5.f-3: For S_2, S_3, S_4 (sec. II.4.d, p. 51), multiply all pairs of classes and find the c's. What are the multiplication tables of the classes? Do they look like group multiplication tables? Find a formula, or algorithm, for the c's for S_n.

Problem IV.5.f-4: Repeat for the geometrical groups (sec. II.2.e, p. 36 — sec. II.2.h, p. 40).

Problem IV.5.f-5: What are the c's for the quaternion group (pb. III.2.e-4, p. 74) [Lomont (1961), p. 24]?

Problem IV.5.f-6: Find an expression for the c's of cyclic groups. For general Abelian groups. Can it be done for dihedral groups (sec. III.2.h, p. 79)?

Problem IV.5.f-7: Is there a formula for the c's in the product of a class with that of its inverses? Suppose the class is ambivalent (pb. IV.5-15, p. 121)?

Problem IV.5.f-8: With transformations that are inverses of each other denoted by j and $-j$, prove that for any finite group, for every class there is another (which may be the same) having as members the inverses of those of the first, with orders (numbers of elements)

$$N_j = N_{-j}. \tag{IV.5.f-2}$$

This gives

$$c^{-k}_{-i-j} = c^{-k}_{-j-i} = c^{k}_{ij}. \qquad \text{(IV.5.f-3)}$$

The product of a class with that of the class of its inverses contains the identity N_j times. That of any other two classes does not contain the identity. Explain. Prove that, with N_j the order of class j, the class constants c^k_{ij} satisfy

$$N_i N_j = \Sigma c^k_{ij} N_k, \qquad \text{(IV.5.f-4)}$$

summed over the classes [Jansen and Boon (1967), p. 74]. Does this give anything interesting if one class is ambivalent? If two, or all three, are (assuming that two, but not three, is possible)? Also,

$$N_k c^k_{ij} = N_{-i} c^{(-i)}_{j(-k)}. \qquad \text{(IV.5.f-5)}$$

Does this give anything for ambivalent classes? How do these follow from the definition of classes, and groups?

IV.6 COSETS

Physically constraining a system reduces the set of transformations on it, giving a subgroup of its symmetry group, and thus a subgroup with a clear physical meaning. Constraints are connected by symmetry transformations; applying *such* a transformation to all elements of the set A_i allowed by the constraint gives set A_j. Thus constraints belong to sets, the members so related. Taking all such constraints gives a set of the sets A_i — each element of the set is itself a set. What is the significance of this set, what does it tell us about the figure (and the group)? Does it have general, or interesting, properties? Is it a subgroup of the group? Is this question important?

Problem IV.6-1: A rod through corners of a square reduces its possible physical transformations. What is the symmetry group now? Is it a subgroup of the group of the square? Above the word "such" was used. Why? For the square, say, is this a real, and necessary, limitation? What are the constraints given by the (symmetry) transformations acting on this? How are the subgroups so obtained related?

IV.6.a The group, and the (co)sets obtained from a subgroup

Having obtained this subgroup, how do we get the group of the square from it? Each transformation of the square can be carried out by moving the rod and then rotating the square about it, or reflecting with it held fixed — each can be written as a product of a transformation of the rod times a transformation with the rod fixed (or in reverse order). Thus we place the rod and so obtain subgroup H. Then we perform a symmetry transformation t_1 (of course, not in the subgroup) on the rod to carry it to a new position, which gives a second set of operators t_1H (which is not a subgroup; it does not contain the identity). Applying all group elements t_i that transform the rod, we get a series of positions for it, and for each such a set.

So in general, the elements of group G can be divided into sets by taking a subgroup H, and operators t_i not in it, with the sets then

$$H, t_1H, t_2H, \ldots, \qquad \text{(IV.6.a-1)}$$

where t_iH means the set of operators found by multiplying each of those of H by t_i, and we take (just) enough t's so that this set of disjoint sets is group G [Baumslag and Chandler (1968), p. 107; Cornwell (1984), p. 28; Grossman and Magnus (1992), p. 82; Hall (1959), p. 10; Hamermesh (1962), p. 20; Kurosh (1960a), p. 60; Landin (1989), p. 107; Ledermann (1953), p. 35; Lyapin, Aizenshtat, and Lesokhin (1972), p. 79; Maxfield and Maxfield (1992), p. 62; Scott (1987), p. 19]. The sets that are members of this set are cosets of the group (sets which together — "co" — make up G). The cosets, except H, not containing the identity, are not groups. The multiplication can be either on the right or the left giving right (Ht) and left (tH) cosets (of H). These cosets seem related in some (fundamental) way to the structure of the group, thus giving information about it. We have to see how. In fact, they lead to the most fundamental information about (finite) groups.

Problem IV.6.a-1: The sets, except H, are not subgroups, and the operation forming them is not a similarity transformation. Is there a connection?

Problem IV.6.a-2: If subgroup H is the identity, what are the cosets? If H is the whole group?

Problem IV.6.a-3: Give examples to show that the left and right cosets are not the same, in general; they are distinct (but equivalent — in what sense?) . How are they related? Are there cases for which the right and left cosets are not distinct? Are the numbers of elements of tH and Ht equal? Explain why the sets of right and left cosets have the same number of members. Also the order of every coset is the same.

Problem IV.6.a-4: Prove that, and explain why, elements t, t' (not in H) can always be chosen such that H, tH and $t'H$ are distinct (that is, these sets are disjoint). What are they if t, t' are in H? This also holds for right cosets. Thus explain why every element of a group can be written as the product of an element of a subgroup, the same subgroup for all group elements, times a group element not in the subgroup — show that a group can always be broken up so that every element is in (one and) only one coset.

Problem IV.6.a-5: Show that for subgroup H, and distinct elements t, t' either cosets tH and $t'H$ are identical (placing what conditions on t, t'?), or they share no element, and $t^{-1}t'$ is not in H. Does this differ from the preceding problem? Let H, tH and $t'H$ be disjoint. Need they all be? Show that $tt'H$ need not be disjoint to both H and $t'H$. Give examples.

Problem IV.6.a-6: For the square, what are the transformations of the rod? Write each transformation of the square as a product of a rotation of the rod followed by one keeping it fixed, then as a product in the reverse order. Using subgroup H obtained with the rod in one position, show that the group is given by the sum of sets obtained by multiplying the elements of H by all transformations that act on the rod. Are there other subgroups that can be obtained by constraining transformations on the square [Cornwell (1984), p. 29; Maxfield and Maxfield (1992), p. 65]? Is there a relationship between these subgroups? Does it seem likely that this way of breaking a group up into subsets is not unique, that there can be different choices of subgroups leading to different cosets? Instead of a rod, can a mirror be used? List the cosets of this group using all subgroups, for both right and left cosets; compare.

Problem IV.6.a-7: Repeat this for the group of the double rectangle (sec. II.2.g, p. 39) and discuss how the constraints differ from those of the square. Find the transformations of the constraining rod (or other object) and write each transformation of the group (uniquely?) as one of these products, again twice (for both orders of the products). How do the transformations with the rod fixed differ for the two objects? Is the number of constraints (subgroups) the same as for the square? Why? Could you tell this by looking at the group table? How about by looking at the figure? Is there a relationship between what you can learn by looking at the group table and looking at the object? Find all cosets, compare with the square, and explain the difference (physically). Repeat this for the water group (sec. II.2.d, p. 35) and the pyramidal group (sec. II.2.e, p. 36), describing how the constraints are carried out. Show that each group element can be written as a product of the two types of transformations. Do this twice, once by first applying the constraint, then transforming, and then in the reverse order. To what extent are these cosets unique? Which are right, and which left. Explain the similarities and differences in the results.

Problem IV.6.a-8: For an atom or molecule imposing constraints using a rod would be inconvenient. But such can arise from, say, electric or magnetic fields (perhaps acting on an impurity atom in a crystal). Readers who wish to think in terms like these rather than rods are encouraged to do so. Think of a few (or many) such examples.

Problem IV.6.a-9: Find the S_3 cosets using the S_2 subgroups [Lyapin, Aizenshtat, and Lesokhin (1972), p. 80, 82, 216, 217]. Repeat using A_3, and A_2 (sec. II.4.d.ii, p. 54). Do the same for S_4, using the S_3, S_2, A_4, A_3 and A_2 subgroups. Is there a general rule for finding all cosets (using all subgroups) for S_n? Is it necessary to implement it as a computer program? Can it be? Do this for both right and left cosets, and see if anything interesting appears in a comparison of the results.

Problem IV.6.a-10: Are cosets of the quaternion group (pb. III.2.e-4, p. 73) [Lyapin, Aizenshtat, and Lesokhin (1972), p. 80, 216] interesting?

Problem IV.6.a-11: What can be said about the cosets of a cyclic group [Baumslag and Chandler (1968), p. 108; Lyapin, Aizenshtat, and Lesokhin (1972), p. 80, 217]? How about a general Abelian group? Is an algorithm, or computer program, necessary to find all of an arbitrary Abelian group?

Problem IV.6.a-12: Check that the number of cosets of the normalizer of element t equals the number of members of the class to which t belongs.

Problem IV.6.a-13: Consider two subgroups H and J, the cosets formed from them, and the intersection of a coset of H and a coset of J. Show that this is a coset formed from the subgroup that is the intersection of H and J. We can say (roughly?) that the intersection of cosets is a coset of the intersection.

IV.6.b The index of a subgroup in a group

The index of subgroup H in group G is the number of cosets formed from H, and this number gives information about the structure of the group, and places restrictions on it. Why?

IV.6.b.i *The order of subgroups and the order of groups: Lagrange's Theorem*

Subgroups cannot have all orders less than that of their groups; they are restricted.

Problem IV.6.b.i-1: Show that the index is the same for right and left cosets, and equals the order g of the group divided by the order of h, of the subgroup — the order of G equals the order of subgroup H times the index of H. Since all cosets have the same number of elements, the order of a group (the number of its elements) divided by the order of any subgroup, is an integer. Thus the order of an element (the power of the element that equals the identity) divides the order of the group. This gives [Baumslag and Chandler (1968), p. 109; Blichfeldt (1917), p. 34; Falicov (1966), p. 12; Grossman and Magnus (1992), p. 82; Hamermesh (1962), p. 20; Maxfield and Maxfield (1992), p. 65]

THEOREM (Lagrange's theorem): The order h of subgroup H divides the order g of group G,

$$g = hk, \tag{IV.6.b.i-1}$$

(their ratio, k, the index of H, is an integer). The converse of this does not hold [Hall (1959), p. 43]; if an integer n divides the order of group G there does not have to be a subgroup of G with order n. Give examples; the subgroups of symmetric groups might be considered. However there are some cases in which there must be subgroups. Prove [Armstrong (1988), p. 68; Maxfield and Maxfield (1992), p. 90]
THEOREM (Cauchy's theorem): If p is a prime divisor of finite group G, then G contains an element (so a subgroup) of order p.

IV.6.b.ii *Groups and subgroups are limited*

The index also provides other limitations and information.

Problem IV.6.b.ii-1: Show that the numbers of even and odd permutations are equal, so the index of A_n in S_n is 2.

Problem IV.6.b.ii-2: Prove that a group of prime order can have only the identity and itself as subgroups. What type of group must one of prime order be?

Problem IV.6.b.ii-3: Take any element t of group G, and subgroup N_t, its normalizer [Jansen and Boon (1967), p. 30, 73; Kurosh (1960a), p. 82]. How many elements are in N_t? The order of the class ξ_t to which t belongs is the number of elements in it. Prove that the index of N_t in G equals the order of ξ_t; also it just depends on the class of t, being the same for all elements in that class. Thus the number of elements in N_t equals the order of the group divided by this index. Is there a (simple) way of explaining this? Given N_t, find N_s, for t and s in the same class. Use this to check the results.

IV.7 INVARIANT SUBGROUPS

Element s can commute with element t — so be invariant under a similarity transformation by t,

$$ts = st \iff s = tst^{-1}. \tag{IV.7-1}$$

And s can commute with all elements of the group (thus be in its center). Is there anything equivalent for subgroups: can one be invariant (go into itself, though perhaps with its elements shuffled) under a similarity transformation, by an element, by all elements, of the group? What type of subgroup is invariant under all, what type of group has such a subgroup, what type does not? There are groups both with, and without, invariant subgroups, giving a fundamental method of classifying groups. Subgroup H is invariant (or normal) if all conjugates of H are equal — the sets obtained by similarity transformations of H, using all elements of G, equal H (the elements, though perhaps in different order, are the same),

$$H = tHt^{-1}, \text{ all } t \text{ of } G. \qquad \text{(IV.7-2)}$$

It is an invariant subgroup (self-conjugate subgroup, normal divisor, normal subgroup) of G (normal in G) [Baumslag and Chandler (1968), p. 111; Burn (1991), p. 96, 100; Cornwell (1984), p. 27; Grossman and Magnus (1992), p. 120; Hamermesh (1962), p. 28; Jansen and Boon (1967), p. 33; Landin (1989), p. 111; Ledermann (1953), p. 99; Lyapin, Aizenshtat, and Lesokhin (1972), p. 86]. Every group contains two invariant subgroups, the identity and itself. Any other is a proper invariant subgroup. How can we tell if a subgroup is invariant, and if a group has a proper invariant subgroup? And why do we want to?

Problem IV.7-1: Show that all left and right cosets are the same (t_iH and Ht_i have the same elements, for all i) for, but only for, a normal subgroup.

Problem IV.7-2: For subgroup H to be normal it is necessary and sufficient that it consists of complete classes of G (if an element in a class of G is in H, all elements of that class are) [Armstrong (1988), p. 79]. Why? It is a union of classes [Burn (1991), p. 157, 164]. Is this and the preceding problem related?

Problem IV.7-3: Another necessary and sufficient condition for a subgroup to be invariant is that the normalizer of the subgroup be the entire group. Is this related to the previous problems?

Problem IV.7-4: Is there a relationship between the invariant subgroups and the center of a group? Is the center normal?

Problem IV.7.-5: Does the product of two subgroups (the set of all products of all elements in one or both subgroups) have to be a subgroup [Scott (1987), p. 33]? Does it matter if one of them is normal? If both are? Is the product normal if one subgroup is? If both are? Is the intersection of two normal subgroups normal? Need the union be? How about one normal subgroup, and one not normal? Explain any differences.

Problem IV.7-6: The definition of a normal subgroup is that the transforms of every element of the subgroup be in it, using all group elements. However take a basis (sec. III.3, p. 82) of the group (giving one of the subgroup; why?) . Show that "element of the subgroup", and of the group, can be replaced by the corresponding "element of the basis" [Armstrong (1988), p. 84].

Problem IV.7-7: Since the order of a subgroup is a divisor of the order of the group (their ratio is an integer) the possible orders of normal subgroups can be obtained by finding all ways the orders of the classes can be added to give a divisor of the order of the group (always including the class of the unit element). Give an example showing that not all subgroups so allowed occur [Burn (1991), p. 94, 99].

IV.7.a Examples of normal subgroups

Before investigating the value of normal subgroups it is worthwhile studying examples.

Problem IV.7.a-1: What are the invariant subgroups of cyclic groups [Baumslag and Chandler (1968), p. 112]? Do all (any) have normal subgroups? Show that any subgroup of a cyclic normal subgroup of arbitrary group G, is also normal in G. Is "cyclic" necessary?

Problem IV.7.a-2: Show that all subgroups of Abelian groups are normal [Baumslag and Chandler (1968), p. 111], and find them (or develop an algorithm to do so). Give a counterexample to show that the converse need not be true.

Problem IV.7.a-3: Every subgroup of index 2 is normal [Burn (1991), p. 150, 154; Landin (1989), p. 112; Ledermann (1953), p. 101]. Why? Check it for S_n, finding such a subgroup for each n; explain why it is obvious.

Problem IV.7.a-4: Give all subgroups of index 2 of S_2, S_3, and S_4. Is there a rule giving all of index 2 for S_n?

Problem IV.7.a-5: Are there subgroups of index 2 of the group of the square (sec. II.2.f, p. 37)? The pyramidal group (sec. II.2.e, p. 36)? Why? Compare these to the group of the double rectangle (sec. II.2.g, p. 39).

Problem IV.7.a- 6: For S_3 show that (1)(2)(3), (123), (132), form subgroup A_3, which is invariant. Find the other subgroups and show that none are normal [Jansen and Boon (1967), p. 34]. For S_2, S_3 and S_4 find all subgroups. Which are invariant? Repeat for the alternating groups (sec. II.4.d.ii, p. 54) [Burn (1991), p. 159, 165]. Is there an S_2 that is a normal subgroup of S_n; of A_n? For S_4, show that both A_3 and A_4 are invariant subgroups. Is A_3 an invariant subgroup of A_4? Why? Is A_{n-k} normal in A_n, for all, for any, k; $n > 4$? Explain why A_n is a normal subgroup of S_n [Baumslag and Chandler (1968), p. 111; Ledermann (1953), p. 101; Maxfield and Maxfield (1992), p. 76]? What are the indices for these cases? Are the answers related? The only normal subgroup of S_n that contains a transposition is S_n [Burn (1991), p. 157, 164]. Why?

Problem IV.7.a-7: The group of the square is an S_4 subgroup. Is it invariant? What is the relevance of that? Answer this for the group of the double rectangle; explain the significance of similarities and differences. Repeat for the pyramidal group.

Problem IV.7.a-8: For D_4 (sec. II.2.f, p. 37) [Burn (1991), p. 157, 164], give the classes and normal subgroups. Can this be done for every dihedral group?

Problem IV.7.a-9: A dihedral group is the product of an order-n cyclic group with one of order 2. Show that the former is invariant [Ledermann (1953), p. 101]. How about the latter? Explain its geometrical significance.

Problem IV.7.a-10: If cyclic group H is a normal subgroup of G, and K is a proper subgroup of H, show that K is normal in G [Baumslag and Chandler (1968), p. 112]. However give an example to show this need not hold if H is not cyclic — thus normality is not transitive (sec. III.4.a, p. 85) [Scott (1987), p. 33]. But if N is a subgroup of both G and H, and is normal in G, it is so in H [Jansen and Boon (1967), p. 73].

Problem IV.7.a-11: Prove that all subgroups of the quaternion group (pb. III.2.e-4, p. 73) are normal [Dixon (1973), p. 4]. All subgroups of an Abelian group are normal, but if all subgroups are normal the group need not be Abelian. This simplest example is a subgroup of every non-Abelian group with all normal subgroups [Kurosh (1960a), p. 67].

Problem IV.7.a-12: Take all conjugate subgroups H_i of group G, starting with an — arbitrary — subgroup H,

$$H_i = t_i^{-1} H t_i, \tag{IV.7.a-1}$$

using all t_i of G [Jansen and Boon (1967), p. 74]. Find the intersection of all H_i, show that it is an invariant subgroup of G, a subgroup of H (why?) , and is the largest normal subgroup of G in H. Is it normal in H? Why? Do the answers differ for H invariant, and not invariant? Try this with a few symmetric groups, and the geometrical groups. Do this give anything interesting for cyclic groups? Abelian ones? Dihedral groups?

Problem IV.7.a-13: If group G has two normal subgroups, H and J, and these have no common element except the identity, show that every element of H commutes with every one of J [Armstrong (1988), p. 84].

Problem IV.7.a-14: Consider p and q, both primes ($p < q$); then in a (every?) group of order pq a subgroup of order q is normal [Hall (1959), p. 34]. What would go wrong if $p = q$? How is this related to pb. III.2.b-4, p. 70?

Problem IV.7.a-15: Can the centralizer of an Abelian subgroup be larger than the subgroup? Does such a centralizer have to be a subgroup? Must it be invariant?

Problem IV.7.a-16: The set of all $n \times n$ matrices with real entries forms a group, for each n, as does its subset of matrices with unit determinant (pb. II.4.g-2, p. 56). Prove that it is a normal subgroup [Lyapin, Aizenshtat, and Lesokhin (1972), p. 87]. Is this true for complex entries? This is a point of some significance for group theory, as we will see, and for physics.

IV.7.b The reasons for normal subgroups

Why physically would a subgroup be invariant? For a square-shaped molecule besides the transformations of interchanging atoms we can also permute electrons within an atom; the general symmetry transformation is a product of a permutation of atoms times a permutation of electrons. These permutations commute; it does not matter if we permute electrons before or after atoms. Thus the group of transformations of the atoms is invariant (as is that of permutations of electrons). If we permuted electrons, then interchanged atoms, then reversed the permutation of the electrons, the result is an interchange of atoms, an element of the subgroup. This is a simple case of a group that is the product of two normal, commuting subgroups.

The group of the square gives a product whose terms do not commute. The subgroup of the identity plus the class of rotations about the line perpendicular to the square is invariant. A reflection followed by one of these rotations about the perpendicular line, followed by the inverse (which, of course, is the same reflection) is a rotation about the line, although not the same rotation. The subgroup is invariant because a transformation of the square, such as by a reflection, acting on the rotation gives one of these rotations, and because the product of two such rotations is another, or the identity. Thus the set of these is a subgroup. And the subgroups of reflections, and that of rotations, do not commute. Physically each of these subgroups is invariant because it contains a different type of transformation than the others in the group, and no group transformation (that leaves the object invariant?) can turn one type into another.

Problem IV.7.b-1: Carry out these operations physically and check this. Can there be (group?) transformations taking one type of operation into another?

Problem IV.7.b-2: Is the sum of the classes for the symmetric group consisting of the identity plus permutations on three objects an (invariant) subgroup? Is it related to the discussion of normal subgroups of the geometrical groups?

Problem IV.7.b-3: In sec. IV.5.a, p. 122, we discussed why elements of a group break into classes, and how those of different classes differ (physically). Here the argument is that a group has a normal subgroup because its elements are different, physically, from the others. Is there a relationship between these views? Why? Does this suggest anything about classes and normal subgroups, and their relationships?

IV.8 MULTIPLYING AND DIVIDING COSETS AND GROUPS

A group is a set of elements plus multiplication rules. We usually think of the elements as transformations, operators, numbers, but they need not be. They might be sets themselves, (sub)groups, classes, cosets, ..., suggesting that from groups we can construct others, with elements subsets of the groups. How can we multiply these, and what is the purpose? For subsets of groups there is a built-in multiplication rule, that inherited from the group, giving the product of elements of the subsets. How does this then give a rule for the product of subsets themselves? And if we can multiply sets, such as groups, can we divide them? What does such multiplication and division tell us about a group, about groups generally, and about the objects described?

The product of sets is that set found by multiplying each element of one by each of the other. Division is somewhat harder, and we have to explore products further before coming to it.

IV.8.a Products of cosets

First we multiply cosets. These are sets, and we have just defined products of sets, so we know how, though perhaps there is ambiguity.

Problem IV.8.a-1: To find the product of cosets t_1H and t_2H we take elements h_a and h_b of H and compute $t_1h_at_2h_b$, giving $t'h'$; the coset to which it belongs is the product. Does it depend on h_a and h_b? Explain why it does not — the product of cosets is unique [Cornwell (1984), p. 31; Hall (1959), p. 27]. Show that it is t_1t_2H. Is this true if H is not normal? Note that we take any two elements of H, not all. Thus each product occurs only once. Compare this to the product of classes (sec. IV.5.f, p. 126) and explain why different definitions are used.

Problem IV.8.a-2: If for subgroup H the product of (every) two cosets is also a coset, prove that H is normal [Jansen and Boon (1967), p. 74]. Is the converse true? Is "every" necessary? Why does the subgroup have to be normal?

IV.8.b Products of groups

Groups also can be multiplied — provided they have the same product rule. What do we get — anything new, or is all the information in the product already available in its terms?

Problem IV.8.b-1: Is the product of groups a group? Is the product of subgroups a subgroup [Grossman and Magnus (1992), p. 128]?

Problem IV.8.b-2: The positive rationals form a group, with multiplication as the product. Their logarithms also do, with addition as the product (pb. III.4.b-5, p. 87). Would we multiply these? Why? Would this give a group?

Problem IV.8.b-3: A special case of the product of groups is the direct product (sec. III.2.b, p. 69; sec. III.5.c, p. 101). The product of G and G' is a direct product if all elements of G commute with all of G'. Let both G and G' be S_2. First let G act on 1,2, G' on 2,3; find the product. Then find the product if G acts on 1,2 while G' acts on 3,4. Compare the product groups. The latter is a direct product, the former is not. Find the normal subgroups of both product groups. Does this suggest any general rule? In what way might the concept of direct product be of use? The group of products of permutations of atoms with permutations of electrons (sec. IV.7.b, p. 136) is a direct product of groups; the group of the square is not a direct product.

Problem IV.8.b-4: Take the direct product of groups G and G' and invariant subgroup K of this product, such that K has no elements (besides the identity) in common with $G \times E$ and $E \times G'$, the subgroups given by the elements of G times the identity of G', and of G' times the identity of G. Show that K is Abelian [Armstrong (1988), p. 84].

Problem IV.8-5: For group G, a direct product of groups H and K, show that the classes are direct products of the classes (what does this mean?) of H and K. How are their numbers related [Joshi (1982), p. 27]?

IV.8.c The (factor) group of the cosets of a normal subgroup

Cosets of subgroup H of G can be multiplied. Objects with a multiplication rule often give groups. Can, need, this be true here; does this product give a group, one whose elements are cosets? How is this group related to H and G? The set of cosets that are constructed using an invariant (normal) subgroup H is itself a group. It is the factor (quotient) group G/H; G "divided" by H [Burn (1991), p. 150, 153; Maxfield and Maxfield (1992), p. 72]. Its elements are sets, and there is a relationship between these sets and the elements of G.

Problem IV.8.c-1: Show that the necessary and sufficient condition for the cosets of H to themselves form a group is that H be normal [Armstrong (1988), p. 79]. What goes wrong if the subgroup is not invariant?

Problem IV.8.c-2: What is the factor group if the normal subgroup is the group itself [Lyapin, Aizenshtat, and Lesokhin (1972), p. 88, 218]?

If it is the identity?

Problem IV.8.c-3: What is the identity of a factor group [Jansen and Boon (1967), p. 36; Lyapin, Aizenshtat, and Lesokhin (1972), p. 88, 218]? What is the inverse of an element — given a coset what is the coset whose product with the first gives the identity set? Give a rule for finding the coset that is the product of two cosets. For factor group $F = G/H$, can, must, F be a subgroup of G? What then is the relationship between F and H? Is F invariant? Is G/F a factor group?

Problem IV.8.c-4: The factor group of G is a group of sets. However show that there corresponds to each of these (co)sets an element of G, and that the product of sets S, T corresponding to elements a, b of G, is related (how?) to (the product) element ab of G. This is reminiscent of an isomorphism or homomorphism. Is it [Grossman and Magnus (1992), p. 126]? Which? In what way? What groups are related by the morphism (sec. III.4, p. 85)? Summarize the relationships between normal subgroups, factor groups, and homomorphisms.

Problem IV.8.c-5: Prove that a factor group of group G is isomorphic to a subgroup of G. However it need not be a subgroup, a important point in crystallography.

Problem IV.8.c-6: If group G has a factor group F by subgroup H, prove that G is a direct product of H and F (so that the notation of group division is not unrealistic). Show that the necessary and sufficient condition for a group to be a direct product of subgroups is that it has a normal subgroup. Are these statements related?

Problem IV.8.c-7: How are the sets of normal subgroups, and of factor groups, related?

IV.8.c.i *Examples of factor groups*

The geometric groups (sec. II.2.e, p. 36 — sec. II.2.h, p. 40) show what factor groups are and why. There are groups consisting of rotations. For these figures we might also have reflections. So this gives a set of transformations, rotations, and products of rotations and reflections, the group dividing thus into two sets. The rotations form an invariant subgroup. The group is then the product of the factor group given by the rotations, and the invariant subgroup, the reflections (or reflections times rotations; either can be taken). There is a factor group because each element can be written as a product of an element of the factor group with one from the invariant subgroup.

Problem IV.8.c.i-1: For the group of the square we have found an invariant subgroup (sec. IV.7.b, p. 136), the rotations, and a coset formed by multiplying the elements of the subgroup by a reflection. Can we consider these sets as elements and multiply them? Using this subgroup, what is the set that is the product of the cosets. Work this out both physically and mathematically. For this group find the factor groups, using all subgroups, constructing their multiplication tables. Do these all give factor groups? Which are invariant? Have we met any before? What is their physical significance? Are the factor groups subgroups? Is there a relationship between their orders and those of the group? Find all groups to which the group of the square is homomorphic. Why is this included in the problem?

Problem IV.8.c.i-2: Why do the rotations form an invariant subgroup? Would they do so if the dimension of space were different?

Problem IV.8.c.i-3: What is the significance of the difference between these results and those for the group of the double rectangle?

Problem IV.8.c.i-4: Repeat these problems for the pyramidal group and for the water group.

Problem IV.8.c.i-5: Do these problems also for the other order-eight groups (sec. III.2.e, p. 73). How much information can you get about these groups, and the differences among them, and about the figures they describe, knowing (just) their invariant subgroups, and knowing (just) their factor groups, and knowing (just) their homomorphisms?

Problem IV.8.c.i-6: Cyclic subgroup H is normal in group G. Show that a (proper?) subgroup of H is itself normal in G.

Problem IV.8.c.i-7: Prove that every subgroup, and factor group, of a cyclic group is cyclic [Dixon (1973), p. 20]. Give examples, especially of factor groups of cyclic groups. Also any finite cyclic group of order n contains a unique (cyclic?) subgroup (up to isomorphism?, automorphism?) of order d, for each d for which n/d is an integer. What happens if it is not an integer? Give examples. What are the normal subgroups of a cyclic group? Construct the multiplication table for all factor groups.

Problem IV.8.c.i-8: Find all factor groups of S_2, S_3, and S_4; prove they are groups. Repeat for the alternating groups. How are these related? Give the group tables for the factor groups. Have these been given names? Are their names surprising? Do this for S_n and A_n (perhaps by computer).

Problem IV.8.c.i-9: A group that has an invariant subgroup is the inhomogeneous rotation group — the set of all rotations and translations, with the translation subgroup normal. Thus with T a translation and R a rotation, $R^{-1}TR$ is a translation, so the translations are invariant, but $T^{-1}RT$ is not a rotation. Try this experimentally. Take an object and rotate, then translate and then rotate it about the same axis, but through the negative of the angle. Notice that its orientation is unchanged, but it is translated. Now translate, rotate, and then translate by the negative distance. Since the axes have been rotated, the inverse translation is in a different direction. Thus the object is both rotated and displaced. So this sequence of operations does not give a rotation — the rotation subgroup is not invariant. The factor group is found by taking an invariant subgroup, here the translations, and generating the cosets, here translations in different directions. Notice that the product of two cosets, here given by the rotations, is a coset. So the cosets form a group, the factor group. Check this experimentally. Also try to form cosets using the (noninvariant) rotation subgroup, and a group using the elements so formed. What goes wrong?

IV.8.c.ii *Properties of factor groups*

Factor groups are important probes of groups. Before studying this it is useful to collect some of their properties.

Problem IV.8.c.ii-1: Group G is a direct product of two groups, H, J; show that H and J are both invariant and that the factor group G/H is (isomorphic to) J. Is it the same as J, or only isomorphic? Also G is homomorphic to both H and J. What is the rule giving the homomorphism (the list of corresponding elements)? Further the number, N_G, of classes of G, is the product of those of H and J,

$$N_G = N_H N_J. \tag{IV.8.c.ii-1}$$

Problem IV.8.c.ii-2: The center Z of group G is always normal. Why? Show that if G is non-Abelian, G/Z cannot be cyclic [Lyapin, Aizenshtat, and Lesokhin (1972), p. 89]. Can it be Abelian?

Problem IV.8.c.ii-3: Consider subgroup H of G and a factor group G/N by another subgroup N. Assume that H/N is a factor group of H (is it always, never?) . Is H/N (isomorphic to) a subgroup of G/N? Find the intersection R of H and N, and H/R. How is H/R related to H/N? To H? Give examples.

Problem IV.8.c.ii-4: Suppose from factor group G/H we can form another factor group $(G/H)/J$. Is $(G/H)/J$ also a factor group of G? If so, denote it by G/K. How are H, J and K related? Give examples.

Problem IV.8.c.ii-5: The factor group of G with respect to invariant subgroup N, G/N, is a group. Thus it can also have subgroups, and normal subgroups. How are these related to the subgroups of G [Jansen and Boon (1967), p. 73, 74]? Show that every normal subgroup of G/N is (isomorphic to?) a normal subgroup of G. Further every normal subgroup of G, that is also a subgroup of the normal subgroup N, is (isomorphic to?) a normal subgroup of G/N. Prove that if N is a normal subgroup of G, and is as well a subgroup of subgroup H of G, then N is also a normal subgroup of H. Further factor group H/N is a subgroup of factor group G/N.

Problem IV.8.c.ii-6: There are elements of groups that can be written in the form

$$c = [s, t] = s^{-1}t^{-1}st, \qquad \text{(IV.8.c.ii-2)}$$

with s and t group elements, and this defines a new product [,] of s and t. For which groups is c always the identity? Find elements of symmetric groups that can be so written. Elements of this form are called commutators (they are unity if s and t commute) [Armstrong (1988), p. 83; Baumslag and Chandler (1968), p. 112; Dixon (1973), p. 30; Hall (1959), p. 138; Jansen and Boon (1967), p. 74; Kurosh (1960a), p. 99; Landin (1989), p. 112; Ledermann (1953), p. 104; Lyapin, Aizenshtat, and Lesokhin (1972), p. 91, 218; Scott (1987), p. 56]. Check that this product [,], is not associative; it does not give a group. Show that if c is a commutator then all members of its class are, as is c^{-1}. Now for group G, find the smallest subgroup (using the group product, not [,]) that contains all the commutator elements of G, the commutator subgroup H_C. It is clearly normal. Form the factor group of G by H_C, G/H_C, show that it is Abelian, and that H_C is the smallest normal subgroup of G giving an Abelian factor group. Give an elementary explanation. Would there be any value to checking examples of these constructions for cyclic groups? Abelian ones? Dihedral groups? What is the commutator subgroup of a symmetric group? How about A_n? Show that it is equal to A_n, $n > 4$. What is the commutator subgroup of the quaternion group (pb. III.2.e-4, p. 74)? Find the commutator subgroups of the geometric groups, and discuss the reasons, group theoretical and particularly geometrical, for elements being commutators, and for the forms of, and the elements in, the commutator and factor groups.

Problem IV.8.c.ii-7: The automorphisms (sec. III.4.e, p. 91) of group G form a group, AUT G. The set of inner automorphisms, those produced by an operator of the group (how?) , is of course (?) a subgroup. Show that it is a normal subgroup (of what?) [Dixon (1973), p. 19; Hall (1959), p. 85; Kurosh (1960a), p. 88]. Also it is isomorphic to $G/Z(G)$, where $Z(G)$ is the center of G. That is G is "divided" by the largest subgroup of elements that commute with all those of G, and the result is the group of inner automorphisms. Why?

IV.8.c.iii *Factor groups and homomorphisms*

Factor groups give information about the structure of groups; one of the ways is by giving their homomorphisms. As we saw these are closely related. We should understand why, and how to use this.

Problem IV.8.c.iii-1: Explain why groups to which group G is homomorphic must be (isomorphic to) subgroups of G, and these are normal. Give examples. What is the relationship between the homomorphisms and the factor groups [Lyapin, Aizenshtat, and Lesokhin (1972), p. 87, 89]? Why? What groups can have homomorphisms? Why?

Problem IV.8.c.iii-2: The kernel of a homomorphism of G is the set of elements of G that are mapped to the identity [Armstrong (1988), p. 86; Burn (1991), p. 149; Cornwell (1984), p. 35; Kurosh (1960a), p. 73; Landin (1989), p. 130; Maxfield and Maxfield (1992), p. 83]. Is it a subgroup? A normal one? How is it related to the normal subgroups of G? To its center? Why?

IV.9 SIMPLE GROUPS

Group G can be constructed from invariant subgroup H (thus from a factor group); H provides knowledge about G, enough so that it can be obtained with only a small amount of other data. And H can similarly be constructed from an invariant subgroup of it. Continuing we come to a group with no normal subgroups (that this is correct and not merely a reasonable guess is implied by the Jordan-Holder theorem [Baumslag and Chandler (1968), p. 163; Hall (1959), p. 126; Scott (1987), p. 35]). Thus G is built from that subgroup. It then seems clear that invariant subgroups, and those groups that have none, play a central role in group theory. A group that has no proper invariant subgroup is called simple. It is on simple groups that much of the theory of (finite) groups rests [Scott (1987)].

Problem IV.9-1: Which cyclic groups are simple [Burn (1991), p. 160, 165]? Explain. Which Abelian groups are [Scott (1987), p. 35]? Why?

Problem IV.9-2: Are S_3 or S_4 simple?

Problem IV.9-3: Show that A_3 is simple (but cyclic). Find the invariant subgroup(s) of A_4, showing that it is not simple (pb. IV.7.a-6, p. 134) [Baumslag and Chandler (1968), p. 174; Kurosh (1960a), p. 71].

Problem IV.9-4: Find all conjugate subgroups, for each subgroup, for the geometric groups (sec. II.2.e, p. 36 — sec. II.2.h, p. 40). Compare results, especially for the group of the double rectangle and the group of the square, and discuss the meaning of the differences. Which of these groups are simple? Some are not because they contain transformations of different, and invariant, types, giving invariant subgroups. Which? Why? Can you give physical (geometrical) proofs that they are (or are not) simple? How about mathematical ones? Find the relationship between their conjugate subgroups and those of S_4.

Problem IV.9-5: Prove that there is only one simple group of order 60; it is A_5 [Armstrong (1988), p. 85; Burn (1991), p. 160, 165; Grossman and Magnus (1992), p. 167; Hall (1959), p. 83; Kurosh (1960a), p. 68; Maxfield and Maxfield (1992), p. 93].

Problem IV.9-6: It is of great importance in several areas of mathematics that A_n, $n > 4$, is simple [Baumslag and Chandler (1968), p. 172; Kurosh (1960a), p. 68; Ledermann (1953), p. 120; Maxfield and Maxfield (1992), p. 95]. All have subgroups, but none invariant.

Problem IV.9-7: Simple groups are by definition subject to a strong condition, so are greatly restricted. There are in fact only a few infinite series, plus some others (the sporadic simple groups). It might be expected then that given the order of a group, if there is any simple one with that order, it is unique. However show that there are two nonisomorphic simple groups of order 20,160 [Scott (1987), p. 314].

IV.9.a Semisimple groups

A semisimple group is a direct product of simple groups (each generator of the group is a product of generators of its subgroups, and the generators of different subgroups commute; the order of the group is therefore the product of the orders of the subgroups making up the direct product) [Lomont (1961), p. 28].

The groups that we consider here are either finite or semisimple. These are both special, so their properties may not hold in general, and it is important not to extend these to other cases without verification [Mirman (1995b)].

Problem IV.9.a-1: A group is semisimple if it has no invariant Abelian subgroup(s). Show that these two definitions are equivalent.

Problem IV.9.a-2: Of the groups considered, symmetric groups, geometrical groups, and so on, some are simple and some not. Of the latter which, if any, are semisimple? Why? What is the significance (physical, geometrical, group theoretical, other) of that?

Problem IV.9.a-3: Prove that a simple group cannot be written as a direct product of its subgroups. A semisimple group is composed of simple groups, but a simple group cannot be broken down into semisimple ones.

IV.9.b The importance of simple groups

We see that there are differences among groups. For some all elements are alike (physically, geometrically, perhaps in other ways) and so cannot be divided into subgroups of different types of elements. Subgroups are transformed into each other by group operations. That is (physically, geometrically), the subgroups are distinguishable but not really dissimilar, differing, say, only as viewed by different observers. Groups having elements, in such a sense all the same, are simple (perhaps in several ways). But we can also consider objects invariant under different types of transformations, like rotations and reflections. Each of these types, by itself, gives a group. The complete group consists of transformations of different types, and products of these. The group is not simple. Thus we see how group theory restricts the transformations of (physical) objects, so objects themselves. There are, for each order, only a finite number of groups, and only a small number, if any, of simple groups.

Normal subgroups and factor groups then tell us that there are a special set of groups, the simple groups, and every (finite) group can be found by combining these. To understand a group, we need understand its invariant subgroups. So for a group, an important question is: Of what simple groups is it composed? To find these we find its invariant subgroups and the factor groups. We can take an invariant subgroup and multiply it by transformations to get the entire group. Thus we can take all simple groups and multiply them in various ways to get all groups, so the simple groups are especially important in group theory. Their compilation, a needed step in understanding them, was completed only recently [Gorenstein (1982, 1983)]. Knowing the simple groups, we can find the others. It is like the prime numbers. Given them (something which has not — yet — been done), all others are their products.

Normal subgroups and factor groups offer reassurance. The set of groups that we have to study does not go on forever. There are certain types (including several infinite series of simple groups) and their combinations. Of course the study of simple groups is far from a simple task, and they can be combined in many different ways. But at least the problem is bounded and we have some hints of tools to help study it.

Chapter V

Representations

V.1 THE PHYSICALLY MOST RELEVANT PART OF GROUP THEORY

A physicist or chemist (and often a mathematician) interested in groups usually is interested in applications. A group itself is abstract, a set of symbols plus rules relating them. What is generally needed are concrete objects, statefunctions, their transformations, quantum numbers, quantities like angular momentum, and so on, and for a mathematician other entities, or often the same, perhaps with different names. How are these related to groups describing systems?

Such physical quantities are not groups, but their representations, specifically the objects that representations consist of, basis states, called in quantum mechanics wavefunctions (or statefunctions, a better term), and the matrices that carry out the group transformations by acting on them. It is not the abstract symbols that we (usually) want, but these matrices and basis states (the objects given by the transformation matrices of a representation acting on any one, say a unit vector). They are often explicit functions of the relevant parameters, such as angles; spherical harmonics are examples. For applications (not only to physics), we have to develop means of classifying and finding the representations of the relevant groups: matrices and basis states and functions of them, and their labels.

Here we study what a representation is, its relationship to its group, and to the physical (and mathematical) quantities and properties of a system described by the group. The purpose of the present chapter is to introduce the concepts and explain their meaning and relevance, putting them into context. Vocabulary is emphasized; it is important to first thoroughly understand the terminology. In the following chapters we develop needed mathematical tools.

V.2 HOW REPRESENTATIONS ARE DISGUISED

Group representations are introduced in courses on elementary quantum mechanics (even in elementary physics), in disguise of course. It is useful to start by removing this disguise — quantum mechanical examples illustrate the concepts and give concrete meaning to the abstract symbols of group theory. They show how quantum objects and procedures fit into the theory and how the usual discussions of such familiar things hide the presence (and value) of representations.

We start with the most familiar representations, those of the rotation groups — for these groups are familiar. However they are continuous groups — these are studied below — rather then finite ones, which these chapters are emphasizing. They are considered only to illustrate the meaning of the terms. Having done that, we give the definitions of representation and associated objects, and then develop techniques for analyzing those of finite groups, leaving continuous ones like rotation groups until later.

Problem V.2-1: Even this elementary discussion is rather advanced, using quantum mechanics. The reader, as he studies this, should think of how the concepts are really introduced much earlier; in particular, the terminology provides hints.

V.2.a Systems with cylindrical symmetry

A system with cylindrical symmetry is one with a potential energy function invariant under rotations around an axis (labeled z); for it the angular part of the wavefunction is

$$\Phi(\phi) = exp(il\phi), \quad l \text{ integral.} \tag{V.2.a-1}$$

This follows immediately from Schrödinger's equation. It does not matter what the system is, this is the angular wavefunction for any with cylindrical symmetry. Why? These wavefunctions are not properties of the equation or the system, but of the symmetry; Schrödinger's equation has no choice but to have these, and only these, as its solutions if it is to describe correctly such systems. If we rotate around z the system is unchanged, thus so must the equation be. And the wavefunction, the solution, seen by one observer must be the same as that seen by a rotated observer, except for the change of angle.

Problem V.2.a-1: Explain why the requirement of the last sentence follows from that of the previous one, and vice-versa.

V.2.a.i *The condition symmetry places on the solutions*

So a rotation acting on any function describing any system with this symmetry gives the same function with only the value of the angle different. What functions have this property? The only ones are $exp(il\phi)$, with l integral. Thus these can, and must, be the solutions of Schrödinger's equation for a potential with cylindrical symmetry (one independent of ϕ). These, times a function of all other variables, are the (only possible) wavefunctions of any such system.

V.2.a.ii *How rotations determine the solutions*

The function $exp(il\phi)$ seen in a system rotated through θ is

$$exp(il\phi + il\theta) = exp(il\theta)exp(il\phi). \qquad \text{(V.2.a.ii-1)}$$

Thus rotation operator $R(\theta)$ — giving a symmetry of the system — corresponds to $exp(il\theta)$. Corresponds means that they have the same effect on $exp(il\phi)$. Note that there is not one function for $R(\theta)$, but one for each (integral) l. Here the symmetry group consists of the operators $R(\theta)$ — purely abstract symbols with a multiplication law — but the effect of these on the statefunctions is given by concrete, explicit, functions. These functions, of clear value in studying the system, the symmetry, and the group, are representation basis states.

Thus the solutions of Schrödinger's equation for a system with cylindrical symmetry, and the functions corresponding to the operators that leave such a system invariant (here rotations around the symmetry axis), are identical. In general, the solutions of Schrödinger's equation for a system with symmetry are the functions determined by the operators that leave the system invariant. These operators, and the statefunctions on which they act, give representations of the symmetry group.

V.2.a.iii *The only possible statefunctions give all representations*

The basis functions (or basis states) of a representation of the two-dimensional rotation group, the set of rotations around an axis, are obtained by the action of $R(\phi)$, represented by the one-dimensional matrix with element $exp(il\phi)$, on a fixed vector, here 1, so are $exp(il\phi)$. These functions form all representations, one for each integer l, of this group. So they are the only possible wavefunctions for a system with this symmetry — these are the only functions that remain the same when rotated, except for the change of angle. That is why they — only — are the solutions of the ϕ part of Schrödinger's equation for a system invariant under this group.

Here, and in general, knowing the symmetry group of a system gives that part of its wavefunction that depends on the variable, like the angle, that the potential is independent of — provided that we know the representations of the group.

So group representations are essential in quantum mechanics, for it must respect, and express, the symmetries of space and of the objects it describes — thus it must describe nature by representations of the relevant groups. More fundamentally, there are different observers, these are related by group transformations, and the formalism must allow, but more be based on, this, so must properly describe the correlations of these observations (sec. I.2.b, p. 4) [Mirman (1995a)].

Problem V.2.a.iii-1: Why must l in $exp(il\phi)$ be integral?

Problem V.2.a.iii-2: Check that trigonometric functions do not have the property that a rotation gives the same function.

Problem V.2.a.iii-3: How is cylindrical symmetry used in these arguments?

Problem V.2.a.iii-4: Show that the function going with $R(-\theta)$ is the inverse of that for $R(\theta)$ and the one going with $R(\theta)R(\theta')$ is the product of the functions for $R(\theta)$ and $R(\theta')$. Explain, both physically and mathematically, why the rotations, $R(\theta)$, form a group.

V.2.b Disguising the three-dimensional rotation group

The two-dimensional rotation group is too simple to be very instructive. Richer, and probably more familiar, systems are those with spherical symmetry, the nonrelativistic hydrogen atom, and the simple harmonic oscillator, say. The three-dimensional rotation group (sec. II.3.a, p. 41), SO(3), is a symmetry group of these.

Schrödinger's equations for systems with this rotational symmetry are solved by separation into equations for the radial coordinate and for the angles. The angular equation does not contain the potential, if and only if, the potential function is independent of the angles, so the equation is same for all spherically symmetric systems. Its solutions then are determined by the symmetry so by the group of transformations that leaves such systems invariant, SO(3). It is this dependence of the solutions on the group that we wish to study, so as to determine the solutions and their properties, and to understand why spherical symmetry requires that these, but only these, functions describe such systems, and why they have the properties that they do.

V.2.b.i *The basis states for spherical symmetric systems*

In elementary quantum mechanics we learn that the states of a spherically symmetric system are labeled by quantum numbers, l, an integer giving the angular momentum, and $l_z(= m)$, an integer giving the angular-momentum z component. These have the same sets of values for all these systems because they really label the rotation group representations and states. Each integer l specifies a representation, and each value of m, $-l \leq m \leq l$, a state of that representation.

The representation basis states, the angular parts of wavefunctions, are called spherical harmonics. For each l there are $2l + 1$, and there are three matrices, representing rotations around the three axes, of dimension $2l+1$, which acting on the basis states give the wavefunctions describing the system rotated around the x, y and z axes, that is they transform the spherical harmonics. These, and matrices (for small l) are given in books on quantum mechanics, and many other subjects, testifying to their importance.

Problem V.2.b.i-1: Explain why the spherical symmetry of a system requires that the statefunctions describing it be the basis states of the representations of the rotation group.

Problem V.2.b.i-2: Why do the representation matrices have dimensions which are odd numbers?

Problem V.2.b.i-3: The (unnormalized) basis states of the $l = 1$ representation are

$$Y_1(\phi) = sin\theta exp(i\phi),\ Y_1{}'(\phi) = cos\theta,\ Y_1{}''(\phi) = sin\theta exp(-i\phi). \tag{V.2.b.i-1}$$

Show that under the transformation

$$\theta \Rightarrow \theta + \theta_1,\ \phi \Rightarrow \phi + \phi_1, \tag{V.2.b.i-2}$$

these three functions go into linear combinations of themselves — and no other functions. Find the coefficients in these linear combinations — the matrix elements. What are the values of m for the functions? Write this set of states as a column vector and check that the matrix acting on the column gives this same transformation.

Problem V.2.b.i-4: What are the basis states (and matrix elements) for $l = 2$? For general l?

Problem V.2.b.i-5: Prove that a rotation acting on spherical harmonic $Y_{lm}(\theta, \phi)$ gives a sum of spherical harmonics, all with the same l — change the angles in the function and show that the resultant function is a sum of terms of $Y_{lm'}$, for a set of m'.

V.2.b.ii *What does finding a representation mean?*

This illustrates what we wish to do: find the labels of the representation (here only one, l) and the states (here only m), then the basis states (the wavefunctions) — here they are functions of the angles — and the matrix elements, the coefficients of the transformations. From these we can determine the other physical variables.

This, of course, is a standard problem of quantum mechanics; it is group representation theory in disguise.

V.2.c Removing the disguise from quantum mechanics

What is the point of this? Schrödinger's equation has an infinite number of solutions, and we can take any linear combinations. Which do we want, and why? The useful states, the ones used in quantum mechanics, are (generally) those solutions that are transformed into each other by rotations; the states of, say, the hydrogen atom are (taken as) the ones with single values of l and m. Why?

If we change the angles (perform a rotation) of a basis function of a single l, the function of the new angles, θ', ϕ' can be written as a sum over basis functions expressed in terms of the old angles, θ, ϕ, with coefficients, $a^l_{mm'}$, which depend on the angles of rotation. And all these functions have same l. These states are mixed by the rotations: a rotation operator applied to a state gives a sum of states of the same l; schematically

$$Y_{lm'}(\theta', \phi') = \sum_m a^l_{mm'} Y_{lm}(\theta, \phi). \qquad \text{(V.2.c-1)}$$

These — for fixed l — all are wavefunctions of the same state of system, but as seen by different observers; they differ because their axes are relatively rotated. Functions of different l are statefunctions of different states of the system (here, different total angular-momentum states); they have different physical properties. To describe correctly an object we must separate statefunctions into sets, all members of each give the same state, though viewed by different observers, different sets (representations) have physically distinct states. For spherical symmetry these sets are labeled by l.

Generally statefunctions of a system are intermixed by the group operators — an operator acting on a state gives a sum of such states. A set of functions so mixed among themselves by the group operators are the basis states (basis vectors, basis functions), and their sums, of one of its representations (basis states are the basis vectors, or coordinate vectors, of the representation vector space). The states of a representation are those states with the property that any operator of the group applied to any state gives a function that can be written as a sum over this set of states (only) — a group operator applied to a basis state equals a sum of basis states of the same representation. These states can be written as a sum of a minimum number (which depends on the representation) and the members of any such minimum set are the basis states, although the term might be used more loosely as referring to any states of a representation. For each system there are several sets (representations) intermixed, with the states of one not mixed with those of another. So rotations acting on a basis state of representation l (a spherical harmonic with label l) give sums of states of — only — representation l.

Different observers see statefunctions obtained from each other by applying group operators, so the ones mixed by these operators all belong to a single set (are the states of a single representation), those not mixed belong to different sets (representations).

A sum of states of different representations describes a system that can be found in a state of each of these representations. One that is a sum over states of the same representation gives, again, the probability of finding the system in each of these states. But (for groups such as these), unlike a sum over representations, there is a coordinate system for which the sum is over only a single state — in which the system will certainly be found in that state. Moreover if the system is invariant under the group, all states of the same representation, but not those of different representations, must have the same energy. Thus organizing the states of a system into representations is necessary in order to find its physical properties.

For spherical symmetry the basis states are labeled by numbers which, as is implied by their names, have physical significance. Since these labels have definite values, they are (given by) eigenvalues of the operators going with the physical quantities, here angular momentum and its z component. Any state with a single angular-momentum value can be taken as a basis state since it is a sum of states all of the same l (being an eigenstate of the angular-momentum operators).

The potential does not appear in the angular equations, so all states of the same m have the same energy. (The equation for r, in which it does appear, is independent of m.) The group provides the labels for the states, and these have physical significance, and it also gives the functions (of say the angles or indices) that are the basis vectors.

Generally in quantum mechanics we need a complete set of states — a set of states such that any (well-behaved) function of the same variables, defined over the same range as these variables, can be expanded in terms of. The collection of all basis states of all representations of a group (of the types considered here) is such a complete set.

Those of the symmetry group of a system (the group whose transformations leave the Hamiltonian unchanged) are the natural choice — at least because all states of a representation have the same energy, as well as the same values of such other physical quantities as total angular momentum. Given these states, and their expectation values, we have completely described the system. But finding the basis states and matrix elements, which give expectation values, is the statement of the problem of constructing the representation of a group. Thus group representation theory is a systematic way of solving quantum-mechanical problems (provided there is a group available). This indicates the role, and usefulness, of group theory in quantum mechanics.

We also want the wavefunctions explicitly, their labels and physical significance, and wish to know how to organize the states into sets, all with the same energy. This provides most of what can be understood about the system; it comes from the representations of the group, this and more. The problem then is given a group of transformations, construct the representations: label and find the basis functions and matrices representing the group elements. Before doing this we must give a (reasonably) rigorous definition of representation.

Problem V.2.c-1: In explaining why the states of the hydrogen atom are the basis states of the rotation-group representations how was spherical symmetry used?

Problem V.2.c-2: Here the appearance of representations in quantum mechanics has been emphasized. But they are introduced much earlier, at the beginning of elementary physics and in mathematics. The reader should note such occurrences and discuss the reasons for them.

V.3 WHAT ARE REPRESENTATIONS?

What does it mean to say a set of matrices forms a representation of the rotation group? By taking some physical object, say this book, and rotating it in various ways we can find experimentally the product of two rotations — the rotation that gives the same final orientation as the two performed in succession (although products are not usually found this way). If we can assign a matrix to each rotation such that the matrices representing (assigned to) the rotations have the same product as the rotations they represent, the set of matrices forms a representation of the rotation group. (This implies, perhaps not quite fairly, that we are not supposed to think of these matrices as the operators, they only represent the operators. Or perhaps we should think of representations as presenting again the group, in a different form; another is a presentation.)

We can now give a definition for any group. A set of matrices forms a representation of a group if to each group element we can assign a matrix and the matrices have the same list of products as the group elements (using matrix multiplication as the combination rule); if

$$AB = C, \tag{V.3-1}$$

then the matrices M representing A, B and C obey

$$M_A M_B = M_C, \tag{V.3-2}$$

for all elements.

Obviously this cannot be done for any set of matrices and finding those for which it can be done, the representation matrices of the group, is a central problem of group theory.

These matrices act on a set of states, column (or row) vectors — the basis states of the representation (the representation is based on how these transform, and the other states are based on these, being linear combinations of them). Of course given the (square) representation matrices we can find a set of basis states, any complete, orthogonal, set of column vectors with the same dimension as the matrices. However often there is more. The states have physical interpretations and are functions of variables, such as angles for angular-momentum states or indices for symmetric-group states. The wavefunction for a square molecule depends on the coordinates of the atoms at each corner. When we ask for the basis states, it is these specific functions of (physically) meaningful variables, these statefunctions, that we want. The basis states are column, or row, matrices, with entries that can be functions (of variables determined by the group) that are mixed among themselves by the representation matrices — a matrix of a representation acting on a state gives linear combinations of the states of that representation. (It is possible to consider functions and operators acting nonlinearly on them [Mirman (1995a,b)], but representation usually implies "linear", and this entire discussion is limited to that.)

A basis vector is a sum of all symbols — some may have zero coefficients — generated from any one by applying all group operators to that one. We might think of continuing and obtaining all symbols produced this way, until a closed set is obtained, but a product of group operators equals a single operator, so all symbols are obtained by applying all group operators to any one. We shall have to see how to find the sets of sums of these symbols (the basis vectors of the different representations), each of which go into themselves under the action of the group.

Problem V.3-1: Prove that matrices, using matrix multiplication, do obey the group axioms, stating which axiom(s) must be considered, and which are satisfied by choice of the set of matrices.

Problem V.3-2: Given a set of matrices forming a representation of a group, which, if any, of the following also forms a representation of that group:

a) the matrices whose elements are complex conjugates of the representation matrices,

b) the inverse matrices,

c) the hermitian conjugate matrices,

d) the transposed matrices?

V.3.a A mathematical view of representations

The motivation for representations, and their usefulness, presented here is physical. But group, and representation, theory are branches of mathematics. So let us summarize this in mathematical terms. A matrix group is a set of matrices obeying the group axioms (a group is merely a set of symbols which can be anything, even matrices). The matrix group forms a representation of the abstract group (the set of symbols with no meanings attached) if the two groups are homomorphic. If the two groups are isomorphic — a distinct matrix is assigned to each group symbol — it is a faithful representation.

Representations are important in physics; they are also important in mathematics. The group symbols have specific meanings in individual contexts in mathematics. Representing these as matrices is helpful, often essential, in studying the systems. And representations are important in group theory. They provide complete information about their groups, and since matrices have properties besides those following from the group axioms, they can be added for example, they can help to extract this information. Representations also provide a tool for studying other aspects of group theory. We shall have to see how.

Problem V.3.a-1: Do representation basis vectors form vector spaces (sec. II.4.j, p. 59)? Why? What postulates about group theory lead to these conclusions?

V.3.b Why groups have representations

Since representations are (usually) the physically relevant aspect, isn't it nice that groups actually do have representations? Of course it is useful to be able to regard the abstract symbols of a group as transformations and to be able to use matrices to find their effect on the functions of physics. Groups have physical significance because we can interpret them this way. But what properties of groups allow them to be so interpreted; why do groups have matrix representations?

The axioms — existence of an identity, of a product, of an inverse, and associativity — state properties also possessed by matrices. So we can use matrices to represent the group transformations. (There are singular matrices; but that there are representations for all, or at least all finite, groups does not require that all matrices belong to representations.)

There is a hole in this logic. That matrices obey the group axioms means that it is possible for them to give representations. But this does not prove that given a group multiplication table there is a set of matrices that satisfy it. The discussion of the regular representation, in the next chapter, shows why the axioms require that the symbols of all (finite) groups can be regarded as transformations; this proves that they — finite groups — can be represented by matrices. (Do not jump to conclusions about other groups.) Thus the group axioms do require that these groups always have representations.

This is the mathematical reason. But it is also significant that (these) physical transformations have matrix representations. If we have an operator without an inverse we might use a matrix to represent it; we do not have to give the inverse matrix a physical meaning. However this would require an auxiliary condition; awkward and a threat to mathematical consistency. There are Hamiltonians that acting on a state produce a rotated state. These Hamiltonians are functions of the operators of the rotation group, and can be written in terms of the representation matrices of the group. Suppose, however, that it were physically impossible to perform the inverse of a rotation. We could write the same Hamiltonian but attach to it a note saying do not perform transformations that are inverse to the allowed ones listed in the attached table — at best mathematically a very inelegant method. There might be other ways of writing a Hamiltonian performing certain transformations but not their inverses, but it is not obvious how, and is not likely to be either mathematically or physically impressive.

Hamiltonians that give transformations belonging to a group are easy to construct, those performing other types are at best doubtful, and really unphysical. Thus the properties that physics requires of transformations leads to their mathematical description by matrices, and to the existence of representations.

Problem V.3.b-1: Write a Hamiltonian that performs a transformation but not its inverse. Write one that performs two transformations but not their product.

Problem V.3.b-2: This analysis is very primitive and there is much more to be considered and said about the reasons physical transformations can be represented as matrices (and indeed why transformations are possible at all). What would physics be like, could there be a consistent universe, and physics, otherwise? Those philosophically-minded may wish to ponder such questions and increase our understanding of why physics leads to (requires?) group representations [Mirman (1995a,b)].

V.3.c Realizations

Physical objects are described by representations. However specifying a representation and its states does not give a full description of a system. For example we can consider wavefunctions describing angular momentum, but it can be the angular momentum of a hydrogen atom, of an oscillator, or of any number of other systems; we can require that states be antisymmetric, these can be states of electrons or of protons. And the representation might be of the SU(2) group describing spin, or that for isospin; these are isomorphic, but have different physical meanings. Also we can realize an operator by a matrix, by a derivative, or in other ways. Thus the z component of angular momentum (whose relationship to the rotation group we consider later) is realizable as

$$L_z = i(x\frac{\partial}{\partial y} - y\frac{\partial}{\partial x}), \qquad \text{(V.3.c-1)}$$

or as various matrices. The eigenvalues are the same, for though of different realizations, they refer to the same representations. Or one realization might refer to the spin of a particle, another to its orbital angular momentum, a third to its isospin. Though the physical meaning is different, the mathematics describing these various physical objects is the same, because these are all different realizations of operators giving the same group, and the group determines the representations, these being independent of the realization.

We consider not only the representation of a group, but its realization (which makes the abstract symbols "real", that is concrete): the actual functions (of relevant variables) that are the basis states, the explicit form of the operators, and their physical meaning. Usually we can ignore this. We do not have to know whether we are dealing with an atom or an oscillator if we want the allowed angular-momentum values. Nor do we need an explicit form for the angular-momentum operators to find these eigenvalues. Usually we can ignore the realization, but not always (often we use it explicitly, but not consciously).

Here we are thinking of the symbols of the group abstractly and regard the operators of, for example, orbital angular momentum, or spin (the Pauli matrices) or isospin, as different realizations of this one group (here SU(2)). We can also take these as different, but isomorphic, groups; we can say that we interpret these symbols differently. These are diverse ways of saying the same thing (and we can take the, say, rotation and isospin groups distinct, but isomorphic, and the orbital angular momentum and Pauli matrices as different realizations of the former — mixtures of terminologies are possible). We prefer (but not always) to use the terminology of realization, it being more explicit.

V.4 TYPES OF REPRESENTATIONS

The definition of a representation is broad, so allows objects of little value to be called representations. Thus we first classify representations, pick out the types that we want to study, and briefly see how others fit in, and why we might not wish to consider them.

V.4.a Must a representation be faithful to its group?

It is easy to find a representation of a group: assign the number 1 to each element — obviously not very useful. Usually we require of a representation more than that the matrices have the same product rules as the group.

A faithful representation, which is ordinarily what we want, is a set of matrices such that distinct matrices are assigned to different elements of the group, so is one-to-one. It is an isomorphism from the group onto a matrix group.

We also get representations by taking homomorphisms from the group into a matrix group; these are not faithful. For them, several different group operators are represented by the same matrix (such as 1). It will usually be assumed that "representation" refers to a faithful representation. However we shall keep an open mind in case we find a useful homomorphism.

Problem V.4.a-1: The "one-to-one" does not follow completely from the assignment of distinct matrices to different elements. Is it possible to assign more than one matrix to an element?

Problem V.4.a-2: Why must every (finite) group have at least one faithful representation? Can it have more?

Problem V.4.a-3: Is it possible that two nonisomorphic groups both have faithful representations that are isomorphic to each other [Grossman and Magnus (1992), p. 105, 182]? Homomorphic? Could they have representations, one faithful the other not, which are isomorphic to each other? How about homomorphic?

Problem V.4.a-4: Can two nonisomorphic groups have isomorphic representations? Can the representations be homomorphic? Under what conditions is a homomorphism possible?

V.4.b Equivalent representations are much the same

Rotation-group representations are labeled by the total angular momentum, the states by its component along z. But the z axis is arbitrary. A transformation to new axes gives other matrices representing the group elements. These sets of representation matrices are physically equivalent; there is an important distinction between representations of different angular momenta, but those for different axes are fundamentally the same. Physically they are identical: the isotropy of space requires that there be no way to distinguish different z axes (what do the last three words mean?). Physically, and mathematically, representations that differ only in their axes are equivalent — they are related by a similarity transformation taking one set of matrices to the other. However this does not depend on whether or not there are axes.

Two representations that can be transformed into each other by a similarity transformation are equivalent; M and M' are representation matrices of the same group element in equivalent representations if

$$M' = SMS^{-1}, \tag{V.4.b-1}$$

where S is independent of the group element — the similarity transformation is the same for all operators. By definition of a representation the multiplication rules must be preserved under similarity transformations, as they are,

$$SAS^{-1}SBS^{-1} = SABS^{-1}. \tag{V.4.b-2}$$

If representation matrices M and M', of the same group element in different representations, cannot be so related for every element, the representations are inequivalent.

Groups can have an infinite number of representations, but they fall into sets such that the members of each can be obtained from the others by a similarity transformation — they are equivalent representations; those of different sets are inequivalent. For a finite group there are a finite number of (sets of) inequivalent representations, with (perhaps) an infinite number equivalent to any one. When we speak of a representation we are (implicitly) referring to a set of equivalent ones, although in writing matrices and basis states explicitly we choose a particular one of a set.

Problem V.4.b-1: Prove that the necessary and sufficient condition that products be preserved under similarity transformation S is that S be independent of the group element. So all, and only, such similarity transformations take a representation to a representation.

Problem V.4.b-2: For an n-dimensional representation there are n basis vectors; these can be taken as column vectors, with all entries 0, except a single 1, this appearing in a different position for each of the n vectors. Check that linear combinations of these n vectors (with any coefficients) are basis states; there are matrices, obeying the group multiplication table, obtainable by a similarity transformation from, but not necessarily distinct from, the original ones, that take these vectors into linear combinations of themselves. Show that for any realization, not only this, the basis states of a representation are linear combinations of the basis states of an equivalent representation — this is the effect of a similarity transformation on the states. Further any n linearly-independent linear combinations of basis vectors, for an n-dimensional representation, are basis vectors of a representation, and these representations are equivalent. It is useful in reading the discussions below to keep in mind the question of how, and when, matrices of equivalent representations differ.

V.4.c Degeneracy and equivalence

A particle with angular momentum can be described using different (equivalent) representations, these distinguished by the axis of quantitization (z axis). If an interaction does not pick out a direction, all states of one representation, and also the states of all equivalent ones, have the same energy; states quantized along different z axes (and also equivalent representations) are degenerate. How are degeneracy and equivalence related? Need one imply the other?

States, or representations, are degenerate if they have the same energy; this means that operators taking one representation, or state, into the other commute with the Hamiltonian, so it is invariant under group transformations and under a similarity transformation. Often equivalent representations and the states belonging to them are degenerate, but not always. The definition of equivalence does not involve a Hamiltonian, nor does a Hamiltonian have to be so invariant.

For the rotation group, if it does not matter what the z axis is, then all states of a single representation have the same energy, and thus so do all states of all representations equivalent to it. But with a magnetic field, states with different m values have different energies. And equivalent representations are distinguished: the "up" states differ in energy. Thus states are equivalent mathematically but distinguished physically: they can be transformed into each other but the Hamiltonian is not invariant under the transformation. The energy of a particle with spin up along B is independent of what axis is called z. But an eigenstate of σ_z has different energy then a $\sigma_{z'}$ eigenstate, thus particles with spin up along different axes have different energy.

Whether representations are equivalent physically — are degenerate (that is have the same energy) — not merely mathematically, depends on whether the Hamiltonian has a term distinguishing them. In discussions of group theory, equivalence means related by a similarity transformation, not invariance of the Hamiltonian.

V.4.d Reducible representations

There is a simple way to construct a representation from other representations; take the direct sum of several sets of representation matrices (from the same or different representations). These representation matrices M are block-diagonal; their diagonal elements are representation matrices and their off-diagonal ones are all zero,

$$M = \begin{pmatrix} A & 0 \\ 0 & B \end{pmatrix}, \qquad \text{(V.4.d-1)}$$

where A and B are matrices. Doing this is generally not useful.

We can now mess things up by performing a similarity transformation to get an equivalent representation with matrices not block-diagonal. Looking at them, could we tell that there is a reverse similarity transformation block-diagonalizing the matrices? How? Are there ever any representation matrices that cannot be block-diagonalized?

A representation whose matrices can be block-diagonalized is a reducible representation (it can be reduced to block-diagonal form). One whose matrices cannot be so reduced, is an irreducible representation. This process of block-diagonalizing the representation matrices is called reduction of the representation. It can be continued, reducing A and B, and so on, until the final blocks are all irreducible representation matrices.

Since we can get a reducible representation from irreducible ones we need only consider these; when we refer to a representation we assume that it is irreducible. So finding the representations of a group means finding its irreducible representations.

An irreducible representation of a group may be reducible under a subgroup (as for an alternating group such as the A_3 subgroup of S_3). That is, the matrices of the subgroup may mix vectors within subsets, but not mix vectors from different subsets. The group elements not in the subgroup mix vectors from different subsets.

Problem V.4.d-1: Check that the M's are representation matrices, if A and B are, and conversely. Also the basis of this representation is a sum of two (in general, several) sets of linear combinations of basis vectors such that those of one set are mixed among themselves by the A's, of the other by the B's, and the two sets are not intermixed. Perform a similarity transformation giving M not block-diagonal.

Problem V.4.d-2: For a representation with matrices $M(T_i)$, consider matrices $M(T_i)^*$, the complex conjugate ones, $M(T_i)^{-1}$, the inverse matrices, and $M(T_i)^+$, the hermitian conjugate matrices,

$$M_{ij}^+ = M_{ji}^*. \tag{V.4.d-2}$$

Show that the M^* matrices form a representation, and if M is irreducible the complex conjugates are, and conversely. Inverse matrices M^{-1} and hermitian conjugates, M^+, do not form representations, except for Abelian groups (that is, if we assign to group elements $a, b, \ldots$, matrices $M_a, \ldots$, forming a representation, then $M_a^{-1}, \ldots$, assigned to $a, \ldots$, is not a representation; of course M_a^{-1} assigned to $a^{-1}, \ldots$, is, the same one). Check this for S_3. What goes wrong with M^{-1} and M^+?

V.4.e Decomposability and reducibility

It may be that a representation cannot be completely reduced; that is we can reduce it to

$$M = \begin{pmatrix} A & 0 \\ C & B \end{pmatrix}, \tag{V.4.e-1}$$

but cannot transform C to zero. So the matrices have block-diagonal elements, plus nonzero elements above (or below) the diagonal, but zero elements below (or above) it. Then the representation is reduced. If we can transform C to zero, it is then decomposed (or completely or fully reduced). A representation that can be completely block-diagonalized is completely reducible (decomposable).

We see later that every representation of a finite group can be completely reduced (decomposed), so for these groups we use reducible with the understanding that it means decomposable.

V.5 EXAMPLES OF REPRESENTATIONS

The meaning of group representations, the reasons they arise, and their physical and mathematical significance and value have now been considered. Before developing methods for finding them we study a few examples.

V.5.a The two-dimensional rotation group, SO(2)

The most intuitive examples are probably SO(2) (sec. V.2.a, p. 147), and SO(3), so we start with SO(2), though it is a continuous group; SO(3) involves too much for now and is considered in a later chapter.

Problem V.5.a-1: Show that the rotation matrix for any representation for a rotation through $\theta + \phi$ equals the product of the matrix for a rotation through θ with the matrix for one through ϕ.

Problem V.5.a-2: For SO(2) the states of each representation are mixed among themselves by rotations around z. There are representations for each integer l and the basis states for representation l are the $l + 1$ functions

$$\Phi_{l,k}(\theta) = cos^k\theta sin^{l-k}\theta, \quad 0 \le k \le l. \tag{V.5.a-1}$$

These are functions (of an angle) and thus, when multiplied by functions of the other parameters like distance, are the wavefunctions of a system with cylindrical symmetry. For a rotation through angle ϕ

$$\theta \Rightarrow \theta + \phi, \tag{V.5.a-2}$$

check that

$$cos^k\theta sin^{l-k}\theta \Rightarrow cos^k(\theta+\phi)sin^{l-k}(\theta+\phi) = \Sigma a(\phi)^l_{kk'}cos^{k'}\theta sin^{l-k'}\theta, \tag{V.5.a-3}$$

so that a state of representation l goes into a sum of terms all of a single l (l labels the representation). Thus these functions are the basis states of representation l. Calculate the values of the a's; these are the matrix elements of representation l between the states labeled by k and k'.

Problem V.5.a-3: SO(2) is Abelian and all irreducible representations of Abelian groups are one dimensional. Why? Thus the states are

$$\psi_l(\theta) = exp(il\theta), \tag{V.5.a-4}$$

and under rotations

$$\psi_l(\theta) \Rightarrow exp(il\theta + il\phi). \tag{V.5.a-5}$$

Calculate the matrix elements for these.

Problem V.5.a-4: Representation l has been given in both $l+1$-dimensional and one-dimensional forms. Find the relationship between the two ways of writing the states. Show that the $(l + 1)$-dimensional matrix can be block diagonalized, with each block of dimension 1, and that there is a relationship between the entry in this block and the matrix element of the one-dimensional representation. This block diagonalization (for other groups not generally to one dimension) is an example of reducing a representation. The representation however is not reduced over the real numbers, but only over the complex ones, these entering explicitly in the basis states.

V.5.b Representations of a finite group, D_3

Although the rotation groups are the most familiar, they are continuous groups. How do we find the representations of a finite group, and what is their physical significance?

The group D_3 (sec. II.4.c, p. 50) is given by the symbols A, B, C, D, F and unit element E, with the requirements (sec. III.2.d, p. 72)

$$A^3 = E, \quad C^2 = E, \quad CA = A^{-1}C,$$

$$B = A^2, \quad D = AC, \quad F = A^2C. \tag{V.5.b-1}$$

From these conditions we get table III.2.d-1, p. 73.

This group has six elements; we can for example regard B as a transformation, or if we do not wish to introduce a new symbol, regard A^2 as a transformation. It may seem that using A^2 reduces the number of transformations. We have to be careful: a symbol that includes letter A is still a different object than A. To say that symbol A^2 is the product of A with itself is the same as saying that B is this product.

Problem V.5.b-1: Verify that these conditions give the multiplication table, and vice-versa, and that these are the same as in pb. II.4.c-5, p. 51).

Problem V.5.b-2: What are the representations of the group? One is obtained by assigning the number 1 to each element. This is clearly not faithful. Another is obtained by assigning 1 to some, -1 to others. Which elements are assigned which values? Is this faithful? What are the basis states?

Problem V.5.b-3: Check that the matrices

$$M(E) = \begin{pmatrix} 1 & 0 & 0 \\ 0 & 1 & 0 \\ 0 & 0 & 1 \end{pmatrix}, \quad M(A) = \begin{pmatrix} 0 & 1 & 0 \\ 0 & 0 & 1 \\ 1 & 0 & 0 \end{pmatrix}, \quad M(B) = \begin{pmatrix} 0 & 0 & 1 \\ 1 & 0 & 0 \\ 0 & 1 & 0 \end{pmatrix},$$

$$M(C) = \begin{pmatrix} 1 & 0 & 0 \\ 0 & 0 & 1 \\ 0 & 1 & 0 \end{pmatrix}, \quad M(D) = \begin{pmatrix} 0 & 0 & 1 \\ 0 & 1 & 0 \\ 1 & 0 & 0 \end{pmatrix}, \quad M(F) = \begin{pmatrix} 0 & 1 & 0 \\ 1 & 0 & 0 \\ 0 & 0 & 1 \end{pmatrix}, \tag{V.5.b-2}$$

have the same multiplication table as D_3. They therefore form a representation of it. Is this faithful? What are the basis states? Suppose that you were given these matrices, but they were not assigned to elements. Could you do so? Is there ambiguity? Why?

Problem V.5.b-4: Try several different similarity transformations on this set of matrices. Check that the resultant matrices also obey the same product rules, so also form representations. These are equivalent to this set. Is this representation irreducible?

Problem V.5.b–5: Give physical interpretations of the representation matrices and basis states, for all three representations.

V.5.c Representations of symmetric groups

Symmetric group S_n (sec. II.4.d, p. 51) provides another example. We consider its action on object $\psi_{123...}$ with n indices, a monomial ("one name", that is term, not a sum of these). This might be a statefunction of an atom with electron 1 in state a, electron 2 in state b, and so on. Physically this is not the correct wavefunction, that must be antisymmetric in the electrons. Such a state would be

$$|a) = \psi_{123...} - \psi_{213...} + \cdots . \qquad \text{(V.5.c–1)}$$

It is obtained by applying symmetric group operator

$$P_a = \varepsilon - (12) + \cdots = (1)(2) - (12) + \cdots , \qquad \text{(V.5.c–2)}$$

to the monomial; ε is the identity operator and (12) the transposition interchanging 1 and 2. As far as the symmetric group cares only the indices are important; the specific function, and its meaning, is irrelevant. So we suppress the function and write just the indices.

For S_2, monomial 12 goes into 21 under (12). The two states mixed by the group operators are

$$|s) = 12 + 21, \text{ and } |a) = 12 - 21, \qquad \text{(V.5.c–3)}$$

with matrix element of ε always 1, that of (12) is 1 for $|s)$, -1 for $|a)$. For S_n there is again the (one-dimensional) completely-symmetric representation — its basis state is the sum of all n permutations of the n indices — and another one-dimensional state half of whose terms have plus signs, the other half minus signs. All matrix elements for the first state are 1, for the second half are 1, half -1.

Problem V.5.c–1: For the completely-antisymmetric state which permutations have matrix element 1, and which -1. Why?

Problem V.5.c–2: For S_3, there are in addition (two) two-dimensional representations. For one, the states are [Schindler and Mirman (1977a), p. 1696]

$$|1) = 3[123 + 213 - \frac{1}{2}(132 + 321 + 231 + 312)], \qquad \text{(V.5.c–4)}$$

$$|2) = 3\sqrt{\frac{3}{2}}(132 - 321 - 231 + 312), \qquad \text{(V.5.c–5)}$$

with (12) having matrix elements 1 and -1, for the first and second. Show that

$$(23)|1\rangle = -\frac{1}{2}|1\rangle + \sqrt{\frac{3}{2}}|2\rangle, \tag{V.5.c-6}$$

$$(23)|2\rangle = \frac{1}{2}|2\rangle + \sqrt{\frac{3}{2}}|1\rangle, \tag{V.5.c-7}$$

so (23) is represented by

$$(23) = \begin{pmatrix} -\frac{1}{2} & \sqrt{\frac{3}{2}} \\ \sqrt{\frac{3}{2}} & \frac{1}{2} \end{pmatrix}. \tag{V.5.c-8}$$

These states form the basis of a representation of S_3; they are mixed among themselves by the permutations. The matrices representing the permutations can then be found (although this may not be the most efficient way) from the action on the states. Find the other two states, show that they are also intermixed, but not mixed with the first two; also find the (23) representation matrix, showing that it is the same. For both sets of states, find the other representation matrices, and check that both representations have the same matrices.

V.5.c.i *Finding the sums forming the states*

In general we construct sums, each term a monomial, a function, or just a symbol, with indices. Each term in the sum has a different permutation of indices, and a different coefficient. These sums, multinomials ("many names", that is terms, in the sum) or vectors, form a vector space (so vector is appropriate; it being used more abstractly and generally). A permutation of symmetric group S_n, with n the number of indices, applied to a multinomial gives a different multinomial (sometimes the same, or as above, its negative).

The problem then is to find a set of multinomials such that when all permutations are applied to any, the resulting multinomials are sums of those in the set; they form a closed set (or space). The sums are thus mixed among themselves by the permutations, and no multinomial outside the set is mixed into it. Each set then consists of the basis vectors of a representation; different sets form bases (set of basis vectors) of different representations.

V.5.c.ii *The definition of, and requirements on, a product*

There is one further thing we require of basis vectors: they should be mutually orthogonal and normalized. So we need a rule for their product. To illustrate, we take for S_n the product of a monomial with itself to be 1, the product of two different monomials — having different orderings of the indices, 0. This agrees with quantum mechanics. If the wavefunction is that describing particle 1 in state A and particle 2 in state B then the norm of the wavefunction, the product of the wave function with itself, is 1. The product of this wavefunction with one for particle 2 in state A and 1 in B is 0; if the system is in the first state, then the probability of finding 2 in A and 1 in B is 0.

Also (as expected) we require, and show later, that states from different representations are mutually orthogonal.

Problem V.5.c.ii-1: What are the normalization factors for the two states of S_2 using this definition of product?

Problem V.5.c.ii-2: Normalize both one-dimensional states of S_3. Is the normalization of the states of the two-dimensional representation of S_3 correct? Show that all these six states are mutually orthogonal.

V.5.d The group of an equilateral triangle

An equilateral triangle is fixed by its three vertices (sec. II.2.c, p. 33). The symmetry group permutes these, thus acting on three objects. It is in fact (isomorphic to) S_3 being of order 6 and noncommutative. While the representations are given in the previous section this interpretation illuminates them [Fässler and Stiefel (1992), p. 13, 15; Jansen and Boon (1967), p. 89]. The symmetry group leaves the center of the triangle fixed. From it the vertices are located with two vectors. Thus there is a two-dimensional representation — giving a geometrical explanation for this representation.

We take the origin at the center of the triangle, the y axis to one vertex, the x axis perpendicular, and denote the rotations (about the origin) leaving it invariant by C_3 (a rotation through $2\pi/3$), and C_3^2, and the reflections through the lines from the origin to the vertices by σ_1, σ_2, σ_3. The representation matrices are then

$$E = \begin{pmatrix} 1 & 0 \\ 0 & 1 \end{pmatrix}, \quad C_3 = \begin{pmatrix} -\frac{1}{2} & \sqrt{\frac{3}{4}} \\ -\sqrt{\frac{3}{4}} & -\frac{1}{2} \end{pmatrix}, \quad C_3^2 = \begin{pmatrix} -\frac{1}{2} & -\sqrt{\frac{3}{4}} \\ \sqrt{\frac{3}{4}} & -\frac{1}{2} \end{pmatrix},$$

$$\sigma_1 = \begin{pmatrix} -1 & 0 \\ 0 & 1 \end{pmatrix}, \quad \sigma_2 = \begin{pmatrix} \frac{1}{2} & \sqrt{\frac{3}{4}} \\ \sqrt{\frac{3}{4}} & -\frac{1}{2} \end{pmatrix}, \sigma_3 = \begin{pmatrix} \frac{1}{2} & -\sqrt{\frac{3}{4}} \\ -\sqrt{\frac{3}{4}} & -\frac{1}{2} \end{pmatrix}. \quad \text{(V.5.d-1)}$$

Problem V.5.d-1: Draw a triangle, with vectors to each vertex (these written in terms of the unit vectors along the axes, e_1, e_2), and verify that these matrices acting on them do leave the triangle invariant, and do perform the transformations used to label them. Would you have expected the entries in these matrices? Why? Show that these obey the

S_3 group-multiplication table. These groups are isomorphic. Explain the isomorphism; justify the particular mapping of the elements of one group onto those of the other. How much freedom is there in choosing this list of corresponding elements? What is the geometrical significance of this freedom? Instead of using vectors along the x and y axes take one along y and the other to a second vertex (these denoted by f_1 and f_2). Find the effect of the transformations on these. Show that the representation matrices are now

$$E = \begin{pmatrix} 1 & 0 \\ 0 & 1 \end{pmatrix}, \quad C_3 = \begin{pmatrix} -1 & 1 \\ -1 & 0 \end{pmatrix}, \quad C_3^2 = \begin{pmatrix} 0 & -1 \\ 1 & -1 \end{pmatrix},$$

$$\sigma_1 = \begin{pmatrix} 1 & -1 \\ 0 & -1 \end{pmatrix}, \quad \sigma_2 = \begin{pmatrix} -1 & 0 \\ -1 & 1 \end{pmatrix}, \quad \sigma_3 = \begin{pmatrix} 0 & 1 \\ 1 & 0 \end{pmatrix}. \qquad \text{(V.5.d-2)}$$

and that these also obey the S_3 multiplication table. Check that they both form faithful representations. Obviously (?) these representations are related; they are equivalent. Explain (including geometrically) the forms of both sets of matrices [Inui, Tanabe, and Onodera (1990), p. 45] and also why the transformation between the vectors is

$$e_1 = -\frac{\sqrt{3}}{3} f_1 - 2\frac{\sqrt{3}}{3} f_2, \quad e_2 = f_2, \qquad \text{(V.5.d-3)}$$

checking that this gives the similarity-transformation matrix

$$S = \begin{pmatrix} -\sqrt{3}/3 & 1 \\ -2\sqrt{3}/3 & 0 \end{pmatrix}. \qquad \text{(V.5.d-4)}$$

Verify that it gives the similarity transformation between one set of representation matrices and the other. It is interesting that the symmetric group has nothing to do with geometry, but geometry determines its properties, or perhaps it is the converse.

Problem V.5.d-2: This gives the two-dimensional representation, but there are two others, both one-dimensional. Why do they arise? In one, the completely symmetric representation, all operators leave the basis vectors invariant. There is a single basis vector, the sum of the squares of the vectors to the vertices, and it is invariant. Justify the statement that this is a basis vector. What about the antisymmetric representation? We notice that for it, the basis vector is invariant under rotations, but its sign is reversed by reflections. This can be checked by comparing with S_3. Thus a vector (along z) perpendicular to a face is invariant under rotations, but reverses sign under reflections, which are equivalent to rotations of π around an axis from the center to a vertex. So the antisymmetric representation comes from marking (but not distinguishing) faces, or distinguishing right-handed from left-handed labeling of the vertices. Draw the triangle and check. What axioms of Euclidean geometry require the existence of these vectors, so the existence of the two (only) one-dimensional representations of the symmetric group — which has nothing to do with geometry? And how do these axioms give the two-dimensional representations?

Problem V.5.d-3: It is worth considering whether similar geometrical interpretations of larger symmetric groups are possible [Fässler and Stiefel (1992), p. 30, 265], and if so, doing the analogous analysis for them.

Chapter VI

The Group as a Representation of Itself

VI.1 THE REGULAR REPRESENTATION

Representation basis states are usually regarded as functions (say wavefunctions) on which group operators act. But it does not matter what meaning the symbols have, or even whether they have any meaning besides being objects so acted on. Thus we can go so far as to take the objects on which the operators act as the operators themselves. That is, we can regard multiplication of operators,

$$T_3 = T_1 T_2, \tag{VI.1-1}$$

as either the product of two operators to give a third, or as T_1 acting on basis vector T_2 to give basis vector T_3.

For S_3, for example,

$$(123) = (12)(23), \tag{VI.1-2}$$

and we can regard operator (12) as acting on vector (23) to produce vector (123).

So what, does it really matter whether we call this multiplication or the action of an operator? If we take the operators as basis vectors, then the group elements themselves form a representation of the group, which seems unusual, so perhaps important. How is this representation, the regular representation, related to the others, and how can it help in finding and understanding their properties? One might guess that taking the group as its own representation — the regular representation — leads to interesting and helpful results. And it is fortunate that it does.

VI.1.a The relationship between group elements and basis vectors

We have realized basis states as functions and also as group operators. Is there a relationship between these views?

For example, S_3: the basis states of the irreducible representations are linear combinations of the six functions obtained by applying all six permutations to any three-indexed function. To specify one of these six functions we either write it explicitly, or state the permutation that gives it when acting on some function, say the one, f_{nt}, with indices in natural order 123,

$$f_{nt} = f_{123}. \tag{VI.1.a-1}$$

Thus we can write either f_{213}, or (12), with the understanding that the latter is to be applied to f_{nt}. So we can regard permutation P' acting on permutation P, as P acting on f_{nt},

$$Pf_{nt} = f_1, \tag{VI.1.a-2}$$

then P' acting on f_1 to give a function with permuted indices,

$$P'f_1 = f_2, \tag{VI.1.a-3}$$

or P' acting on P to give permutation $P'P$ which acting on f_{nt} gives f_2,

$$(P'P)f_{nt} = f_2. \tag{VI.1.a-4}$$

We can take either the permutation operators or the permuted functions that they produce. The result is the same.

This is true not only for the symmetric group but in general — the operators acting on a fixed function (or symbol) give a set of functions. These, or sums of them, are taken as the basis vectors, or we can regard as basis vectors not the functions (which we can suppress), but the operators that produce each — sums of the group elements themselves. The two sets completely correspond. These are key points in the development of representation theory.

VI.1.b Basis vectors of any, and of the regular, representation

All basis vectors are sums of those of the regular representation, which means that all irreducible representations are contained in the regular representation — thus its importance. Why does the regular representation contain every irreducible representation (that is, it is reducible, and when reduced, all irreducible representations appear in its decomposition)? Because it contains all group operators, so all symbols produced by these, so any sum of the symbols (any representation basis state) is expressible as a sum over regular representation basis states — the group operators. A group operator acting on such a sum changes each symbol to another, and the sum to another such sum. Hence irreducible-representation bases are sums, mixed among themselves, of symbols, each given by a group operator acting on a fixed symbol, thus sums of group operators (regular-representation basis vectors) acting on this symbol.

Basis vectors of any representation are sums of those of the regular representation, therefore conversely, its basis vectors are linear combinations of the basis vectors of every (inequivalent, irreducible) representation of the group. This implies that the regular representation is reducible; finding these linear combinations reduces the regular representation. When completely reduced it contains all representations of the group, thus providing a way of obtaining these, and much information about them.

Applying all operators of the group to any one gives a set of monomials, and these form a complete set — there is no way to get an expression linear in these objects except by taking a sum over the monomials. Note "these objects"; the wavefunction for two electrons, one in state a, the other in b, is a linear combination of the symmetric and antisymmetric states (although only the latter is allowed), but a wavefunction for two protons cannot be written as a sum over electron states. A function can be written as a sum over monomials, but of the same realization.

The dimension of the regular representation is g, the order of the group — the number of its operators. If the regular representation contains all (inequivalent, irreducible) representations there must be, fortunately, a bound on their number, it determined by the number of group operators.

So we have to obtain all representations of the group by reducing the regular representation — finding all basis vectors. There are no other independent sums, so no other representations. Then we find the matrices, say from the action of the group elements on the vectors.

Problem VI.1.b-1: For example for S_2 the basis vectors are $f_{12} + f_{21}$ and $f_{12} - f_{21}$; there cannot be another tensor independent of these two because they contain all permutations. Check this for S_3.

Problem VI.1.b-2: Although it is generally kept secret, the algebra of basis vectors, and thus of the regular representation, involves sums of group transformations. However the group axioms do not include a definition of sums of group elements. Is this introduction of such sums a new postulate about them, does it follow from the group axioms, might it be a hidden assumption that is part of the concept of basis vector, or perhaps it is brought in when we regard groups as transformation groups, or is it something else, or introduced elsewhere?

VI.1.c There is always a symmetric representation

Having shown that every representation is contained in the regular representation it would be unfortunate if it turned out to be irreducible. To show that it is reducible, we give a representation that is always present. Every finite group has a one-dimensional, completely symmetric, representation. Its basis state is the sum of all operators, with all coefficients 1; any operator applied to this sum gives the sum with terms simply rearranged.

Problem VI.1.c-1: This should be simple to prove. What are the matrix elements for this representation? How does it follow from the group axioms and the definition of representation?

Problem VI.1.c-2: In addition, for symmetric groups, the multinomial $f_{123...} - f_{213...}$ gives the completely antisymmetric representation; permutations leave it invariant or change its sign. This is a second representation obtained in the reduction of the regular representation. Prove and explain this. Do alternating groups have completely antisymmetric representations? Why? Give examples.

Problem VI.1.c-3: It is not obvious *a priori* that the regular representation is reducible, and especially that the remainder, after removing the completely symmetric one, is still so. That it is, says something about the structure of groups, which we should understand. From the discussion of classes (sec. IV.5, p. 119) explain why the regular representation is reducible. Why do all groups have a completely symmetric representation?

VI.2 NUMBER OF APPEARANCES WITHIN THE REGULAR REPRESENTATION

The basis states of every representation are sums of those of the regular representation. Can there be different sums that all transform the same way — different sums that form the same basis vector of the same irreducible representation? There are; the number of appearances of representation r within the regular representation (the number of linear combinations of regular-representation states that transform as a given state of r) equals the dimension of r — if its dimension is d_r, it appears d_r times in the reduction of the regular representation.

The completely symmetric representation is one-dimensional; it appears once. The S_3 two-dimensional representation (sec. V.5.c, p. 165) appears twice; there are four linear combinations of regular-representation states that are basis states of this representation, two mixed among themselves by the group transformations, the other two also mixed among themselves (but not mixed with the first pair).

VI.2.a Why the number of appearances equals the dimension

Why does a representation appear several times, and why does the number of its appearances equal its dimension?

When the operators of a group act on the operators themselves to give the regular representation they can do so in two ways, from the left and from the right. Thus for basis vector (23) we have (12)(23) and (23)(12), and these are different. Apply operators on, say, the left to get the regular representation and decompose it: find the sums mixed (only) among themselves by the operators — applied from the left. Since the representations are irreducible every state of representation r appears at least once in a sum resulting from the application of an operator to some state of r; irreducible representation means that the set of states does not break into subsets each invariant under operators — applied from the left. Applying to irreducible-representation basis states f_i group operators T_α, from the left, gives

$$T_\alpha f_i = \Sigma \alpha_{ij} f_j, \qquad \text{(VI.2.a-1)}$$

where α_{ij} is the (ij) matrix element of transformation α.

However the representations are reducible. That is, we can break each set into subsets, each invariant, having states mixed among themselves by the group operators — applied from the right. So

$$f_k T_\alpha = \Sigma \alpha_{kl} f_l. \qquad \text{(VI.2.a-2)}$$

How are these states related? First let us introduce a better notation using two indices on f, which was not previously needed. Then,

$$T_\alpha f_{ik} = \Sigma \alpha_{ij} f_{jk}, \qquad \text{(VI.2.a-3)}$$

$$f_{ik} T_\alpha = \Sigma \alpha_{kl} f_{il}. \qquad \text{(VI.2.a-4)}$$

Notice which indices vary, and which do not, in the two cases.

If this representation has dimension d, so all indices go from 1 to d, then we have a set of d^2 states, f_{ik}, with action from the left mixing among themselves the states from each of the d sets labeled by k, and action from the right likewise mixing among themselves the members of the d sets labeled by i. Thus each representation, of the group operators applied from the left, appears d times, and these d sets are mixed by the operators applied from the right.

The states of the two two-dimensional representations of S_3 provide an example. Applying a permutation on the left to a state gives a sum of the two states of the same representation. Applying the permutation on the right gives a sum of the two corresponding states from each of the representations. Thus any sum of two corresponding states goes into a sum of the same linear combinations of the two states from the two representations, under permutation from the left. These sums are basis states of representations irreducible under operators from the left; but the proper linear combinations of them must be taken to get basis states of representations irreducible from the right.

Problem VI.2.a-1: Check these statements starting with the states in pb. V.5.c-2, p. 165, then for all S_n.

Problem VI.2.a-2: Explain why this shows that each representation appears (what does this mean?) the same number of times as its dimension.

VI.2.b Representation dimensions and group order

That the number of appearances of a representation in the decomposition of the regular representation equals its dimension gives information about representation dimensions, among other uses.

Problem VI.2.b-1: Obviously

$$\Sigma_r d_r^2 = g, \tag{VI.2.b-1}$$

the order of the group, where d_r is the dimension of representation r, and the sum is over all inequivalent, irreducible representations.

Problem VI.2.b-2: As all groups have a symmetric representation at least one d_r is 1. Thus show that the dimensions (in parentheses) are for S_2, (1,1); S_3, (1,1,2); S_4, (1,1,2,3,3); the pyramidal group, (1,1,2); the group of the square, (1,1,1,1,2). How about the group of the double rectangle? For S_3 there are two one-dimensional representations, the completely symmetric and the completely antisymmetric ones. Each appears once. There is also a two-dimensional representation which appears twice. This gives a total of six basis states, the dimension of the regular representation; $g = 6$ for S_3. Check that likewise the dimensions and orders of all these groups agree.

VI.3 THE MATRICES OF THE REGULAR REPRESENTATION

A representation is given by its basis states and the matrices that transform them. What are the representation matrices of the regular representation, and how do they decompose when it is reduced to a sum of all irreducible representations?

Problem VI.3-1: Show that the representation matrices of the regular representation consist of 1's and 0's only [Chen (1989), p. 30; Jones (1990), p. 67]. How many 1's are in each row? In each column? Why? Describe these representation matrices. How are they related to the group multiplication table? Why do the axioms, and definitions, give this?

Problem VI.3-2: The matrices of the regular representation can be reduced (block-diagonalized). Show that since each basis state, under multiplication from the left, belongs to a set of the same representations under multiplication from the right, each block (matrix of an irreducible representation) appears the same number of times as its dimension. Verify eq. VI.2.b-1, p. 175, from this.

Problem VI.3-3: For the S_3 six-dimensional regular representation, the matrices, in block form, contain the two one-dimensional representations (one-dimensional blocks) each once, and the two-dimensional representation twice (there are two identical two-dimensional blocks). This gives a six-by-six matrix before and after decomposition. However the states before and after are different sums. Verify this by writing the regular representation matrices and decomposing.

VI.4 EQUIVALENCE CLASSES

The decomposition of the regular representation gives sets of equivalent representations and the number in each set equals the dimension of any member; we call these sets equivalence classes. The regular representation breaks into equivalence classes, and each of these into (mutually) equivalent representations.

Problem VI.4-1: What are the equivalence classes of S_3?

Problem VI.4-2: Show that S_4 has two equivalence classes of the same dimension, which are inequivalent to each other, and each contains three mutually equivalent representations.

VI.4.a Classes and equivalence classes

We are using class in two different senses: class of elements of the group (sec. IV.5, p. 119), and equivalence classes of representations in the regular representation. These names should be carefully distinguished for they have nothing to do with one another, except for one point. The number of equivalence classes of representations of a finite group (that is the number of sets of inequivalent representations) equals the number of classes of the group elements.

Let us see why. Elements in each class can be transformed into each other by similarity transformations (by definition) but ones from different classes cannot; the number of sets of elements that cannot be so transformed by similarity transformations is the number of classes. But the number of linear combinations of regular-representation basis vectors (group elements) that cannot be transformed into each other by similarity transformations equals the number of — inequivalent — irreducible representations this representation breaks into, that is the number of inequivalent irreducible representations of the group. So these two numbers are equal.

There is an interesting relationship between the definition of class, and of equivalence class of representations. Group elements of the same class can be transformed into each other by similarity transformations; representations of the same equivalence class can be transformed into each other by similarity transformations. Distinct classes and inequivalent classes of representations are two different ways of dividing group elements into sets. It is not surprising that their numbers are equal.

Problem VI.4.a-1: Are only their numbers related, or could there be a closer relationship between the two types of classes?

Problem VI.4.a-2: Check this for the (at least, smaller) symmetric groups. Find the numbers of members of the classes. The numbers of classes, and of equivalent classes, are equal. Are the sets of numbers of members of the two types of classes equal?

VI.4.b Labeling representations and states

Since a representation appears several times in the decomposition of the regular representation we need to distinguish this set of equivalent representations, an equivalence class, from the individual representations. For S_3 the two two-dimensional representations equivalent to each other form an equivalence class, as do the one-dimensional representations, giving three classes. To specify a representation then we give the group, the equivalence class, and the representation in the class. A final label is needed to give a state.

For the rotation group we give l and then m_l, for the representation and the state. For finite groups giving the equivalence class labels, the analog of l, is not sufficient (the information lost by dropping a rotation-group label must be left to later consideration).

VI.5 EXAMPLES OF THE REGULAR REPRESENTATION AND ITS REDUCTION

To see how useful the regular representation is in providing the irreducible representations and information about them, and how useless it is for groups of large order, it is helpful to try examples.

Problem VI.5-1: Find the matrices and states of the regular representation of S_3. Reduce it and find all S_3 representations. Repeat for S_4 [Fässler and Stiefel (1992), p. 30, 265]. Anyone not convinced of the difficulties of doing this to find the representations of a group can try S_5.

Problem VI.5-2: Find the matrices and states of the regular representation of A_3. Reduce it and find all representations. How are these related to those of S_3? Repeat for A_4 and A_5.

Problem VI.5-3: Find the matrices and states of the regular representations of the water group and the pyramidal group. Reduce them and find all representations of these groups.

Problem VI.5-4: Find and decompose the regular representations of the group of the square and the group of the double rectangle. Compare and explain the results.

Problem VI.5-5: These are subgroups of S_4. Identify the operators of the subgroups with those of the group (pb. IV.3.d.iv-4, p. 115) and find their representations from those of S_4. How are the regular representations related?

Problem VI.5-6: Every group is a subgroup of a symmetric group. Thus its representations can be obtained from those of a symmetric group. For every subgroup of S_2, S_3 and S_4 find the inclusion rules, the rules telling which transformations of the subgroup correspond to which elements of the larger group. From these find the representations of all subgroups, which are generally reducible (aren't they?). Reduce them and so find all irreducible representations of all S_3 and S_4 subgroups. If this works repeat for S_5. Can it be done for S_n?

Problem VI.5-7: Find the regular representation, and from its decomposition, all representations, of the dihedral groups (sec. III.2.h, p. 79), starting with the smallest [Elliott and Dawber (1987), p. 45; Fässler and Stiefel (1992), p. 24; Jones (1990), p. 57, 262; Tung (1985), p. 28, 52; Aivazis (1991), p. 11]. Explain the form of the matrices.

Problem VI.5-8: If all operators of a group commute, what are the dimensions of its representations? Do the representation matrices for the regular representation differ in any significant way from those of a non-Abelian group? Can you tell by looking at these matrices whether the group is Abelian?

Problem VI.5-9: It would seem that since all elements of an Abelian group commute it is easy to state its representations. Do this for all cyclic groups [Baumslag and Chandler (1968), p. 216]. Then give a rule or algorithm for the representations of a general Abelian group, knowing its group table [James and Liebeck (1993), p. 81]. Implement it as a computer program. Compute the representations for a reasonable number of such groups. The representations of Abelian groups are relatively simple, but if a group has an Abelian subgroup, its representations are complicated. That is a reason we concentrate on simple groups.

Problem VI.5-10: Find an algorithm for constructing the regular representation of any finite group, and for reducing it into the set of irreducible representations. Implement it as a computer program.

VI.6 THEOREM OF THE REGULAR REPRESENTATION

The results of this chapter are important enough to summarize and to give that summary a pretentious name. Thus we get the
THEOREM OF THE REGULAR REPRESENTATION: The regular representation of a finite group contains every representation of the group (that is, the basis states of every representation are sums of the group operators, or, to put it another way, these operators acting on a function), and it contains each representation several times (several different sums give the same representation — the same representation matrices). The number of times it contains a representation is equal to the dimension of that representation. This means that the representation matrices of the regular representation can be decomposed (block-diagonalized), with the blocks being the representation matrices of each irreducible representation, and the number of times each block appears equals the dimension of the block.

So using the regular representation we can find all representations of the group — for any finite group.

Chapter VII

Properties of Representations

VII.1 REPRESENTATIONS AND FUNCTIONS ON THEM

Representations are essential, certainly for applications, but also because they, that is their matrices and states, illuminate their groups, and groups in general. These have properties following from being constituents of group representations, or which we require of them. Here we develop ways of finding and using them, collect some of these properties, quantities, and functions, and try to understand their reasons and rationales, and how to obtain and apply them. And we try to learn what they teach us about groups and physics.

Quantities and functions over the representations and states provide a foundation for this. A function is an assignment of (collections of) values to points. Here the points are the group elements, at present finite in number (so discrete). Thus we have functions on the group. Any set of values can be assigned to any points. But certain functions arise naturally; they are properties of the group. Knowing them we obtain information about it.

What functions, scalars, vectors, matrices, are there naturally on groups and representations, what are their properties, and why are there such? How do these arise from, and reflect, the underlying properties of groups and representations? And how do they help us?

Since most people seem to be more comfortable seeing lots of symbols and equations in what purports to be a book on mathematical physics, we emphasize this here more than previously. However we also try to explain what the equations mean. We leave it to the reader to decide whether the words and the equations say the same thing.

VII.2 UNITARY REPRESENTATIONS

A representation whose matrices are all unitary (sec. II.3.d, p. 43) is a unitary representation. Unitary is better for various reasons; one is that generally the transformations of quantum mechanics are required to be unitary, so that the norm (the unit) and product of states (thus probabilities) are invariant. Also requiring unitarity leads to restrictions which can be helpful (unitary matrices have properties that arbitrary ones need not have).

By unitary representation we mean unitary with respect to a product (that has been defined). A representation may be unitary for one product but not for another. Of course the unitarity of a matrix is independent of products. However representation matrices differ in different representations and may be unitary in one representation while not in an equivalent one.

If a representation is equivalent to a unitary one we use the unitary form. We take all representations unitary, which may seem like a strong restriction, but (here) is not.

Problem VII.2-1: Prove that the rows, and likewise the columns, of a unitary matrix are orthonormal (using what definition of a product, and with what justification for it?) [Hamermesh (1962), p. 91; Gel'fand (1989), p. 104]. Show that this is not true for an arbitrary matrix. Would a matrix be interesting if they were only orthogonal? Give an example. Could the rows be orthonormal (or orthogonal) and the columns not? If so, would this be interesting? Could there be a physical interpretation? Are these possible for group-representation matrices?

Problem VII.2-2: Can a unitary matrix (always) be diagonalized? What are the conditions on its diagonal elements? Can a general matrix [Gel'fand (1989), p. 106]? Why? This is another reason unitary representations are nice, and why it may be easier to get information from them, and use them in applications (not only physical).

Problem VII.2-3: Write down the general 2×2 unitary matrix. Also give the general nonunitary one. Find the similarity transformation from one to the other. Under what conditions is such possible? Are there differences between a similarity transformation connecting two unitary matrices (or representations), and one connecting unitary and nonunitary ones? How about for two nonunitary ones? Similarity transformations can have physical meaning; they might be transformations between coordinate systems, say. Which is more likely to have a physical interpretation, such a transformation between two unitary matrices, or one between a unitary and a nonunitary one? Explain (being careful not to restrict transformations to ones like rotations).

Problem VII.2-4: What do we mean by saying that an operator is unitary? Show that matrix M is unitary, for a product denoted by {}, if

$$\{M\psi, M\upsilon\} = \{\psi, \upsilon\}. \qquad \text{(VII.2-1)}$$

Is this necessary for M to be unitary? Why? Since M is independent of the product, why should this depend on the definition of the product? Does it? What would happen if we considered the same matrix, but a different product? Could we? This equation is extended to apply to any operator, to give the definition of a unitary operator. Why is this extension reasonable (or should it be necessary?)? So if operator T is unitary with respect to a product (ψ_i, ψ_j), then

$$(T\psi_i, T\psi_j) = (\psi_i, \psi_j); \qquad \text{(VII.2-2)}$$

the product of the transformed vectors equals that of the untransformed ones. What are the matrix elements of T acting on ψ_i?

VII.2.a Representations of finite groups can be taken unitary

Are all representations equivalent to unitary ones? In general no, but all representations of finite groups are equivalent to unitary representations [Inui, Tanabe, and Onodera (1990), p. 77; Hamermesh (1962), p. 92]. For finite groups, representations are (almost always) assumed unitary, as here — which is not a restriction.

To prove this we first demonstrate that there is a product that gives all matrices of a representation unitary. We take an arbitrary product (ψ, υ) of basis vectors ψ and υ. For it, the representation may not be unitary. However a product $\{\psi, \upsilon\}$ can be defined for which operator, T_t, for each group element t, is unitary. With g the order of the group, the number of elements, we define

$$\{\psi, \upsilon\} = (1/g) \sum (T_t\psi, T_t\upsilon), \qquad \text{(VII.2.a-1)}$$

summed over all group elements. To show that this is unitary we take T_s, for arbitrary s, and get

$$\{T_s\psi, T_s\upsilon\} = (1/g) \sum (T_tT_s\psi, T_tT_s\upsilon) = (1/g) \sum (T_{ts}\psi, T_{ts}\upsilon) = \{\psi, \upsilon\}, \qquad \text{(VII.2.a-2)}$$

as the sum over the products of all elements with a fixed one is the sum over all elements. Thus T_s is unitary for product {} — but not for ().

Problem VII.2.a-1: Would this last statement make sense if T were matrix M whose entries are the products? Can they be? Is the $(1/g)$ coefficient necessary? What type of product would be obtained if it were omitted? Would it be arbitrary? How are the matrix elements of T for the two products related? Is there a similarity transformation (independent of the group element) connecting the two sets of matrices?

Problem VII.2.a-2: If the representation is unitary using product (), would there be a difference with {}? Why? In what way?

VII.2.a.i *All representations are unitary for all products*

So for every finite group there is a product that gives a unitary representation. Of course this product may not be useful, it may not have any meaningful physical or mathematical interpretation (at least within the context that we are interested in). Thus given a product, perhaps picked for physical reasons (force and distance to give work, or quantum mechanical statefunctions, for example) and given any representation of a finite group, can we find a representation to which it is equivalent, using the product (required by, say, the application)?

VII.2.a.ii *Every representation is equivalent to a unitary one*

In fact every representation R of a finite group is equivalent to a unitary representation, for any product. Take a set of basis vectors, ψ, for the unitary representation, with product $\{\}$. The vectors in representation R are η; their product is (). Define a T (which has to be shown to exist) relating them,

$$\psi_i = T\eta_i. \tag{VII.2.a.ii-1}$$

Using their products, ψ and η can always be taken as orthonormal sets,

$$(\eta_i, \eta_j) = \delta_{ij}, \quad \{\psi_s, \psi_t\} = \delta_{st}, \tag{VII.2.a.ii-2}$$

giving, with arbitrary vectors

$$v = \sum a_i\eta_i, \quad w = \sum b_j\eta_j, \tag{VII.2.a.ii-3}$$

a relationship between the products,

$$\{Tv, Tw\} = \sum a_i^* b_j\{T\eta_i, T\eta_j\} = \sum a_i^* b_j\{\psi_i, \psi_j\}$$

$$= \sum a_i^* b_i = \sum a_i^* b_j(\eta_i, \eta_j) = (v, w). \tag{VII.2.a.ii-4}$$

Then having the representation with matrices M_s, which is unitary for product $\{\}$, we define matrices N, which we want unitary for (),

$$N_s = T^{-1}M_sT \tag{VII.2.a.ii-5}$$

and get from eq. VII.2.a.ii–4 used twice, and eq. VII.2–1, p. 182,

$$(N_s v, N_s w) = \{TT^{-1}M_sTv, TT^{-1}M_sTw\} = (T^{-1}M_sTv, T^{-1}M_sTw)$$

$$= \{M_sTv, M_sTw\} = \{Tv, Tw\} = (v, w); \qquad \text{(VII.2.a.ii–6)}$$

the equivalent representation given by N_s is unitary for ().

Thus in summary, given a product, so vectors, and a representation following from these, which need not be unitary, there is also a product for which the operators are unitary, and a set of vectors that are orthonormal with respect to it. The transformation between the two sets of vectors takes the representation to an equivalent one that is unitary for the original product.

This holds for finite groups. For an infinite number of elements there are obvious problems that the sum over the elements can cause; this, which is indeed more complicated (and perhaps interesting), is not considered here.

Problem VII.2.a.ii–1: Why is there always such a transformation T (with what conditions on it?)? Why can we take both sets orthonormal? Is the Gram-Schmidt orthonormalization procedure relevant [Courant and Hilbert (1955), p. 4, 50]?

Problem VII.2.a.ii–2: Given a set of representation matrices M, find (the) T that transforms it to an equivalent unitary representation.

Problem VII.2.a.ii–3: There might be a slight loophole in these arguments. A product is required to be

a) positive definite:

$$(u, v) \geq 0, \quad (u, u) = 0 \Rightarrow u = 0, \qquad \text{(VII.2.a.ii–7)}$$

b) linear:

$$(u, av + bw) = a(u, v) + b(u, w), \quad (au + bw, v) = \ldots; \qquad \text{(VII.2.a.ii–8)}$$

why? Check that all products here, and throughout the book, satisfy.

Problem VII.2.a.ii–4: There are other proofs that all representations of finite groups are equivalent to unitary ones (perhaps giving the transformation) [Falicov (1966), p. 22; Elliott and Dawber (1987), p. 53]. Can they be obtained from this, and conversely? Do these provide, or require, further information or assumptions?

Problem VII.2.a.ii–5: Two sets of representation matrices have been given (sec. V.5.d, p. 167) for the symmetry group of the equilateral triangle (S_3). Show that one set is unitary, the other not. One representation is obtained using orthogonal coordinate vectors, the other using nonorthogonal ones. Is this relevant to it not being unitary? How? Is there a similarity transformation between these? What is its significance? Can it be obtained without knowing the matrices?

VII.2.b Unitary representations are decomposable

That a representation is unitary is an important — and often necessary — restriction on it. And it has major consequences. We usually require (or hope) that the representations we use are fully reduced. But is this always possible? Unitary representations — usually the most important — can always be fully reduced.

Problem VII.2.b-1: Prove that for a matrix of the form $\begin{pmatrix} A & 0 \\ C & B \end{pmatrix}$ to be unitary, $C = 0$, where A, B, C are numbers. Does it matter if C is above the diagonal? Show that this also holds if A, B, C are matrices. Are there conditions on these matrices? Thus every unitary matrix is equivalent to a block-diagonal one. This means that every representation equivalent to a unitary representation can be decomposed [Hamermesh (1962), p. 98] — why? Check the converse; every decomposed representation is (equivalent to?) a unitary one.

VII.2.c Decomposability of finite group representations

Every representation of a finite group can be decomposed — every representation is equivalent to a unitary representation, and every unitary representation is completely reducible. If a representation is not completely reduced there is a similarity transformation that transforms it to an equivalent decomposed representation (this is often called Maschke's theorem) [Boerner (1963), p. 48; Elliott and Dawber (1987), p. 53; Jones (1990), p. 56; Ledermann (1987), p. 21, 24; Lomont (1961), p. 49; Scott (1987), p. 320].

This result is important enough to be given a pretentious name: THEOREM OF COMPLETE REDUCIBILITY: Every representation of every finite group is equivalent to a (unitary) completely reduced one — every representation can be decomposed.

VII.2.d Why finite group representations can be taken unitary

The unitarity of the representations, for finite groups, is essential in many ways, certainly for their application to quantum mechanics. Why does it hold? What properties of groups, and representations, require it?

Given a — finite — set of vectors, here the representation basis vectors, we can orthonormalize them for any (reasonable) product [Courant and Hilbert (1955), p. 4, 50]. The group transforms them to a new set of vectors, in the same space, but these may not be orthonormal. However these again can be orthonormalized. That is, there is a transformation T, independent of the group elements, that takes them to an orthonormalized set. The similarity transformation of the group elements by T gives group elements taking an orthonormalized set of vectors to another orthonormalized set — and such a transformation is unitary. Thus these (transformed) matrices form a unitary representation of the group.

Problem VII.2.d-1: Show that these transformed matrices do take an orthonormal set to another orthonormal set. Explain why a transformation between orthonormalized vectors is unitary (and conversely?) [Gel'fand (1989), p. 105]. Can a matrix that takes an orthonormal set to one that is not, be unitary? What is the difference in these arguments for vectors that are real (have real components), and ones that are complex?

Problem VII.2.d-2: Very little was used here, beyond the fact, which is true for finite groups, that representation spaces are finite (why?). What other assumptions, if any, are used (or needed)? Are the assumptions all the same in the different (versions of the) proofs appearing, at least in part, in almost every relevant book? Which of these, if there are differences besides presentation, are the most general? Why? Which can be best extended, if that is possible, to nonfinite groups? Can they be extended to all such groups, or only to certain classes? Why?

VII.3 SCHUR'S LEMMAS

The basic objects in representation theory are the (unitary) irreducible representations; given a complete set of these, for any (finite) group, any other representation is either equivalent to one, or a direct sum of (ones equivalent to) them. But given representations, how do we know what we have? Suppose we have two; they may look very different, but may not actually be so, they may be equivalent. How do we tell? And if we have a representation we (usually) wish to reduce it. But how do we know if the representation is reducible, or already completely reduced? And to find the (decomposed) representations we must be able to recognize them, and tell which are equivalent, or we will be finding representations forever.

The methods for answering these questions are based on two of the most important (of course) theorems of group theory, though these are called lemmas [Fässler and Stiefel (1992), p. 33; Hamermesh (1962), p.

98].

VII.3.a Schur's first lemma

How do we tell if representations are equivalent? We have SCHUR'S FIRST LEMMA: If matrix S intertwines all matrices of two inequivalent, irreducible representations with matrices $m(t)$, $M(t)$, that is

$$Sm(t) = M(t)S, \tag{VII.3.a-1}$$

for all t, then S (the same for all t) is the zero matrix,

$$S = 0. \tag{VII.3.a-2}$$

Thus if there is a nonzero matrix that intertwines all elements, the representations are equivalent. If we can show that there is none, they are inequivalent.

This is quite obvious (making it easy to overlook the problem); taking the inverse of S gives

$$Sm(t)S^{-1} = M(t), \tag{VII.3.a-3}$$

which is the definition of equivalent representations. The only way they cannot be equivalent is if $S = 0$ — unless S, though not 0, is singular, it has no inverse.

There are two cases; first the two representations have different dimensions. Then for each t, summing over repeated indices,

$$\sum m(t)_{ij}S_{ja} = \sum S_{ib}M(t)_{ba}, \tag{VII.3.a-4}$$

multiplying on the left by basis vector ψ of m, summing, defining

$$\phi_b = \sum \psi_i S_{ib}, \tag{VII.3.a-5}$$

and taking group operator $m(t)$ to give basis vectors ϕ from ψ, with, of course, the number of linearly independent ϕ's not greater than that of the ψ's, we get, with the transformed vectors denoted by primes,

$$\sum \psi_i m(t)_{ij}S_{ja} = \sum \psi_i S_{ib}M(t)_{ba} = \sum \phi_b M(t)_{ba} = \sum \psi'{}_j S_{ja} = \phi_a{}'. \tag{VII.3.a-6}$$

But this means that the functions ϕ_b form the basis of a representation of M reducing it — the ranges of the indices on ψ and ϕ are not the same, thus we have expressed a set of basis states in terms of a smaller one, which reduces the representation, contrary to the assumption that M is irreducible. Thus if the representations have different dimensions there can be no nonzero matrix connecting them. (It is not really surprising that representations of different dimension are inequivalent.)

Problem VII.3.a–1: In this proof, which representation must have the larger dimension, m or M? Explain why expressing one set of basis states in terms of a smaller set means that the representation is being reduced. How does this show $S = 0$?

Problem VII.3.a–2: If the two representations have the same dimension we have to show that S cannot be singular. Again we consider the functions

$$\phi_j = \sum \psi_i S_{ij} \tag{VII.3.a-7}$$

which must be linearly independent or else representation m would be reducible. Why? But S singular means that they are not linearly independent. Why? Show that if the vectors are not linearly independent the matrices of m are reducible. How is this related to the case of different dimensions? Show that the linear independence of these functions means that matrix S has an inverse, and conversely. Thus if a nonzero S intertwines all matrices of two irreducible representations, they are equivalent.

VII.3.b Schur's second lemma

To find if a representation is irreducible we have
SCHUR'S SECOND LEMMA: If a (single) matrix commutes with all elements of an irreducible representation,

$$Sm(t) = m(t)S, \tag{VII.3.b-1}$$

where t runs over all elements of the group, then S is a multiple of the identity, and conversely — essentially the only matrix that commutes with all matrices of an irreducible representation is the unit matrix. Thus if it can be shown that only multiples of the identity commute with all elements of a representation, it is irreducible. If there is a commuting matrix not proportional to the identity, the representation is reducible. (If it were not irreducible it could be block-diagonalized and S could have different multipliers for different blocks.)

To show this, diagonalize S and assume that it is not proportional to the unit matrix, say $\begin{pmatrix} A & 0 \\ 0 & B \end{pmatrix}$, where matrices A and B are different. For an irreducible representation there is a group transformation t from any vector in the space, say one with nonzero components only in rows corresponding to A, to any other, say with nonzero components only in rows corresponding to B,

$$m(t)S\psi_1 = Sm(t)\psi_1 = m(t)A\psi_1 = Am(t)\psi_1 = A\psi_2 = S\psi_2 = B\psi_2. \tag{VII.3.b-2}$$

So $A = B$. There is always a t that mixes the states and we can find two states that are mixed since the representation is irreducible. Thus S is proportional to the unit matrix.

Problem VII.3.b-1: The slight problem is that it may not be possible to (block) diagonalize S. However certain matrices can always be diagonalized. Why can this be made to apply to S?

Problem VII.3.b-2: Explain why

$$Sm(t) = m(t)S, \tag{VII.3.b-3}$$

for all t requires that the set of eigenvectors η of S,

$$S\eta = \lambda\eta, \tag{VII.3.b-4}$$

all with the same eigenvalue λ, be invariant under the group. So this invariant space forms a representation space, and if it is not the entire space then the representation is reduced on it, contrary to the assumption of irreducibility. Would it matter if $\lambda = 0$? Thus S is either 0, or the unit matrix. This proof does not require diagonalization of S. How are the two proofs related? Does this show that S can be diagonalized? Is it possible that S has no eigenvalues?

Problem VII.3.b-3: Generalize eq. VII.3.b-2 to more than two states. Check that the argument goes backwards: if there is an S not the unit matrix commuting with all matrices, the representation is reducible. What would go wrong if S were to commute with only some of the matrices?

Problem VII.3.b-4: It was stated several times that the representations of an Abelian group are all one-dimensional. Prove this using (which?) Schur's Lemma. This shows that the lemmas are actually useful and we shall see more evidence of this below.

VII.3.c Why Schur's lemmas?

Why do Schur's lemmas hold? Are they properties of groups, representations, or vector spaces? The first one is almost a definition, that of equivalence of representations. There is a possible problem, but that actually did not arise. The reason is that the basis vectors (of irreducible representations?) are linearly independent.

Why is a matrix, S, that commutes with all of an irreducible representation, a (multiple of the) unit matrix (multiplying a matrix by a number does not affect whether it commutes)? An S not (proportional to) the identity would transform basis vectors into others, and this set, because of the commutation, would be an invariant subspace, which does not exist for an irreducible representation. Also, two matrices that commute can be simultaneously diagonalized. If S commutes with all matrices of a representation then all would be diagonalizable, which is not true in general (for a reducible representation they can be simultaneously *block* diagonalized).

Problem VII.3.c–1: Why are basis vectors linearly independent? How is this related to the definition of a representation?

Problem VII.3.c–2: While the Schur lemmas are of (fundamental) importance in group theory, do they need all the assumptions of group theory? Might they hold for other, or broader, objects [Boerner (1963), p. 19; Fässler and Stiefel (1992), p. 34, 39; Grove and Benson (1985), p. 98, 101; Lomont (1961), p. 7]? Which? Why? Are they properties of groups, representations, matrices, or all? Which properties are needed for which of the lemmas? Might there be objects for which a part of the statement of a lemma holds, but not the converse? Do they hold for an infinite-dimensional space, for a discrete number of basis vectors? How about a space in which the basis vectors are (or should it be, need be?) labeled by a continuous (set of) label(s)?

Problem VII.3.c–3: Schur's lemmas are discussed in (almost) every book on finite groups. These discussions, and proofs, differ. It would be worthwhile to go through (several of) them and see if the differences are merely in notation and style, or, and to what extent, they reflect different underlying mathematical (and physical) assumptions. Might different proofs hold for different structures? Which of the proofs are closest to physics (say, quantum mechanics)? Why? How can the proofs here be obtained from the others, and conversely?

VII.4 CHARACTERS

Representations are (given by) matrices, but these are not just collections of numbers. Being representation matrices they are limited — few sets of matrices can represent a group — so have characteristics that bound, define, and sometimes identify them. There are numbers, often useful in finding and naming such matrices, that give information about their properties. Why are there such, how are they related to the matrices, and the group, how are they determined and used?

Problem VII.4–1: The sum of the diagonal elements of a matrix,

$$T = \sum_{i} M_{ii} \tag{VII.4–1}$$

is called its trace. Why should this be of interest? Prove that the trace is invariant under similarity transformations. It is this invariance that makes the trace of value in group theory (and elsewhere). The matrix of a group operator differs for different representations — but if these are equivalent, their traces are the same. Also, clearly, traces are class functions — for each representation, they are the same for all matrices in a class. Are these two independent results, that traces are the same for all equivalent representation, and for all class members?

VII.4.a Traces are characters and determine the representations

The traces of the representation matrices are called the characters of that representation [James and Liebeck (1993); Ledermann (1987); Littlewood (1958)]. Since they are invariant under similarity transformations, the characters of equivalent representations,

$$M = SmS^{-1}, \tag{VII.4.a-1}$$

are the same. And the characters of inequivalent representations are different (there is at least one character, the trace of at least one representation matrix, that differs between the two representations). Thus the characters of a representation distinguish inequivalent representations giving a method for telling if two representations are, or are not, equivalent. Representations (that are equivalent) are labeled by their set of characters, which also tell much about them (characters characterize the representation).

That equivalent representations have the same set of characters has been shown in pb. VII.4-1, and we must show the converse — inequivalent representations differ in their character sets. However this has to wait until we learn more about the properties of characters, which we now try to do.

The terminology of representation matrices is often used for the characters. Thus the characters of an irreducible representation are called irreducible (also simple), of a reducible representation, reducible (also compound), and of a faithful representation, faithful.

Problem VII.4.a-1: The character of the identity equals the dimension of the representation. Two representations of the same dimension can be inequivalent; the character of the identity is the same but they differ in at least one other. If they differ in the character of the identity they are inequivalent. Does this follow from the relevant part of Schur's lemmas, can it be used to prove that, or does it make it irrelevant?

Problem VII.4.a-2: Explain why the characters of an Abelian group are the representation matrices.

Problem VII.4.a-3: The trace of a product of matrices is the same whatever order they are written in [Jones (1990), p. 44], as can easily be proven:

$$Tr(ABC\ldots) = Tr(B\ldots C\ldots A) = \ldots . \qquad \text{(VII.4.a-2)}$$

From this show that characters are the same for equivalent representations.

Problem VII.4.a-4: Group G is the direct product of groups H and K. Find the characters of G in terms of those of H and K.

Problem VII.4.a-5: Obtain the characters for the representations in sec. V.5.d, p. 167. Are they the same?

Problem VII.4.a-6: Find examples to show that groups with the same character tables (thus the same set of dimensions for their representations) need not be isomorphic [Dixon (1973), p. 66, 157; Joshi (1982), p. 107; Ledermann (1987), p. 62]. It is interesting that characters distinguish representations, but not groups.

Problem VII.4.a-7: If, for a faithful representation, the degree (dimension of the representation matrices) is less than the smallest prime dividing the order of the group, then the group is Abelian [Dixon (1973), p. 67, 158]. Prove this, and check it for some small-order groups, finding examples of Abelian groups (are there, must there be, any?) that have such representations; also noting that if the group is not Abelian, then the degrees are greater than, or equal to, this prime.

VII.4.b Characters are class functions

Characters are invariant under similarity transformations so they are class functions — every member of a class has the same character, for the representation; we denote the characters by ζ_c, where c runs over the classes. In general the characters of different classes are different, but not always. For example, for the completely symmetric representation the character of all classes is 1.

Problem VII.4.b-1: If a representation is unitary, what restriction is placed on its characters — and therefore on the characters of any finite group? For the characters of ambivalent classes (pb. IV.5-15, p. 121), what information can be given?

Problem VII.4.b-2: Show that for a finite group, the number of real irreducible characters (the irreducible-representation characters that are real numbers) equals the number of ambivalent classes [Dixon (1973), p. 67, 157]. Also if a finite group has odd order, the only real irreducible character is that of the identity. Have we considered any odd-order groups? Check this for them. Further if it has odd order n, and k classes, then $k = n$ (mod 16). Verify this for any groups to which it applies. Character tables are given in Burns (1977), p. 379; Elliott and Dawber (1987), p. 265; Falicov (1966), p. 36; Hamermesh (1962), p. 115;

Heine (1993), p. 448; Jones (1990), p. 68; Ledermann (1987), p. 64, 208; Wilson, Decius and Cross (1980), p. 312.

VII.5 ORTHOGONALITY THEOREMS

Representation matrices and characters carry indices, suggesting that they can be treated in certain (and fundamental) ways as vectors. Can they be? Can they be multiplied as if they are? If so, are there, can (should) we impose, orthogonality conditions on them? Such conditions place constraints, and supply information. These objects can be taken as (collections of) vectors — and orthonormalized (in a sense). Why should this be, what does it have to do with groups, and how is it useful? Here we study how products of these vectors are defined, the extent to which they are orthogonal, and the information provided about them and representations [Hamermesh (1962), p. 101].

Problem VII.5-1: When we regard matrix elements and characters as vectors, is this just a convenient terminology because they carry indices, or do they transform as vectors? The term "basis vector" also has implications. Is this a proper use of language?

VII.5.a Orthogonality of the representation matrices

To study irreducible-representation matrices M (of dimension n) of a group (with g elements) we return to Schur's lemma. We define matrix

$$S = \sum_t M(t)XM(t^{-1}), \tag{VII.5.a-1}$$

summed over the group elements, where X is an arbitrary matrix (that is we are not telling yet what our choice for it is).

Problem VII.5.a-1: Prove that S satisfies the condition of the relevant lemma (it commutes with all representation matrices M). Would S commute if the sum were not over all elements? Is X arbitrary — does this result depend on it? This shows that there is actually a matrix that can be used to test for irreducibility, and gives it. So S is a multiple of the identity, and we write the constant as λ_{ij}, i, j arbitrary, and choose all entries in X to be 0 except

$$X_{ij} = 1, \text{ so } X_{ab} = \delta_{ai}\delta_{bj}. \tag{VII.5.a-2}$$

Hence check that

$$\sum_t M(t)_{ki}M(t^{-1})_{jm} = \lambda_{ij}\delta_{km}, \tag{VII.5.a-3}$$

and for a unitarity representation

$$\sum_t M_{ki}(t)M^*_{mj}(t) = \lambda_{ij}\delta_{km}; \tag{VII.5.a-4}$$

* denotes the complex conjugate of the element. Evaluate λ by setting $k = m$ and sum (over what?), which should give,

$$\sum_t M_{ki}(t)M^*_{mj}(t) = (g/n)\delta_{ij}\delta_{km}, \tag{VII.5.a-5}$$

the orthogonality condition for the matrices of a single representation. For a nonunitary representation

$$\sum_t M_{ki}(t)M_{jm}(t^{-1}) = (g/n)\delta_{ij}\delta_{km}. \tag{VII.5.a-6}$$

Problem VII.5.a-2: Show that

$$S = \sum_t M^1(t)XM^2(t^{-1}), \tag{VII.5.a-7}$$

satisfies the conditions of the relevant Schur's lemma so must be 0 for representations M^1, M^2 inequivalent. Is this true for any X? This provides a way of determining if two representations are equivalent for it gives the matrix required for the test. If the two representations have different dimensions then this matrix product requires that which matrix not be square? Does this seem reasonable? Explain. With this, and the orthogonality condition for a single representation, show that

$$\sum_t M^r_{ki}(t)M^s_{mj}{}^*(t) = (g/n)\delta_{ij}\delta_{km}\delta^{rs}, \tag{VII.5.a-8}$$

$$\sum_t M^r_{ki}(t)M^s_{jm}(t^{-1}) = (g/n)\delta_{ij}\delta_{km}\delta^{rs}; \tag{VII.5.a-9}$$

r, s label the representations. These are the orthogonality conditions for unitary and nonunitary matrices (the second holds for the unitary case also if we do not use the unitarity). Check that there is another simple way of finding the multiplicative factor on the right. Take $i = j$ and $k = m$, sum over k (which is canceled on the right-hand side by the n); each term gives the diagonal element of the identity and there are g terms.

Problem VII.5.a-3: Explain why these orthogonality conditions for the representation matrices mean that the rows, and also the columns, of the matrices, are sets of orthogonal vectors. The components of the vectors are labeled by the group elements. What labels the vectors?

These vectors are normalized, except that the normalization factor is not 1, but g/n, for a representation of dimension n. Could we arrange it so that the normalization factor is 1? Would it be a good idea?

Problem VII.5.a-4: For two vectors to be orthogonal means that their product is defined. State explicitly what the product is here.

Problem VII.5.a-5: Verify, when representation matrices can be easily obtained (sec. VI.5, p. 178), the orthogonality conditions for S_2, S_3 (sec. II.4.d, p. 51; sec. V.5.d, p. 167) the water group (sec. II.2.d, p. 35), and the geometrical groups (sec. II.2.e, p. 36 — sec. II.2.h, p. 40). In other cases, after the discussion in the next two chapters of the symmetric groups, of which these are subgroups, return to this problem, using the results there.

VII.5.b The reason for the orthogonality of the matrices

Why must these orthogonality conditions hold? Each matrix element is a component of a vector, labeled by its representation and row and column position (by the row and column in the regular-representation matrix), these labeled by the group element. That vectors are orthogonal (or at least can be orthogonalized) means that they are linearly independent (orthogonality and linear independence are closely related). If the vectors were not, either they would not give a representation (contrary to assumption) or the transformations would not be independent, which we assumed implicitly (as we had to) in the group axioms. Linear independence gives orthogonality.

Problem VII.5.b-1: This statement about the labeling skims over one point. Give it precisely. Explain why it is correct.

Problem VII.5.b-2: Why are orthogonality and linear independence closely related?

Problem VII.5.b-3: Is there an (implicit) assumption in the axioms that the symbols are independent, or is that part of a definition?

Problem VII.5.b-4: What would go wrong in the definition (or construction) of a representation if basis vectors were not linearly independent? Suppose that there were a nonlinear relation connecting vectors (if that is possible). Would it make a difference?

Problem VII.5.b-5: The rows and columns of a unitary matrix are orthonormal. Can this (and linear independence?) be used to show that all representations of a finite group are equivalent to unitary ones? What might go wrong for continuous groups?

VII.5.c Orthogonality of the characters

Like columns and rows of matrices, characters have indices so can be regarded as vectors — and with orthogonality conditions; these follow from those for the representation matrices. Taking the characters as (collections of) vectors we find that those of different representations are orthogonal, of the same one orthonormal (with norm g). The components of the character vectors are labeled by the classes, so the numbers of components and of classes are equal.

Problem VII.5.c-1: In the orthogonality conditions for the matrices set equal the two indices on each matrix element and sum over both pairs. Show that this gives for unitary representations r and s,

$$\sum \zeta_c^r \zeta_c^{*s} g_c = g\delta^{rs}; \tag{VII.5.c-1}$$

the sum is over classes (remember) and g_c is the number of elements in class c. This is the orthonormality condition for the characters ζ (character vectors of inequivalent representations are orthogonal). This is for unitary representations. Why? What would it be if this were not imposed? This holds only for irreducible representations; why?

Problem VII.5.c.-2: Let $M^r(t_i)$ be the representation matrices for any representation, except the completely-symmetric one. Then

$$\sum M^r(t_i) = 0, \tag{VII.5.c-2}$$

summed over all group elements, and

$$\sum g_k \zeta_k^r = 0; \tag{VII.5.c-3}$$

the sum of characters ζ_k^r is over all classes, these with order g_k. The equations can be obtained independently, also either gives the other. So summing over elements,

$$\sum \zeta_k^s \zeta_k^r = 0, \tag{VII.5.c-4}$$

if ζ^r is not of the completely-symmetric representation.

Problem VII.5.c-3: Show that no character vector is 0. Explain how this follows from the axioms and definitions (of what?). Also character vectors are linearly independent [Scott (1987), p. 326].

Problem VII.5.c-4: State the formal definition of the product of two character vectors.

VII.5.c.i *The character orthogonality condition for classes*

Since characters carry two indices (for the representation and for the class) we expect two orthogonality relations. For irreducible n_r-dimensional representation r, with matrices M^r, the functions

$$S_c^r = \sum M^r(t_c), \tag{VII.5.c.i-1}$$

summed over all elements t_c of class c, commute with all matrices N of the representation; if $S = NSN^{-1}$ (the sum over all elements of a class is invariant under a similarity transformation) then $NS = SN$. Then by Schur's lemma, for class c, with character ζ_c, and g_c elements,

$$S_c^r = \lambda_c^r I_n, \tag{VII.5.c.i-2}$$

with I_n the $n \times n$ unit matrix. Taking the trace gives

$$n_r \lambda_c^r = g_c \zeta_c^r \text{ (no sum).} \tag{VII.5.c.i-3}$$

So

$$\lambda_c^r = g_c \frac{\zeta_c^r}{n_r} = g_c \frac{\zeta_c^r}{\zeta_1^r}, \tag{VII.5.c.i-4}$$

where ζ_1^r is the character of the identity (class).

We now multiply S_c^r and S_k^r. As the S's are sums of elements, the product is a sum of products, which is then a sum of group elements, and this is a sum of sums (from different classes), each over (of course) all elements of a class. However (sec. IV.5.f, p. 126), a class can appear several times, say a_{ck}^l, for class l. We then get

$$S_c^r S_k^r = \sum M^r(t_c t_k) = \sum a_{ck}^l S_l^r, \tag{VII.5.c.i-5}$$

where the first sum runs over all elements t_c of class c, and t_k of class k, the second over the classes. This gives

$$\lambda_c^r \lambda_k^r = \sum a_{ck}^l \lambda_l^r, \tag{VII.5.c.i-6}$$

and

$$\frac{g_c g_k \zeta_c^r \zeta_k^r}{n_r^2} = \sum_l a_{ck}^l g_l \frac{\zeta_l^r}{n_r} = \sum_l a_{ck}^l g_k \frac{\zeta_l^r}{\zeta_1^r}. \tag{VII.5.c.i-7}$$

Summing over the representations gives

$$g_c g_k \sum_r \zeta_c^r \zeta_k^r = \sum_l \sum_r a_{ck}^l g_l n_r \zeta_l^r = \sum_l a_{ck}^l g_l \sum_r \zeta_1^r \zeta_l^r$$

$$= \sum_l a_{ck}^l g_l g \delta_{l1} = g_k g \delta_{-ck}, \tag{VII.5.c.i-8}$$

since n_r is the character of the class of the unit element. The only classes that have the unit element in their product are $(c,-c)$, the inverses of each other (pb. IV.5.f-8, p. 127), the classes whose elements are the inverses of those of the other class, and the unit element appears g_c times, so

$$a_{c,-c}^1 = g_c. \tag{VII.5.c.i-9}$$

Thus we get

$$\sum_r \zeta_c^r \zeta_k^r = \frac{g}{g_c}\delta_{-ck}, \qquad \text{(VII.5.c.i-10)}$$

the second orthogonality relation for the characters — character vectors of the classes, summed over all irreducible representations, are orthonormal, with norm g/g_c. For unitary representations

$$\sum_r \zeta_c^r \zeta_k^{*r} = \frac{g}{g_c}\delta_{ck}, \qquad \text{(VII.5.c.i-11)}$$

Thus character vectors, of different (irreducible) representations, and also of different classes, are orthogonal. That

$$\sum_r \zeta^r \zeta^r = \sum n_r^2 = g, \qquad \text{(VII.5.c.i-12)}$$

also follows from representation r appearing ζ^r times in the regular representation which has g elements.

(A sum for group operators can be defined giving the group algebra, the algebra of these sums [Bhagavantam and Venkatarayudu (1951), p. 228; Boerner (1963), p. 53; Burrow (1993), p. 58; Chen (1989), p. 32; Hamermesh (1962), p. 106; Scott (1987), p. 329; Tung (1985), p. 307]. However this is unnecessary here since we need only sums of representation matrices, whose definition comes from that of matrices, and it is clear that the sum of group operators can be defined to agree with that of their representation matrices. This can also be used to obtain these results.)

VII.5.c.ii *The necessary and sufficient condition for equivalence of representations*

So we get the converse of the result of pb. VII.4-1, p. 190, giving that the necessary and sufficient condition for two representations to be equivalent is that their character sets be identical; the character sets of inequivalent representations differ. There we saw that if representations are equivalent, they have the same set of characters. Since we have found here that the character vectors of inequivalent representations are orthogonal, if the character vectors are the same, the representations are equivalent. This gives a rule for determining if two representations are equivalent. If so, all their characters are the same, if not they differ in at least one.

Problem VII.5.c.ii-1: There is a related result [Jansen and Boon (1967), p. 184]. Can different classes have the same character in all irreducible representations? Prove that they cannot. Thus we can distinguish two classes in a way similar to that for irreducible representations — they differ in their characters for at least one representation. Is there a reason for this similarity?

Problem VII.5.c.ii–2: Two representations with matrices M and N are equivalent if

$$t^{-1}M_i t = N_i, \qquad \text{(VII.5.c.ii–1)}$$

where t is the same for all group elements. However show that if

$$v_i^{-1}M_i v_i = N_i, \qquad \text{(VII.5.c.ii–2)}$$

(there is a transformation, but one that depends on the element) then there is one t that is the same for all elements [Ledermann (1987), p. 65, 209]. Thus the representations are equivalent.

Problem VII.5.c.ii–3: Verify the orthogonality conditions for the characters of s_2, s_3, (s_4, s_5, and s_n, if you are ambitious), the water group, the pyramidal group, the group of the square, and the group of the double rectangle (pb. VII.5.a–5, p. 195).

Problem VII.5.c.ii–4: The characters follow immediately from the representation matrices. Are there simpler methods for finding the characters [Elliott and Dawber (1987), p. 45; Inui, Tanabe, and Onodera (1990), p. 54; Joshi (1982), p. 87; Lomont (1961), p. 312]?

VII.6 APPLICATIONS OF CHARACTERS

Characters not only have interesting characteristics, but useful ones. Next we turn to the use of them as tools for studying and obtaining representations and their properties.

VII.6.a Reduction of representations using characters

Reducible representations are direct sums of irreducible ones, so their characters are sums of those of the irreducible representations. Given a reducible representation we (usually) want to find the irreducible representations of which it is composed — which appear, and how many times each. For this, we can use the characters.

Suppose we have a reducible representation with matrices $M(t)$, and a set of irreducible representation matrices $M^r(t)$, with characters ζ_c^r, with r the representation, c the class, and t the transformation. Then

$$M(t) = \sum a_r M^r(t), \qquad \text{(VII.6.a–1)}$$

where a_r is the number of times representation r appears, and is the desired quantity. Summing the diagonal elements,

$$\zeta_c = \sum a_r \zeta_c^r, \qquad \text{(VII.6.a–2)}$$

for each class. If we know the characters of the group (we do if we know the irreducible representations), and if we know the characters of the reducible representation, we can use these equations to find the a's, giving the number of times each irreducible representation appears. (Of course we might still want the similarity transformation reducing the representation.)

The characters are orthonormal and we use this to get the a's. With g_c the number of elements in class c, we multiply by $\zeta_c^{s*} g_c$, and sum over the classes,

$$\sum_c \zeta_c \zeta_c^{s*} g_c = \sum_r a_r \sum_c \zeta_c^r \zeta_c^{s*} g_c = \sum_r a_r g \delta^{sr} = g a_s, \qquad \text{(VII.6.a-3)}$$

giving

$$a_s = (1/g) \sum_c \zeta_c \zeta_c^{s*} g_c. \qquad \text{(VII.6.a-4)}$$

So the number of times irreducible representation s appears in the decomposition is given by the product of the two character vectors, of the reducible and irreducible representations, with the terms in the sum over the classes weighted by the number of elements in each (which is the same as summing over the group elements); the normalization is given by dividing by the number of group elements g. Thus the characters of the irreducible representations determine which appear in a reducible representation, and how many times.

Problem VII.6.a-1: Prove that the characters of a reducible representation are sums of those of its constituents.

Problem VII.6.a-2: What is the regular representation of a cyclic group (pb. VI.5-9, p. 179)? How can it be reduced? Verify that each representation of cyclic group C_3 occurs in its regular representation the same number of times as its dimension. Do this in general, for any cyclic group, then for any group.

Problem VII.6.a-3: All groups have at least one irreducible representation, the completely-symmetric one. Symmetric groups have two (sec. V.5.c, p. 165). What determines the number of these? Prove that the number of one-dimensional representations equals the index, G/H_C, of the commutator subgroup H_C in G (pb. IV.8.c.ii-6, p. 142) [Jansen and Boon (1967), p. 184]. Why should this be? Is it true for Abelian groups, dihedral groups, symmetric groups? Check it for some others.

VII.6.b How to tell if a representation is completely reduced

The number of times an irreducible representation appears in the decomposition of itself is one. Does this trivial point tell us anything?

Eq. VII.6.a–2, p. 199, for the character of a reducible representation, and its complex conjugate, gives for class c,

$$\zeta_c = \sum a_r \zeta_c^r, \quad \zeta_c^* = \sum a_r^* \zeta_c^{*r}; \tag{VII.6.b–1}$$

we multiply these, sum over the classes (times g_c), and get

$$\sum_c g_c \zeta_c \zeta_c^* = \sum_c g_c \sum a_r \zeta_c^r a_s^* \zeta_c^{*s} = \sum_{sr} a_r a_s^* \sum_c g_c \zeta_c^r \zeta_c^{*s}$$

$$= \sum_{sr} a_r a_s^* g \delta^{sr} = g \sum a_s^2 = \sum_c g_c |\zeta_c^r|^2, \tag{VII.6.b–2}$$

using the orthogonality relations. But for an irreducible representation (r), the only nonzero term in the sum is $a_r = 1$, so

$$\sum_c g_c |\zeta_c^r|^2 = g. \tag{VII.6.b–3}$$

Thus a representation is irreducible, if the absolute square of its characters, summed over the classes, weighted with the number of elements in the class (or summed over all elements), equals the total number of elements, else is reducible. Given the characters of a representation we have a simple test to see if it is irreducible.

Problem VII.6.b–1: These formulas were found for unitary representations. Why? What are they in general?

Problem VII.6.b–2: Even if $\sum a_s^2 \neq 1$, it may be possible to obtain information. If $\sum a_s^2 = 2$, the representation is reducible, but contains two different irreducible ones, and once each [Hamermesh (1962), p. 105]. Explain this, and also describe the reducible representation for $\sum a_s^2 = 3, 4, \ldots$.

Problem VII.6.b–3: For two reducible representations with characters ζ and ϕ, that contain irreducible representation r, a_r and b_r times, show that [Hamermesh (1962), p. 106]

$$\sum_c \frac{g_c}{g} \zeta_c \phi_c^* = \sum_r a_r b_r. \tag{VII.6.b–4}$$

This can be used to tell how many irreducible representations they have in common. Show that if

$$\sum_r a_r b_r = 1, \tag{VII.6.b–5}$$

they have one irreducible representation in common, and this appears once in each. What can be said if $\sum_r a_r b_r = 0, 2, 3, \ldots$?

Problem VII.6.b–4: For the regular representation of any group, determine the characters [Falicov (1966), p. 45; Joshi (1982), p. 87]. Also find $\sum a_s^2$.

Problem VII.6.b–5: Verify these equations for some of the simpler representations of the groups we have discussed.

Problem VII.6.b–6: S_3 has three irreducible characters. Why? Find these. Repeat for S_4, and A_3 and A_4 [Dixon (1973), p. 66, 156]. Try it for S_n and A_n [Hamermesh (1962), sections 7-1-6, p. 182]. Check the orthogonality conditions. What is the relationship between the characters of S_n and A_n, for $n = 2, 3, 4, \ldots, k$?

Problem VII.6.b–7: What are the characters of a cyclic group? Do the orthogonality conditions hold? For an Abelian group that is a direct product of cyclic groups show that the irreducible characters have degree 1, and find them [Dixon (1973), p. 66, 157]. What can be said in general about the characters of Abelian groups [Falicov (1966), p. 38]? Can these questions be answered for dihedral groups, or only for low-order ones?

Problem VII.6.b–8: There are several conditions (Schur's included) for a representation to be irreducible. How are they related? Can they be derived from each other? Are the derivations different views of the same thing, or do they use different properties of representations?

VII.6.c Equality of number of classes and representations

In sec. VI.4.a, p. 176, we discussed why the number of inequivalent irreducible representations of a finite group equals the number of its classes. This can also be shown using the orthogonality relations. There is a set of vectors, the characters, one for each irreducible representation, and each with components labeled by the classes; thus these form a space whose dimension equals the number of classes. The maximum number of independent vectors in a space is its dimension. So the number of inequivalent irreducible representations can be no greater than the number of classes. Are they equal? The orthogonality of these vectors, labeled by classes, with components labeled by representations (eq. VII.5.c.i–10, p. 198), shows that the number of classes is no greater than the number of such representations.

Hence the number of inequivalent irreducible representations equals the number of classes.

Problem VII.6.c–1: Why do the orthogonality relations give an upper bound rather than an equality?

Problem VII.6.c–2: What is the relationship between these proofs of equality of the number of classes and inequivalent representations? Can one be obtained from the other? Are the assumptions used the same? Can orthogonality relations be derived using these proofs? Explain.

Problem VII.6.c–3: The number of distinct irreducible character vectors of a finite group equals the number of its classes. Why? This is an example of the symmetry between classes and irreducible representations. How far does this symmetry extend? Why?

VII.6.d Characters of the regular representation

The character-orthogonality theorems give an expression for the number of times each irreducible representation is contained in a reducible one. There is one reducible representation of particular importance, the regular representation.

Problem VII.6.d–1: For the regular representation, show that the characters of all elements are 0, except for the identity (of course), whose character equals the number of elements g [Hamermesh (1962), p. 107]. Explain the result. Apply this, and show that the number of times, a_r, that representation r appears is (with ζ_1 the character of the identity)

$$a_r = \zeta_1^r = n_r, \tag{VII.6.d-1}$$

where n_r is the dimension of representation r — the number of times a representation appears equals its dimension, as we know. Also

$$g = \sum (n_r)^2, \tag{VII.6.d-2}$$

the sum of the squares of the dimensions of all irreducible representations equals the number of group elements (sec. VI.2.b, p. 175). How are these results, found in different ways, related?

VII.6.d.i *Number of elements and representation dimension*

The regular representation contains all irreducible representations so when we reduce it — block-diagonalize its matrices — we find that its characters equal sums of the characters of each irreducible representation in the decomposition (of the group), each multiplied by the number of times the representation appears.

With $R(t)$ the matrices of the regular representation, ζ_c its character for class c, and a_r the number of times representation r, with matrices M^r, appears, we have for element t,

$$R(t) = \sum a_r M^r(t), \tag{VII.6.d.i-1}$$

summed over all representations in the decomposition. Taking the trace we get,

$$\zeta_c = \sum a_r \zeta_c^r. \tag{VII.6.d.i-2}$$

Multiplying by the complex conjugate of ζ and by g_c and summing over the classes, we obtain for the regular representation

$$\sum \zeta_c \zeta_c^* g_c = g \sum a_c^2 = g^2. \qquad \text{(VII.6.d.i-3)}$$

Then for the class containing (only) the identity

$$\zeta_e = \sum a_r \zeta_e^r. \qquad \text{(VII.6.d.i-4)}$$

The character of the regular representation for the identity is g, while that for r is n^r, its dimension. So, as we expect,

$$g = \sum a_r n^r. \qquad \text{(VII.6.d.i-5)}$$

Problem VII.6.d.i-1: Show by the same argument, taking instead of the regular representation some irreducible representation, that

$$\sum g_c |\zeta_c^r|^2 = g, \qquad \text{(VII.6.d.i-6)}$$

as also shown in eq. VII.6.b-3, p. 201.

Problem VII.6.d.i-2: A transformation on a set of objects permutes some of them, perhaps leaving others fixed. So a permutation on n indices, an S_n operation, might affect only k. A transformation of the square, or the tetrahedron, permutes their vertices, but not necessarily all. Give other examples. Consider an n-dimensional representation of S_n that has for each matrix a single 1 in each row and in each column, and all other matrix elements 0. Find it for S_3. Show that in this representation the character of transformation t is the number of objects (such as indices) left fixed by it [Ledermann (1987), p. 18; Wilson, Decius and Cross (1980), p. 102]. Apply this to the identity. What is the character of a permutation of S_n that affects n objects? Does this agree with previous results? Every group is a subgroup of a symmetric group; does the relationship between the character of a transformation and the number of objects left fixed by it, hold for all finite groups? Why? State, and check, it for the geometrical groups (sec. II.2.e, p. 36 - sec. II.2.h, p. 40). Does it have meaning for Abelian groups? Cyclic groups? Give a geometrical interpretation for dihedral groups.

Problem VII.6.d.i-3: Find the characters for each class of the regular representation for S_2, S_3, S_4, the water group, the pyramidal group, the group of the square and the group of the double rectangle. Compare the last two. Compute the number of times each representation appears in the reduction of the regular representation for these groups. Do these numbers equal the dimensions of the representations?

Problem VII.6.d.i-4: These results can used to find the dimensions of the representations (for small g). Find the dimensions for S_3 and

S_4, and A_3 and A_4. What is the difference between the symmetric and alternating groups? Why? S_3 can also be realized as the symmetry group of the equilateral triangle. Its dimensions, for this realization, were discussed in sec. V.5.d, p. 167. Using this, explain the dimensions of A_3, and why they differ from those of S_3. While this is feasible for small S_n, a computer program that finds characters for any n would be interesting.

Problem VII.6.d.i-5: Prove that every finite group has at least one faithful representation (irreducibility not implied) [Ledermann (1987), p. 65, 209; Lomont (1961), p. 68]. Is this related to the orthogonality and independence of character vectors? Explain why this follows from the group axioms and the definition of representations. Why do these arguments not require that all representations be faithful? For which groups is it irreducible? Do these properties say anything about possible homomorphisms? Are the homomorphisms and the nonfaithful representations related? What does this say about the representations of simple groups? How about those of semisimple groups?

VII.6.d.ii *How simplicity of a group affects its character table*

What is the difference between the character tables of simple and nonsimple groups? The latter contains an invariant subgroup, one consisting of complete classes. We can write the matrices of the regular representation (and the group multiplication table) such that the rows and columns labeled by the elements of a subgroup are the leftmost columns and the highest rows. Thus these break into blocks, with the upper left-hand block going with the subgroup. The character table has its rows and columns labeled by the classes. Can a similar thing be done for it, can the upper left-hand block go with the subgroup? Yes if, but only if, the group is not simple (and the subgroup consists of complete classes). Otherwise the subgroup cannot be distinguished in the character table. Beyond this, since an invariant subgroup consists of complete classes, there is a correspondence (though not one-to-one) between the columns of the tables of the group and the invariant subgroup. If the characters of the latter are known, they can thus be used to find those of the group.

Problem VII.6.d.ii-1: Show that an invariant subgroup goes into itself under an inner automorphism (sec. III.4.e, p. 91). Is this true for an outer one also? Is there a difference between the automorphism group of a simple group and a nonsimple one? How does an automorphism group affect the character table? Does it distinguish these two classes of groups? The group multiplication table, and the character table, give bases of representations of the automorphism group, of different dimensions. Are these (all) irreducible? Are there differences between the representations for the two types of groups?

VII.6.e Uniqueness of the reduction of a representation

In the discussion of reducing representations there was one unstated assumption (at least; the reader should go back and see how many others he can find). There are many similarity transformations and so many ways of reducing a representation.

Problem VII.6.e-1: The reader should now be prepared to prove that (up to equivalence) the decomposition is unique (except for ordering of representations); in the decomposition of a reducible representation, the irreducible representations, and the number of times each appears, is the same for all ways of performing the decomposition [Boerner (1963), p. 51; Hall (1959), p. 252]. This is for finite groups; these are, in many ways, rather simple. Continuous representations can be more subtle. We leave it to the reader to notice during his study of group theory as he comes across more and more esoteric groups whether, and when, uniqueness of the reduction fails. Why is the word "reduction" used in the previous sentence?

VII.7 COMPUTATION OF CHARACTERS

How do we find characters [Elliott and Dawber (1987), p. 67; Falicov (1966), p. 36; Inui, Tanabe, and Onodera (1990), p. 71; Jones (1990), p. 68; Joshi (1982), p. 81; Lomont (1961), p. 61; Tinkham (1964), p. 28]? In general this is not trivial so we do not discuss general (realistic) methods. But it is worth mentioning some basic points (giving a method for groups of small order).

For a group there is, besides a multiplication table, a character table, with columns and rows labeled not by elements, but by classes, and representations, and clearly square. Each table is headed by the classes, multiplied by the number of elements in the class; the rows are labeled by the representations. The first column, headed by the identity, contains the dimensions of the representations. The first row, that of the identity (completely symmetric) representation, has all entries one. Rows, and likewise columns, are mutually orthonormal (with the proper weight factors).

The numbers of representations and of classes are equal, and we have

$$g = \sum (n_r)^2, \qquad \text{(VII.7-1)}$$

for the dimensions, as well as the character-orthogonality relations.

There is one one-dimensional representation (at least), and every representation has a unit matrix (whose character is immediate).

Problem VII.7-1: There is another relation we can use [Falicov (1966), p. 37], this coming from the multiplication of classes (sec. IV.5.f, p. 126), with decomposition coefficients a^l_{ck}. Show that for each representation the characters ζ_c of class c, with g_c members, obey

$$g_c\zeta_c g_k\zeta_k = d\sum a^l_{ck} g_l\zeta_l, \tag{VII.7-2}$$

where d is the dimension of the representation. Check this using the identity class and also the identity representation. Also (pb. VII.4.b-2, p. 192) the number of real characters is equal to the number of ambivalent classes (ones containing the inverses of the — of course, all — elements in them). The rest depends on luck (in principle); we shall not consider the methods for special groups to increase the ease of success.

Problem VII.7-2: Prove that for one-dimensional representations the characters are the roots of unity. What determines which roots?

Problem VII.7-3: A square table, such as the character table, can be regarded as a matrix. Prove that for the characters this matrix is nonsingular [Ledermann (1987), p. 64, 208]. What is the reason for this?

Problem VII.7-4: Why are all characters of symmetric groups real (in fact rational) [Ledermann (1987), p. 164]? Why must these groups satisfy the conditions giving this result? However the alternating (and other) subgroups do not have all real characters [Ledermann (1987), p. 161]. Explain. If the characters of a group are real, how could those of a subgroup not be?

Problem VII.7-5: Are the irreducible characters of group G and those of its factor group(s) related (sec. IV.8.c, p. 138) [Ledermann (1987), p. 79]? Can this be used in finding the group characters? In particular, prove that for elements u and u' of (invariant) subgroup H, and t of G, with

$$u' = tut^{-1}, \tag{VII.7-3}$$

$$\zeta_{u'} = \zeta_u, \tag{VII.7-4}$$

where ζ is the character. Thus knowing the character of each element of the (invariant) subgroup, we know the character of each element of the class *of the group* to which it belongs. Why must H be invariant?

VII.7.a Abelian groups and their characters

For these groups, the number of representations equals the number of classes equals the number of group operations; all representations are thus one-dimensional. The characters and representation matrices are identical. For unitary representations all characters have absolute value 1; they differ only in phases. So the character (representation matrix) of an element and its inverse are complex conjugates. The character of the identity is 1 in all representations.

VII.7.a.i *Characters of cyclic and dihedral groups*

For cyclic groups the characters are the k'th roots of unity, where k is the order of the group [Falicov (1966), p. 40; Jansen and Boon (1967), p. 119; Jones (1990), p. 69; Ledermann (1987), p. 53].

Problem VII.7.a.i-1: Why? Write down the character table for cyclic groups with $k = 1,2,3,4,5$. The character tables of other Abelian groups are more interesting than those of cyclic groups [Falicov (1966), p. 38; Ledermann (1987), p. 54].

Problem VII.7.a.i-2: It should be obvious that the character table of the order-two group is

	e	t
S	1	1
A	1	-1

Table VII.7-1: Character Table for the Order-Two Group

For order three, the table is, with

$$\omega = exp(2\pi i/3) = -\frac{1}{2} + i\frac{\sqrt{3}}{2}, \qquad \text{(VII.7.a.i-1)}$$

	e	t	t'
S	1	1	1
A	1	ω	ω^2
R	1	ω^2	ω

Table VII.7-2: Character Table for the Order-Three Group

The satisfaction of the orthonormality conditions for these groups is obvious, once the relationship between t and t' is seen. What is the significance of the representations?

Problem VII.7.a.i-3: Verify the character table (with the representations arbitrarily labeled) for the four-group (sec. III.2.c, p. 71), and orthonormality relations [Ledermann (1987), p. 56]. Identify the representations. Are any mutually (complex) conjugate? Does this have factor groups (with character tables)? It should be

	e	a	b	ab
S	1	1	1	1
Γ_1	1	−1	1	−1
Γ_2	1	1	−1	−1
Γ_3	1	−1	−1	1 ,

Table VII.7–3: Character Table for the Four-Group

Problem VII.7.a.i–4: We get the character table for an order-six group,

	e	a	a^2	a^3	a^4	a^5
S	1	1	1	1	1	1
A	1	−1	1	−1	1	−1
Γ_1	1	ω	ω^2	1	ω	ω^2
Γ_2	1	ω^2	ω	1	ω^2	ω
Γ_3	1	$-\omega$	ω^2	−1	ω	$-\omega^2$
Γ_4	1	$-\omega^2$	ω	−1	ω^2	$-\omega$,

Table VII.7–4: Character Table for an Order-Six Group

with ω the same as for the order-three group. Why should the cube, rather than the sixth, root of unity enter? Identify the representations. Are any mutually conjugate? Check orthonormality. Does this have factor groups (with character tables)? Which order-six group is this?

Problem VII.7.a.i–5: Dihedral group D_2 has four elements so four classes and four representations. The character and group tables are identical. The first row and first column both have all elements 1. As every element has order 2, the table can be arranged so that the unit element lies along the diagonal, and has character 1. Why? Show that this makes the table symmetric around the diagonal. Give the table. Identify the symbols with the physical transformations. Describe the representations. Check that all orthonormality conditions are satisfied. Find any factor groups and their character tables. The entries in this table are all real, while some of those in the above tables are complex. Is there any reason for this? Other dihedral groups are not Abelian [Jones (1990), p. 71, 76, 265; Tung (1985), p. 53; Aivazis (1991), p. 23]. Find a general rule for their characters.

VII.7.a.ii *Description of the characters of Abelian groups*

The representations (characters) of an Abelian group (sec. III.2.b, p. 69) can be described in general [Dixon (1973), p. 66, 157; Ledermann (1987), p. 52]. For a cyclic group of order k, these are the k k'th roots of unity,

$$c_n(k) = exp(2\pi in/k). \quad \text{(VII.7.a-1)}$$

An Abelian group of order g is the direct product of cyclic groups of prime order,

$$g = p^a q^b \ldots. \quad \text{(VII.7.a-2)}$$

Thus its characters are of the form

$$c_{nm\ldots} = exp\{2\pi i\frac{n_1 + n_2 + \cdots}{p} + \frac{m_1 + m_2 + \cdots}{q} + \cdots\}. \quad \text{(VII.7.a-3)}$$

Problem VII.7.a.ii-1: State this more explicitly. Check it for the character tables above. Prove it. The numbers n_1, n_2, ..., m_1, ..., and the characters they give, $\zeta_{nm\ldots}$, both form groups, one with product addition, the other multiplication. How are these related?

Problem VII.7.a.ii-2: It is simple to show that the matrix

$$a = \begin{pmatrix} 0 & 1 \\ -1 & -1 \end{pmatrix} \quad \text{(VII.7.a-4)}$$

gives a representation of C_3, the cyclic group with three elements [Ledermann (1987), p. 34, 206]. Prove it, and explain why this representation is irreducible if only real numbers are used (over the field of real numbers), but reducible (and should be reduced) to two one-dimensional representations using complex numbers (over the complex number field). This emphasizes that whether a representation is reducible or not (pb. V.5.a-4, p. 163) can depend on the numbers (real, complex, or for the more adventurous, others, say quaternions) we use — as does other aspects of the theory. We always (unless stated) use complex numbers. The mathematically-minded may find it interesting to check which of the results do not hold if only real numbers are allowed. Might some results not hold (or might some results hold) if only rational numbers are permitted? How about integers? Would anything interesting be obtained if quaternions could be used? These are questions we shall carefully avoid, but some readers may not wish to.

VII.7.b Characters of symmetric groups

For S_3 (sec. II.4.d, p. 51; sec. V.5.d, p. 167) there are six transformations, and three representations, two one-dimensional (symmetric and antisymmetric) each appearing once in the decomposition of the regular representation, and one two-dimensional representation appearing twice. There are thus three classes, so a 3×3 character table. We can see that

$$g = \sum n_r^2 = 1^2 + 1^2 + 2^2 = 6 \quad \text{(VII.7.b-1)}$$

is satisfied (fortunately).

Problem VII.7.b–1: For every group there is a one-dimensional representation. Show that for S_3 this equation requires the other two representations to have dimensions 1 and 2; the given ones are the only possibilities [Falicov (1966), p. 37]. Could there be a different number of representations? The characters of all classes of the symmetric representation are 1, and those of the antisymmetric representation have square 1. Why? However they cannot all be 1 since the character vectors of these two representations are orthogonal. The character of the identity is 1. So the character of one of the other two classes is -1. Show that it is the class of the transpositions. This leaves the characters of the two-dimensional representation. That of the identity is 2. The characters of the other two classes are determined by the requirement of orthonormality of the character vector with those of the other two representations. Check that the character table is

	ζ_1	$2\zeta_2$	$3\zeta_3$
S	1	1	1
A	1	1	-1
T	2	-1	0 .

Table VII.7.b–1: Character Table for S_3

Verify that all the orthonormality conditions are satisfied. Find the factor groups and their character tables.

Problem VII.7.b–2: The character table of A_3 should be clear from this?

Problem VII.7.b–3: Repeat this for S_4, and A_4, and for as many other of these groups as possible [Dixon (1973), p. 66, 156; Hamermesh (1962), p. 276; Ledermann (1987), p. 50, 61, 75, 79, 106, 137, 140; Tung (1985), p. 53; Aivazis (1991), p. 22]. Find the character table for A_5, and use it to prove that A_5 is simple (pb. IV.9–6, p. 144) — S_5 is not, having A_5 as an invariant subgroup.

Problem VII.7.b–4: The quaternion group (pb. III.2.e–4, p. 74) is a subgroup of a symmetric group. Find its character table [Dixon (1973), p. 66, 157; Jones (1990), p. 76, 265; Ledermann (1987), p. 62], and compare it to that of the smallest symmetric group of which it is a subgroup. It is the same as that of what other (nonisomorphic) group?

VII.7.c Characters for the group of the square

For this group (sec. II.2.f, p. 37) it is interesting to consider whether the properties that we have found are related to the character table

[Falicov (1966), p. 62; Jansen and Boon (1967), p. 113; Joshi (1982), p. 79].

Problem VII.7.c-1: Verify the orthonormality relations from the table. Identify the classes and representations. Can any properties previously found be seen from the table? The table is

	ζ_e	$2\zeta_a$	$2\zeta_b$	ζ_c	$2\zeta_d$
S	1	1	1	1	1
Γ_1	1	−1	−1	1	1
Γ_2	1	1	−1	1	−1
Γ_3	1	−1	1	1	−1
Γ_4	2	0	0	−2	0 .

Table VII.7.c-1: Character Table for the Group of the Square

VII.7.d Characters of the tetrahedral group

The tetrahedral group (sec. III.2.f, p. 76) has twelve elements, four classes, so four irreducible representations [Falicov (1966), p. 41, 67; Fässler and Stiefel (1992), p. 215; Jones (1990), p. 77, 265; Tung (1985), p. 53; Aivazis (1991), p. 20].

Problem VII.7.d-1: Check the number of classes. Also from the relationship between the square of the dimensions and the group order, show that three representations have dimension 1, and one has dimension 3. There is one normal subgroup of order 4, so index 3. Find it. The factor group is the third-order cyclic group; its characters (representations) are given in table VII.7-2, p. 208. Using this and the orthonormality conditions, we get the character table. Given line two do you expect line three? It should be simple to find line two from one property of the elements in the class. Line four then follows using orthonormality. These should all be checked. Identify the classes and representations. Show that, with ω the (a?) cube root of 1, the table is

	ζ_1	$4\zeta_a$	$4\zeta_b$	$3\zeta_c$
S	1	1	1	1
Γ_1	1	ω	ω^2	1
Γ_2	1	ω^2	ω	1
Γ_3	3	0	0	−1 ,

Table VII.7.d-1: Character Table for the Tetrahedral Group

VII.7.e Examples of character tables

Character tables of specific groups illustrate the concept, and how characters are found, and the difficulties in finding them in general. Character tables have been given here to provide exercises to (hopefully) increase understanding of characters and groups. Tables for many small finite groups are widely available, but perhaps under different names [Elliott and Dawber (1987), p. 265; Falicov (1966), p. 37; Hamermesh (1962), p. 125; Inui, Tanabe, and Onodera (1990), p. 363; James and Liebeck (1993); Ledermann (1987), p. 205; Lomont (1961), p. 78; Tinkham (1964), p. 323].

Problem VII.7.e-1: Relate each character table here to one in any of the referenced tables, and verify that they are the same.

Problem VII.7.e-2: These character tables may also be useful in finding, and understanding, automorphism groups (sec. III.4.e.ii, p. 93).

Problem VII.7.e-3: In these chapters we have raised various questions, indicating what we wished to know about the topics under discussion, and why. It would be useful to review these, and see which (if any) have been answered, and to what extent. Ones that have not been should be answered now, which the reader, having gone through these chapters, presumably should be able to do. What have they taught us about groups, and about physics?

Chapter VIII

The Symmetric Group and its Representations

VIII.1 SYMMETRIC GROUPS

Perhaps the most important finite groups are the symmetric groups. Every finite group is a subgroup (sec. IV.3.d.ii, p. 113); in addition, they provide tools used in studying groups in general, including continuous ones, and representations. And their representations illustrate many properties of the theory. Thus, and because there is much material available, we study them in depth [Baumslag and Chandler (1968), p. 56, 167; Boerner (1963), p. 28, 102; Burnside (1955), p. 168; Chen (1989), p. 117; Duffey (1992), p. 213; Elliott and Dawber (1987), p. 425; Fässler and Stiefel (1992), p. 136; Grossman and Magnus (1992), p. 141; Hall (1959), p. 53; Hamermesh (1962), p. 182; Inui, Tanabe, and Onodera (1990), p. 333; Landin (1989), p. 74; Ledermann (1953), p. 62; Lomont (1961), p. 258; Lyapin, Aizenshtat, and Lesokhin (1972), p. 62; Maxfield and Maxfield (1992), p. 49; Miller (1972), p. 116; Murnaghan (1963), p. 91, 168; Robinson (1961); Rutherford (1948); Sagan (1991); Schensted (1976), p. 119; Scott (1987), p. 8, 298; Tung (1985), p. 64; Weyl (1931), p. 281; Weyl (1946), p. 115; Wigner (1959), p. 124; Wybourne (1970); Wybourne (1974), p. 372]. These have been introduced, and discussed many times, above (sec. II.4.d, p. 51; pb. III.3-3, p. 82; pb. III.4.e.i-6, p. 93; sec. IV.3.d, p. 112; sec. V.5.c, p. 165; pb. VII.6.d.i-4, p. 204; pb. VII.7-4, p. 207; sec. VII.7.b, p. 210), and we start with the assumption that the reader has looked at previous chapters.

VIII.2 PROPERTIES OF SYMMETRIC GROUPS

The representations of a group consist of states, often explicit functions, here sums of terms with different permutations of indices, and matrices acting on them. To find these we need labels, ways of referring to the equivalence classes, representations, states, matrix elements, and then properties of these states and matrices [Miller (1972), p. 116; Schensted (1976), p. 124]. First we must collect some properties of the groups.

How are symmetric groups generated? What are the subgroups, invariant subgroups, factor groups (sec. IV.8.c, p. 138)? Which are simple? How are the representations realized? Is there freedom in answering these questions? These must be considered before the groups can be further analyzed.

VIII.2.a Cycle structure, subgroups and simplicity

Symmetric groups have invariant subgroups, the alternating groups, so are not simple. The alternating groups are simple (sec. IV.9, p. 143), except A_4, which has an invariant subgroup [Armstrong (1988), p. 85; Hall (1959), p. 61; Kurosh (1960a), p. 68]; this is related to there being no formula for the solution of an algebraic equation of degree greater than 4 (sec. IV.3.d, p. 112). How is simplicity, or its lack, related the the class and cycle structure of the groups?

Problem VIII.2.a-1: Symmetric group S_3 has elements of cycle length 1,2,3. Check that all of each cycle length are in the same class, ones of different order are in different classes, and that the number in the classes are respectively 1,3,2. The three elements of order 1 and 3 form a subgroup, which is simple, and is the only (invariant) subgroup. For S_4 the elements have orders 1,2,3,4. Those of different order are in different classes (of course). However there are two classes for order 2. There is one element of order 1, eight of order 3, six of order 4; for order 2 there are six transpositions, and three permutations in the "double transposition" class, giving twenty-four. We label classes using their orders, with a prime for the second order-two class,

$$L_1 = 1, \quad L_2 = 6, \quad L_2{}' = 3, \quad L_3 = 8, \quad L_4 = 6. \qquad \text{(VIII.2.a-1)}$$

Note that L_4 is the order of S_3; why? How far can we get in finding the subgroups using the requirement that the order of a subgroup divides that of the group (sec. IV.6.b.i, p. 131)? The (invariant) group A_4 has classes L_1, $L_2{}'$, L_3, giving twelve elements, half of 24. Verify. An invariant subgroup consists of complete classes. Thus we have a number-theoretic problem. How many ways can we add these numbers (always including 1) to get a divisor of 24 (which are 1,2,3,4,6,8,12)? They give the possible normal subgroups, but the candidates must be checked to show that they are normal; which are? Verify (numerically and by checking the multiplication rules) that L_1 and L_2', but not L_1 and L_3, give a normal subgroup of A_4. What are the subgroups of S_4 and A_4? For any element, all powers of it must be in the subgroup. We get nine (isomorphic) order-two subgroups by taking the identity and any element of L_2 or L_2', four of order-three by taking an element of L_3, and so on. Check and finish this. Which subgroups are simple?

Problem VIII.2.a-2: Repeat this for S_5 and S_6.

Problem VIII.2.a-3: Find the classes of S_5 and A_5, show that A_5 is the only proper normal subgroup of S_5, and is simple [Dixon (1973), p. 12, 83].

Problem VIII.2.a-4: Show that the only proper subgroup of S_n of index less than n is A_n (which has index 2), for $n > 4$ [Dixon (1973), p. 14, 86]. Why should smaller symmetric groups be different?

Problem VIII.2.a-5: Find the normalizer (sec. IV.4, p. 116) of the cyclic group $(12\ldots n)$ in S_n [Dixon (1973), p. 12, 83].

Problem VIII.2.a-6: Write a computer program to find all subgroups, all normal subgroups, all simple subgroups and all simple, normal subgroups of S_n. Compare with the list of all simple groups [Gorenstein (1982; 1983)].

Problem VIII.2.a-7: For S_6 find an outer automorphism [Hall (1959), p. 90]. This interchanges the two classes of order 3. Do any other symmetric groups have (similar) outer automorphisms? Is there a reason?

VIII.2.b Realizations of the symmetric group

Symmetric group S_n consists of all permutations of n objects. What objects? These can be students in a classroom, pages in this book (which would not leave it invariant, but neither would permutations of students leave a classroom invariant), atoms in a molecule, electrons in anything, and so on. S_n is the group of permutation transformations; there is no implication of invariance. There is an important difference physically (and in other ways) between students and electrons. Students can be distinguished (though some teachers are not able to), electrons can not be. For purposes of the formalism we must be able to label objects, though physics may require that allowable statefunctions not distinguish them. Thus we call the objects 1 and 2 (or Isaac and Albert if we wish), but a statefunction must be symmetric or antisymmetric; experimentally there is no way of telling which particle is which.

We can realize (sec. V.3.c, p. 157) the group in many ways, say by describing students or describing electrons. These can be far different physically but need have no mathematical distinction. Here we do not realize the symmetric group with physical objects, rather use a most important discovery in theoretical physics — subscripts. We consider a function (or symbol, it does not matter) with indices, a monomial, and ones with permutations of these indices. Statefunctions are sums of symbols with different arrangements of indices, each with coefficients, these sums called multinomials, or tensors. To find representation matrices we do not need a realization. But it helps for concreteness, and these functions can be interpreted (often appropriately) as quantum mechanical statefunctions.

Basis states of a representation are then taken as multinomials, or equivalently as sums of permutations (with coefficients) acting on a fixed monomial [Hamermesh (1962), p. 243], the standard monomial (usually the one with indices in natural order $123\ldots$, or $abc\ldots$, the natural monomial). It does not matter for the group what symbol the subscripts are attached to, so there is usually no reason to write it. Thus we often write just indices, or what is equivalent, permutations.

VIII.2.c Transpositions

The simplest permutations are transpositions, those that transpose two symbols. They are the simplest and most important; the theory of representation matrices is built on them. Why are transpositions important? Any permutation can be written as a product of transpositions, and especially as a product of neighboring transpositions (which interchange neighboring numerals like 1,2 or 5,6, or, depending on our view, which interchange numerals in neighboring positions — neighboring indices).

Any S_n permutation can be written as a product of, at most, $n - 1$ neighboring transpositions (which is all there are). Consider the indices placed on a line and we wish to move (permute) them so to arrange them differently. We can do this by a series of transpositions of neighboring indices. For example, the index that goes to the first position on the line can be moved there by such a series, then the one going to the second position moved, without affecting the one now the first, and so on. The result is stronger: the transpositions a permutation is a product of need not be neighboring [Baumslag and Chandler (1968), p. 168; Mirman (1991e)].

Problem VIII.2.c-1: Restate the argument, taking transpositions to interchange indices with neighboring values (like 3,4). Write the permutation from 23658714 to 56427381 in terms of transpositions using both views (on neighboring values, and on neighboring positions). How do they compare? Why? Pick another set of $n - 1$ transpositions, not (all) neighboring, and do this using them. Give a rigorous proof for the expression of any permutation in terms of transformations.

Problem VIII.2.c-2: To see that any $n-1$ transpositions can be used, check that the $n-1$ transpositions $(ij), (jk), \dots$, have all numerals in two transpositions except for two numerals in one, and do not form disjoint sets (having no numerals in common). Every numeral of S_n is in at least one (and at most two) of these; there are $2(n-1)$ positions in which to insert them, two are filled by the two numerals appearing once, so the other $2(n-2)$ are filled by $n-2$ numerals, for a total of n. So (ij), in particular neighboring transposition $(i\; i-1)$, can be written as a product of transpositions from a set, say of neighboring ones, $(ij), (jk), \dots$, multiplied by one of the set, then multiplied by the product, but in reverse order:

$$(i\; i-1) = \dots (jk)(ij)(mn)(ij)(jk) \dots . \qquad \text{(VIII.2.c-1)}$$

Thus any S_n permutation can be written as a product of these transpositions since any neighboring one can. Verify these statements. Also for (12), and for permutation (3526), write this product using the set $(45) \dots$. In addition show that, in general, any transposition can be written as NtN; with N neighboring and t a transposition. And t itself can be given in this form. Transposition (ij) can be written as the product of all, and only, neighboring transpositions on numerals between i and j. Does this argument depend on whether transpositions are defined as acting on numerals or on positions?

Problem VIII.2.c-3: Write these permutations (notice that this is independent of S_n) as products of neighboring transpositions: $(13), (123), (132), (14), (24), (12)(34), (13)(24), (14)(23), (1234), (2134), (3526)$. Write them as products of transpositions none of which are neighboring. Write $(13), (14), (24), (15)$ in the form NtN.

Problem VIII.2.c-4: Is it possible to write a permutation as a product of two different sets of neighboring transpositions? How different can they be? Give examples.

Problem VIII.2.c-5: S_n is generated by transpositions. Explain how they give a presentation (pb. III.3-3, p. 82). Show that A_n is generated by the $n-2$ three-cycles $(123), (124), \dots (12n)$ [Armstrong (1988), p. 30; Dixon (1973), p. 13, 84]. Find another presentation. Is there an analogous one for S_n [Lyapin, Aizenshtat, and Lesokhin (1972), p. 64, 188]? Which of these is better? Why? Are there others that might be advantageous?

Problem VIII.2.c-6: Any permutation can be expressed as a product of disjoint cycles. Show that its order is the least common multiple of their lengths [Lyapin, Aizenshtat, and Lesokhin (1972), p. 63]. Can this be done by taking transpositions as the cycles? Give examples.

VIII.2.d Representations and the regular representation

A complete set of permutations of indices (or a set of symbols with all differently permuted indices) forms the bases of the regular representation. All permutations applied to one (permutation or symbol) gives all others; the result is a closed space — every permutation (of indices) is a member of the set.

The regular representation is reducible and we want to reduce it, finding sets of sums (each set with the smallest number of sums) of the different permutations (of indices) mixed among (and only among) themselves, to get the irreducible-representation basis states. Thus we need a procedure for generating all these sums, showing that they form the representation basis spaces, and are irreducible. They must completely cover the space of the regular representation; any sum of permutations of indices can be written as sums of these basis states.

For example T_{12} is a sum of the states of the completely-symmetric and completely-antisymmetric representations: $(T_{12} + T_{21})$ and $(T_{12} - T_{21})$.

VIII.2.e Categories of representations

Each equivalence class contains a set of equivalent representations; all have the same matrices, with the states different. However there are equivalent representations to these equivalent representations; each set of matrices can have (perhaps infinitely) many equivalent forms, of which three have been described and named (by the Rev. Alfred Young, an English country clergyman, who did much work on the theory of the symmetric groups).

The first is the natural representation (category) [Boerner (1963), p. 114; Rutherford (1948), p. 53; Sagan (1991), p. 74; Schensted (1976), p. 145]. Naturally this has no simple properties. Furthermore it seems unnatural in having no physical applications (known to this author). We ignore it.

The second category is the semi-normal [Rutherford (1948), p. 23]. It has the advantage that the matrix entries are rational numbers and the disadvantage that the representation matrices are not unitary and the matrices of the transpositions are not symmetric ($t_{ij} \neq t_{ji}$).

Fortunately, and in accord with the theorem that every finite-group representation is equivalent to a unitary one, there is the orthogonal representation (category) [Rutherford (1948), p. 49]. This has the disadvantage that many matrix entries are not rational, being square roots for the transpositions, and in general products and sums of these. But it has the advantage that the matrices are, as one might suspect, orthogonal. And since the entries are real numbers they are unitary. Further the transposition matrices are, of course, symmetric.

The orthogonal and semi-normal categories are closely related. The states and matrices of one can be obtained from the other by multiplication with functions, called tableau functions. These have been discussed elsewhere and seem not to shed much light on properties of representations so we do not consider them and refer to the literature should they be needed [Rutherford (1948), p. 47; Schindler and Mirman (1977a), eqs. III-14–16; (1977b), table II].

For most of the discussion it does not matter whether we consider the orthogonal or semi-normal representation categories. However if it does matter, as in questions of unitarity, we consider only representations of the orthogonal category.

Problem VIII.2.e–1: Show that an orthogonal matrix with real entries is unitary; state and prove the converse. (There are orthogonal matrices with complex entries [Mirman (1995a)].)

Problem VIII.2.e–2: Is there a relationship between matrices being orthogonal and being symmetric? Are the matrices of the other permutations symmetric?

VIII.3 LABELS

It would not do much good to construct states if we did not have a way of referring to them, nor matrices if we could not label rows and columns. So we need labels, names (given by numbers) for the states and matrix elements. There are three, giving the equivalence class, the representation and the state. Often representation and equivalence class are not distinguished because only representation matrices are required, and these (for each representation category) are the same for all representations of an equivalence class. However states, quantum mechanical statefunctions, of different representations of the same class differ, and must be distinguished. The method of labeling equivalence classes, representations and states was developed by Rev. Young and is useful because it is intuitive and pictorial. Rev. Young also did much work on the symmetric-group representation theory giving the rules for finding the matrices. Thus the labels of the classes are called Young frames, those of representations and states, Young tableaux. The terminology varies with authors; here we use frames and tableaux to be consistent with our usage (making it easier to take over our results [Mirman (1987a,b, 1991e); Schindler and Mirman (1977a,b,c, 1978a,b); Soto and Mirman (1981a,b,c, 1982)], increasing access to our work for further details, also making access to work of others unfortunately slightly more difficult).

For conciseness we often refer to an equivalence class as a frame, and representations and states as tableaux (so care is needed; this term can be ambiguous).

VIII.3.a Young frames

The equivalence classes of S_n are labeled by (standard) Young frames each of n boxes arranged in rows, with the length of the rows non-increasing down — no row is longer than a row above it. With the numbers giving the row lengths in order starting from the top, rows $(1,1,1), (2,1,0)\{=(2,1)\}, (3,0,0)\{=(3)\}, (3,2)$ and $(3,3)$ give standard frames, but not $(2,3)$. There is a one-to-one correspondence between standard Young frames (all arrangements of n boxes with non-increasing rows) and equivalence classes.

For S_3 there are three frames, (3) has all boxes in the same row, (2,1) has two in the first and one in the second, and (1,1,1) has one column. But

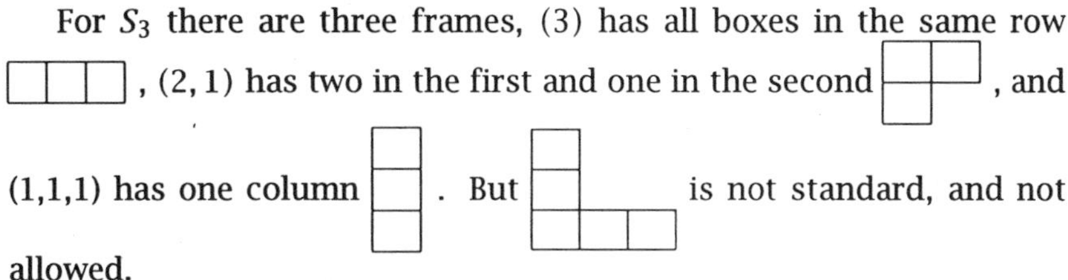

is not standard, and not allowed.

Problem VIII.3.a-1: Draw the frames for S_2 and for S_4 and S_5. The number of classes, so of inequivalent representations, of S_n equals the number of ways of partitioning n (dividing n numerals into sets in all different ways) [Lomont (1961), p. 259]. Do these agree with number of standard frames?

VIII.3.b Young tableaux

Standard Young tableaux of S_n are found for each frame by putting numerals $1 \ldots n$ into boxes such that the numbers are increasing to the right and down; the upper left-hand box always has 1. For example

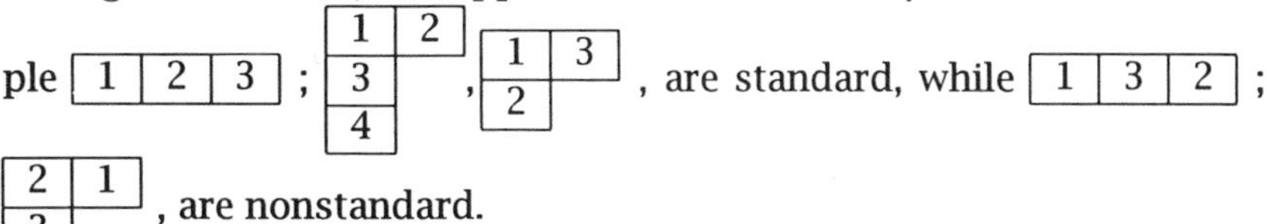

are standard, while ... are nonstandard.

The number of standard frames equals the number of equivalence classes, and the number of standard tableaux equals the number of representations in each class, and the number of states of each representation (these numbers being equal, fortunately — the number of times each representation appears in an equivalence class in the decomposition of the regular representation equals its dimension); nonstandard frames and tableaux are not needed and do not give labels.

Tableaux label representations (with one tableau) and states (with a second tableau) of the equivalence class given by the frame. There is a one-to-one relationship between the set of standard tableaux and the representations of each class, and another one-to-one relationship between this set and the states of each representation. Not only are the numbers equal, but, as we see, many properties of the states can be read from the tableaux.

The two-dimensional representation appears twice, labeled by the two Young tableaux with two rows, and the two states of each are again labeled by another pair of the same two tableaux.

Problem VIII.3.b–1: Write all Young tableaux for S_2, S_3, S_4, S_5 [Tung (1985), p. 78; Aivazis (1991), p. 35]. Count the number of tableaux for each frame — the dimension of the representation. The sum of the dimensions squared should be the order of the group.

VIII.3.c Labeling states

A state then is labeled by a frame (giving the equivalence class), and two tableaux of that frame (for the representation and the state). Of course knowing a tableau we know the frame but it is convenient to give it explicitly. We eventually introduce symbols to represent the states and these have as subscripts two tableaux. The left one labels the state, the right the representation. A permutation mixes either states or representations, depending on which side it is applied to; it cannot mix both at once. Applied from the left it mixes states of the same representation, from the right corresponding states of all representations of a frame are mixed.

Problem VIII.3.c–1: This follows from the discussion of pb. VI.2.a–1, p. 175, and the convention for products of permutations (sec. II.4.d.i, p. 54). Why? Check that if we used the opposite convention, left and right would be interchanged.

Problem VIII.3.c–2: The frames and tableaux not only label the representations and states of S_n, but give the S_{n-1} representations in any S_n representation, and the S_{n-1} representation to which an S_n state belongs. Explain why the frame of the S_{n-1} representation of a state is obtained by removing the box containing n, in the state's S_n tableau (which, right or left?); the resultant tableau is that of the S_{n-1} state. Why does this always give a standard S_{n-1} tableau? So every S_{n-1} representation is contained in an S_n state zero or one times, no more. By writing all tableaux of a frame and then removing the boxes containing

n, we get all S_{n-1} frames for that representation.

VIII.3.d Conjugate frames

The conjugate frame of a frame is found by interchanging rows and columns. If we draw a line from the upper left-hand corner of the frame to the lower right, the conjugate is the one given by reflection in this line. Thus row lengths (5,3,2,1) are the column lengths of the conjugate frame, so its row lengths are (4,3,2,1,1), and conversely.

Problem VIII.3.d-1: Draw these two frames and verify that they are conjugates.

Problem VIII.3.d-2: Can a frame be its own conjugate? Find all conjugates for S_2, S_3, S_4 and S_5.

Problem VIII.3.d-3: Show that every standard frame has a conjugate which is also standard.

Problem VIII.3.d-4: Check that the number of ways of inserting $1\ldots n$ into a frame (its dimension) equals that of its conjugate.

Problem VIII.3.d-5: What is the conjugate of the frame of a single row? What are the dimensions of these two frames? Try this for $S_2,\ldots,S_5$, and then it is a simple combinatorial problem to do it in general.

VIII.3.e Ordering frames and tableaux

Frames and tableaux would not be convenient labels if we always had to draw pictures. What we want as labels are numbers, that is ordinals, so we must order the frames and tableaux. Here we give the convention and then usually refer not to frames and tableaux, but to their ordinals.

VIII.3.e.i *Ordering frames*

The convention is that of last-letter ordering. Two frames are ordered by the left-most column in which they differ; the first is the one having the longest column (the one that puts the "last letter", numeral n, lowest). The completely antisymmetric representation is always frame 1, the completely symmetric one is the last frame.

Problem VIII.3.e.i-1: So for S_3, the row lengths of the frames in order are (1,1,1), (2,1,0), (3,0,0); these are

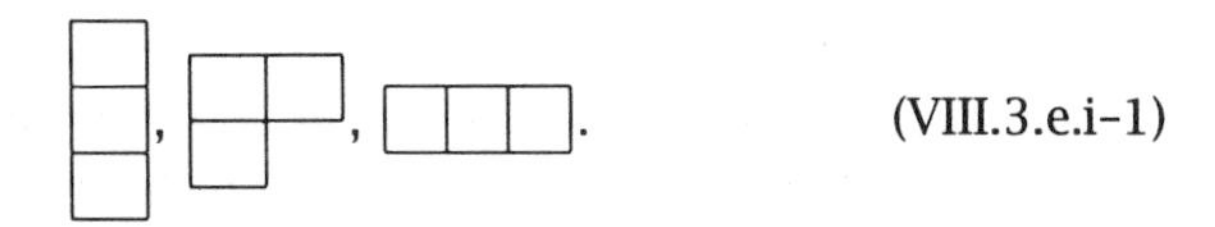

(VIII.3.e.i-1)

Draw in order the frames of S_2, S_4 and S_5.

Problem VIII.3.e.i-2: Show that this rule gives a complete ordering of the frames; there are none whose order is undecided. Prove that it gives a total ordering; if frame 1 < frame 2 and frame 2 < frame 3, then frame 1 < frame 3.

VIII.3.e.ii *Ordering tableaux*

For the tableaux we start with the "last letter", numeral n; if two tableaux differ in the row in which it appears, the one in which it lies lowest comes first. For n in the same row in both, the one in which $n-1$ lies lowest comes first, if this is in the same row also, we consider $n-2$, and so on until the tableaux are ordered.

Problem VIII.3.e.ii-1: For S_3,

$$\begin{array}{|c|c|}\hline 1 & 2 \\ \hline 3 \\ \cline{1-1} \end{array} < \begin{array}{|c|c|}\hline 1 & 3 \\ \hline 2 \\ \cline{1-1} \end{array}. \qquad \text{(VIII.3.e.ii-1)}$$

Order all tableaux of S_2, S_3, S_4 and S_5.

Problem VIII.3.e.ii-2: Show that this gives a complete ordering of the tableaux; there are no tableaux whose order is undetermined. Prove that it gives a total ordering; if tableaux 1 < tableaux 2 and tableaux 2 < tableaux 3, then tableaux 1 < tableaux 3. Is ordering an equivalence relation (sec. III.4.a, p. 85)?

Problem VIII.3.e.ii-3: Check that the first tableau of every frame is ordered lexicographically: reading from left to right across the rows in turn from top to bottom (as we read the words in a book), we get numerals $123\ldots n$ in order.

VIII.3.f How states are related to their labels

Now that we have the three labels for the states, a frame and two tableaux of that frame (these are the labels not the states) we have to construct the states — the multinomials labeled by the tableaux [Tung (1985), p. 64; Weyl (1931), p. 281; Weyl (1946), p. 115]. We need a procedure for generating these and the matrices, a procedure because the number of symmetric groups is infinite so we cannot give all matrices and states. The solution here means, not explicit functions, but an algorithm [Soto and Mirman (1981a,b,c)].

A multinomial realizing a state is a sum of terms with different permutations of indices. The question is how are the coefficients related to the labeling tableaux. One way of constructing states is to take the numerals in the boxes as the indices on the symbols. For numerals in the same row in the tableaux we symmetrize; we construct a multinomial that goes into itself upon interchange of these numerals. Numerals in the same column are antisymmetrized; the multinomial becomes its negative when two are interchanged. Thus to construct the basis state given by a tableau we first symmetrize all indices in each row, then take the resultant sum and antisymmetrize the indices in each column. After a tensor symmetric in a pair of indices is antisymmetrized, the resultant sum is no longer fully symmetric.

For tableau $\begin{array}{|c|c|l}\hline 1 & 2 & \cdots \\ \hline 3 & \multicolumn{2}{|l}{\cdots} \\ \cline{1-1} \multicolumn{1}{c}{\cdots} \end{array}$, we first symmetrize with respect to (12), giving $T_{123...} + T_{213...} + \cdots$; this we antisymmetrize with respect to (13) to get the state

$$|\ ) = T_{123...} + T_{213...} + \cdots - T_{321...} - T_{231...} - \cdots . \qquad \text{(VIII.3.f-1)}$$

This is no longer symmetric in 1,2, as we see for

$$\begin{array}{|c|c|}\hline 1 & 2 \\ \hline 3 \\ \cline{1-1}\end{array} = 123 + 213 - 321 - 231. \qquad \text{(VIII.3.f-2)}$$

The frame with all boxes in the same row labels the completely symmetric representation, the one with all in the same column the completely antisymmetric one. These are one-dimensional; there is only one way of inserting the numerals in these frames to get a standard tableau. For (2,1) of S_3, there are two ways of inserting numerals 1,2,3, so this labels a two-dimensional representation; there are two of these.

Problem VIII.3.f-1: We can also antisymmetrize according to the positions of indices on each term. What would we get? Would it matter?

Problem VIII.3.f-2: What is the effect of (12) on $\begin{array}{|c|c|}\hline 1 & 2 \\ \hline 3 \\ \cline{1-1}\end{array}$?

VIII.3.g The problems with these states

From these labels we can find many properties of the states. But these are labels, not states, and statements about the latter must be justified, not simply read off. Also these are not the states we use. The problem is that multinomials with these symmetry properties (which correspond simply to those of the tableaux) are not orthogonal. We use ones that are.

There is another problem. States are labeled by pairs of tableaux, one giving the representation of the frame, the other the state (so for the (2,1) frame of S_3 there are four states). However with this prescription, the method of constructing these states is not completely defined. That used for the orthogonal states is. Take

$$|1,1) = \begin{array}{|c|c|}\hline 1 & 2 \\ \hline 3 \\ \cline{1-1}\end{array}, \begin{array}{|c|c|}\hline 1 & 2 \\ \hline 3 \\ \cline{1-1}\end{array} = 123 + 213 - 321 - 231, \qquad \text{(VIII.3.g-1)}$$

$$|1,2) = \begin{array}{|c|c|}\hline 1 & 2 \\ \hline 3 & \multicolumn{1}{c}{} \\ \cline{1-1}\end{array}, \begin{array}{|c|c|}\hline 1 & 3 \\ \hline 2 & \multicolumn{1}{c}{} \\ \cline{1-1}\end{array} = 132 + 312 - 231 - 321, \qquad \text{(VIII.3.g-2)}$$

$$|2,1) = \begin{array}{|c|c|}\hline 1 & 3 \\ \hline 3 & \multicolumn{1}{c}{} \\ \cline{1-1}\end{array}, \begin{array}{|c|c|}\hline 1 & 2 \\ \hline 3 & \multicolumn{1}{c}{} \\ \cline{1-1}\end{array} = 132 + 231 - 312 - 213, \qquad \text{(VIII.3.g-3)}$$

$$|2,2) = \begin{array}{|c|c|}\hline 1 & 3 \\ \hline 2 & \multicolumn{1}{c}{} \\ \cline{1-1}\end{array}, \begin{array}{|c|c|}\hline 1 & 3 \\ \hline 2 & \multicolumn{1}{c}{} \\ \cline{1-1}\end{array} = 123 + 321 - 213 - 312. \qquad \text{(VIII.3.g-4)}$$

State $|1,2)$ is obtained from $|1,1)$ and $|2,2)$ from $|2,1)$ by permuting values of numerals, $|2,1)$ is obtained from $|1,1)$ by permuting the positions of numerals. However there is no clear prescription for this. Thus we have to consider the orthogonal (category of) states.

Problem VIII.3.g-1: Construct all states for S_2, S_3 and S_4 following this prescription. Find their products (sec. V.5.c.ii, p. 166). Are they orthogonal? Normalize them.

VIII.4 CRITERIA FOR CORRECTNESS OF THE LABELS

Why is this labeling scheme correct [Boerner (1963), p. 105; Miller (1972), p. 121; Rutherford (1948), p. 11; Schensted (1976), p. 134; Tung (1985), p. 68; Weyl (1931), p. 362]? The requirements are that different standard frames give inequivalent equivalence classes (the representations in each are mutually equivalent but the classes are inequivalent — representations in different ones are inequivalent), all such classes are given, the standard tableaux label all states of all representations (all states of the regular representation), different right tableaux give different representations and all, and different left tableaux give different states and all.

Labels would not be useful if one gave more than one irreducible representation or the states it gave (according to the rules) were not a complete basis of a representation, or included states of different representations. Of course we can use labels arbitrarily (provided we have the right number) but what is valuable is that states and labels are simply related, and we can read properties of representations and states from their labels; these, like those we want for any group, are related in a natural way to the objects they label (so labels, and their assignments, are not arbitrary). Thus do they have the required properties? In particular, we have given rules relating labels and states, here by explicit construction, for the orthogonal states, as we will see, less directly, and must show that the resultant states are correct.

Problem VIII.4-1: These paragraphs are somewhat redundant (why?), but the emphasis is worthwhile.

VIII.4.a Number of frames and number of equivalence classes

One way to show completeness is to count. The numbers of equivalence classes and of classes of group elements are equal (sec. VII.6.c, p. 202). If all equivalence classes have been labeled, and only once, the number of standard frames must equal the number of classes. Does it?

Problem VIII.4.a-1: Show that the number of classes of group elements equals the number of partitions of n — the number of ways that integers $1, \ldots, n$ can be assigned to subsets (say by putting various sets of parentheses around the numerals in strings), with the ordering of subsets, and the identity and ordering of numerals in each subset, being irrelevant. Verify that the number of standard frames equals the number of ways that integers $1, \ldots, n$ can be divided into subsets. Thus the number of standard frames is sufficient, and no more, to label all equivalence classes.

Problem VIII.4.a-2: This explains why all, and only, standard frames label all equivalence classes. Each class is given by a partition into cycles of the n S_n numerals. Cycle lengths can be taken as row lengths of a frame. Why? They can be ordered with the largest at the left and nonincreasing in size to the right; this gives all classes. This is the same as ordering row lengths with (by convention) the longest at the top and nonincreasing down; all standard frames give all classes. Also, permutations mix numerals but do not affect cycle lengths (which as seen, leads to equivalence classes being labeled by the sets of cycle lengths), or their ordering, so do not mix frames; different frames give different classes, and they give all. Verify these.

VIII.4.b Frames give different classes

The numbers of frames are correct. Let us see how the action of the permutations gives these results. Why does a permutation applied to a state of one frame not give a sum which includes states of others — thus different frames actually do give different equivalence classes?

Problem VIII.4.b-1: To illustrate, take the completely symmetric representation and one having a state symmetric in 1,2 and antisymmetric in 1,3. Any permutation applied to the symmetric state gives the same state. For other representations, there are permutations (here (13)) on a state that gives its negative. So the completely symmetric state cannot be written as a sum over other states, and vice versa. Thus, for $|1,1)$, $|1,2)$, (eqs. VIII.3.g-1, VIII.3.g-2, p. 225) let

$$\begin{array}{|c|c|c|}\hline 1&2&3\\\hline\end{array} = a\begin{array}{|c|c|}\hline 1&2\\\hline 3\\\cline{1-1}\end{array} + b\begin{array}{|c|c|}\hline 1&3\\\hline 2\\\cline{1-1}\end{array}. \qquad \text{(VIII.4.b-1)}$$

Apply (12) and (13) and show $a = b = 0$.

Problem VIII.4.b-2: For S_4 similarly show that if

$$(34)\begin{array}{|c|c|}\hline 1&2\\\hline 3\\\cline{1-1} 4\\\cline{1-1}\end{array} = (34)\{a\begin{array}{|c|c|}\hline 1&2\\\hline 3&4\\\hline\end{array} + b\begin{array}{|c|c|}\hline 1&3\\\hline 2&4\\\hline\end{array}\}$$

$$= -\begin{array}{|c|c|}\hline 1&2\\\hline 3\\\cline{1-1} 4\\\cline{1-1}\end{array} = ?, \qquad \text{(VIII.4.b-2)}$$

then again $a = b = 0$. Show that states of frame (2,2) cannot be written in terms of those of other S_4 equivalence classes.

Problem VIII.4.b-3: Two frames have some leftmost column in whose length (number of boxes) they differ. This requires that the number of antisymmetric indices differ. Consider frames

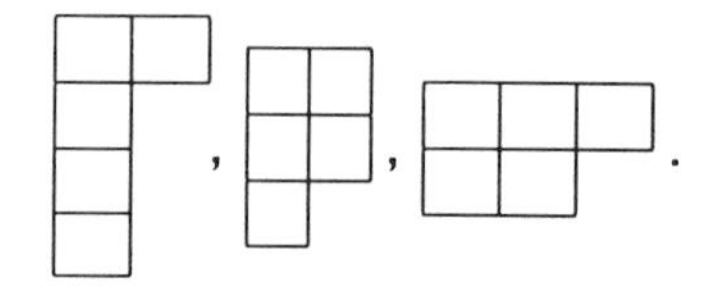

The first has four antisymmetric indices; there are six transpositions (only transpositions need be considered — why?) that take it to its negative. The others have three and one such indices for the leftmost column. Show that if these frames had columns to the left of those drawn, all the same length, or rows above those in the diagram, again of the same length, the number of antisymmetric indices would be greatest for the left frame, least for the right. Prove, that since the column lengths are nonincreasing to the right (why?), for (any) two frames the number of antisymmetric indices of the frame with the shorter leftmost column is less than that of the other, whatever the relative lengths of the columns to the right of it.

Problem VIII.4.b-4: The result of these arguments is that states of different frames are linearly independent. Suppose that a state of one frame were written as a sum that includes states from other frames. Show that it is always possible to find transpositions under which the state goes into its negative, but not all states in the sum do (or for which these go into their negatives but not the state being expanded), so all coefficients of these states are 0. Different frames label different equivalence classes.

VIII.4.c Tableaux correctly label representations and states

The arguments for independence of states of representations labeled by different (right) tableaux of the same frame, and of states of the same representation labeled by different (left) tableaux, are similar. For a state written as a sum over other states, there are enough transpositions (usually more than one are needed) to require that all coefficients in the sum be 0 (using the different symmetries of the states following from their being labeled by different tableaux).

Problem VIII.4.c-1: Prove this.

Problem VIII.4.c-2: However there is a difference between frames and tableaux. If we apply a permutation to a state, the resultant sum does not include states of different frames, but it does include different states. Why? Why does the argument that the coefficients are 0 is, or appear to be, different for frames than for tableaux? Moreover both representations and states are labeled by tableaux, yet a permutation mixes states labeled by different tableaux but not representations that are also labeled by different tableaux. The properties of the two labeling tableaux do not differ. How do representations and states differ in this regard? To answer these questions we must look more closely at the effects of permutations on states and representations. What have we left out?

VIII.4.c.i *Independence of tableaux*

That tableaux give distinct representations and states is clear from their symmetry properties. A proof is given elsewhere [Schindler and Mirman (1977a)]. However it is useful to be, at least visually, more explicit. The argument that no standard multinomial can be written as a sum over others is the same as for frames. For one neighboring transposition (on the left) at least (failure for one is sufficient), one multinomial goes into itself, the other not. This gives all coefficients in the sum 0. Thus to get all states of a representation all standard tableaux are needed.

The argument for representations, when permutations are applied from the right, is the same (sec. VIII.3.c, p. 222).

So states of different frames, states of different representations and states of the same representation are linearly independent. Permutations do not mix frames, acting from the left they do not mix representations, nor states when acting on the right. Thus each frame labels an equivalence class, and each (right) tableau a representation.

It remains to show that all representations of each class, and all states of each representation, are given by the standard tableaux (so the states labeled by all these tableaux are complete — they are independent linear combinations of all regular-representation states; multinomials, not so labeled, are not needed, and not independent).

Problem VIII.4.c.i-1: The reason permutations mix states (or representations) but not frames is that they cannot change the symmetry properties of a tableau (if a tableau contains k antisymmetric indices this is unaffected by the permutation), but they do change numerals in the boxes, so permuting the indices, mixing tableaux. Explain these. Would these statements have been affected if we considered symmetric indices, instead of antisymmetric?

VIII.4.c.ii *Why the standard tableaux label the states*

Why do all, and only, standard tableaux label states? Each S_n state is also a state of a frame of S_{n-1}, one found by removing the box containing n. However this gives a frame only if n is at the right of its row and the bottom of its column — no holes are allowed. The argument continues with $n-1$, and so on, giving that each numeral must be at the end of its row and column, once larger ones have been removed. This is true for standard tableaux — only. And all standard tableaux must be used: if one labeled a state expressible as a linear combination of other standard states then the states of the corresponding S_{n-1} frame could be written as sums of those of other S_{n-1} frames, which cannot be.

Problem VIII.4.c.ii-1: Why is the S_{n-1} frame found by removing the box containing n? Must this be so, or is it a matter of definition?

Problem VIII.4.c.ii-2: There are other proofs, given in the references, that only standard tableaux are needed, and possible. Do these directly, or indirectly, use the requirements about the decomposition into representations of smaller groups? How? Can it be avoided? How do they compare with the discussion here? Which is the most rigorous?

VIII.4.c.iii *What about nonstandard multinomials?*

To see why the set of states labeled by standard frames and tableaux (standard states) is complete, we consider a mythical state not so labeled and show that it can be written as a sum over standard states (so these are a complete basis set). Such a mythical state would be labeled by a nonstandard frame or tableau. Can there be such? The tensor realizing the state is found by symmetrizing the natural monomial, but using the operations given by a nonstandard, instead of a standard, tableau. Such a tensor can be written, but it can be expanded as a sum of standard states, so is not an independent basis state.

If a frame is not standard — there is (at least) one row longer than that above it — we can put into it not the numerals $1\dots n$, but ab..., or any n symbols whatever, provided that they are ordered. Thus in the nonstandard frame we simply reorder the entries so that the objects in the lower row come before those in the upper. So for row 1 we take $2 < 3 < 1$ instead of $1 < 2 < 3$. With this new order the tableau, so the frame, is standard. This reordering is a permutation. A state is given by a symmetrizing operator acting on a monomial, the natural monomial. But a nonstandard state is given by a standard symmetrizing operator acting on some other monomial, thus on a permutation acting on the natural monomial. The permutation can be taken to act on the symmetrizing operator, giving a sum of such (standard) operators, acting on the natural monomial, so a sum of standard states.

However we cannot perform this reordering to change one standard frame into another. So $\begin{array}{|c|c|}\hline \ & \ \\ \hline\end{array}$ requires the two numerals to be symmetric, while they are required to be antisymmetric for $\begin{array}{|c|}\hline \ \\ \hline \ \\ \hline\end{array}$. There is no way to relabel indices to change a symmetric sum into an antisymmetric one.

Problem VIII.4.c.iii–1: Check that $\begin{array}{|c|cc}\hline 1 & \multicolumn{2}{c}{} \\ \hline 2 & \multicolumn{1}{c|}{3} \\ \hline\end{array}$ and $\begin{array}{|c|c|}\hline 1 & 2 \\ \hline 3 & \multicolumn{1}{c}{} \\ \cline{1-1}\end{array}$ are equivalent by writing the multinomials (which must be defined) and reordering.

Problem VIII.4.c.iii–2: Show that the three standard frames of S_3 cannot be changed into each other by relabeling. Show that no standard frame of S_n can be changed into another standard frame by relabeling. Nonstandard frames can be converted to standard ones by reordering, but a standard one cannot be changed to another standard one.

Problem VIII.4.c.iii–3: Construct all nonstandard frames of S_2, S_3 and S_4 and show they can be converted into standard frames by relabeling. Show that every nonstandard frame of S_n can be changed into a standard one by relabeling (reordering the indices on the monomial).

Problem VIII.4.c.iii–4: Show that for S_4, the completely symmetric, and completely antisymmetric, frames plus the three with tableaux

$$\begin{array}{|c|c|}\hline 1 & 2 \\ \hline 3 & \multicolumn{1}{c}{} \\ \cline{1-1} 4 & \multicolumn{1}{c}{} \\ \cline{1-1}\end{array}, \quad \begin{array}{|c|c|}\hline 1 & 3 \\ \hline 2 & \multicolumn{1}{c}{} \\ \cline{1-1} 4 & \multicolumn{1}{c}{} \\ \cline{1-1}\end{array}, \quad \begin{array}{|c|c|}\hline 1 & 4 \\ \hline 2 & \multicolumn{1}{c}{} \\ \cline{1-1} 3 & \multicolumn{1}{c}{} \\ \cline{1-1}\end{array}; \quad \begin{array}{|c|c|}\hline 1 & 2 \\ \hline 3 & 4 \\ \hline\end{array}, \quad \begin{array}{|c|c|}\hline 1 & 3 \\ \hline 2 & 4 \\ \hline\end{array};$$

$$\begin{array}{|c|c|c|}\hline 1 & 2 & 3 \\ \hline 4 & \multicolumn{2}{c}{} \\ \cline{1-1}\end{array}, \quad \begin{array}{|c|c|c|}\hline 1 & 2 & 4 \\ \hline 3 & \multicolumn{2}{c}{} \\ \cline{1-1}\end{array}, \quad \begin{array}{|c|c|c|}\hline 1 & 3 & 4 \\ \hline 2 & \multicolumn{2}{c}{} \\ \cline{1-1}\end{array}, \qquad \text{(VIII.4.c.iii–1)}$$

provide a complete set, and any nonstandard frame and tableaux can be written in terms of them.

Problem VIII.4.c.iii-5: Does the set of all arrangements of n boxes, the set of standard plus nonstandard frames, give all multinomials? Each multinomial, standard or nonstandard, contains all n! permutations of the indices. Can we take combinations of these to give the individual monomials (so any multinomial)? Prove that if we consider every sum of permutations labeled by a frame there are more sums than permutations. Prove that there are linear combinations of the standard multinomials that equal any monomial.

VIII.4.c.iv *What does relabeling do?*

What is the effect of relabeling? Nonstandard states are obtained using standard operators except that the numerals are changed (but the coefficients in the sums are not); permutations acting on the monomial give the same state with just its indices relabeled — the state is constructed from a permuted monomial, which is given by a permutation acting on the standard monomial (with indices in natural order $123\ldots$). On this acts the sum of permutations to give the (nonstandard) state. But a permutation acting on a sum of permutations just gives a sum of permutations, each labeled by a standard tableau because a permutation acting on a standard state gives a sum of standard states. These now act on the standard monomial, so each sum gives a standard state. Thus the relabeling produces from the nonstandard state a sum of standard states.

These frames provide all equivalence classes and since the nonstandard ones can be converted into standard ones by relabeling, the (states labeled by) standard frames and tableaux provide a basis in terms of which any multinomial can be written.

Problem VIII.4.c.iv-1: Why do frames give irreducible representations (for each representation-labeling tableau of a frame, all the states are mixed by the permutations)? If a representation were reducible there would be a subset of states mixed among themselves, but not with others. But given any pair of (state-labeling) tableaux of a frame it is clear (why?) that there is a permutation rearranging the numerals of one to give the other. There can be no invariant subsets. Check this.

VIII.5 STATES OF THE ORTHOGONAL CATEGORY

Although the states constructed in sec. VIII.3.g, p. 225, from the frames and tableaux, are visually attractive, having properties easily seen from their labels, they are not orthogonal — and this is more important. Nor are they unambiguously defined. Here we describe states that are, though they are somewhat removed from the pictorial properties of their labels; these have the same labels but different coefficients in the multinomials. After that we give the rules for finding the matrix elements, thus completely describing the representations of the orthogonal category.

The states are sums of all monomials generated by all permutations acting on a fixed one, the standard monomial. Equivalently they can be sums of permutations which act on the standard monomial; these sums of permutations we call basis state operators. The monomial they act on is not of interest here and is usually suppressed (though it might be interesting if it is a quantum-mechanical statefunction, and also, say, a function of coordinates). These sums, which we require orthonormal, are described by giving a rule for the coefficients of the permutations.

VIII.5.a Basis state operators

The states are constructed by having an operator — a basis state operator, a sum of permutations, ξ^{α}_{ij}, act on the natural monomial. It is labeled by α, the ordinal of the frame, and i and j, the ordinals of (the tableaux labeling) the state and representation respectively, and is

$$\xi^{\alpha}_{ij} = \frac{1}{\theta^{\alpha}} \sum_{\sigma} \mu^{\alpha}_{ij}(\sigma)\sigma, \qquad \text{(VIII.5.a-1)}$$

where the sum is over all permutations σ of S_n, and $\mu^{\alpha}_{ij}(\sigma)$ is the ij matrix element of σ. The normalization factor is

$$\theta^{\alpha} = \frac{n!}{f^{\alpha}}, \qquad \text{(VIII.5.a-2)}$$

with f^{α} the dimension of any representation of frame α. This is the required sum over permutations σ, with the coefficients being the matrix elements. The representation category is determined by the choice of matrices. It is necessary to show that these are the sums with the properties wanted (and the only ones).

Problem VIII.5.a-1: Prove that the ξ's are idempotent [Schensted (1976), p. 72, 134, 164; Tung (1985), p. 78; Aivazis (1991), p. 38]

$$\xi^{\alpha}_{ij}\xi^{\alpha'}_{i'j'} = \delta^{\alpha\alpha'}\delta_{ji'}\xi^{\alpha}_{ij'}. \qquad \text{(VIII.5.a-3)}$$

(Idempotent apparently comes from Latin with the meaning of same or, of course, identity, and power; the — second — power of the operator is the same operator, $T^2 = T$. Here the definition is stretched slightly, as we can see from the subscripts.) If an operator is idempotent, it is also called a projection (and conversely) [Boerner (1963), p. 7, 58]; why?

VIII.5.a.i *Basis state operators can be taken as states*

States are given by basis state operators acting on the (single) monomial with indices in natural order (123 ...), so are sums of monomials with indices in permuted orders. If the monomials are suppressed, the basis state operators, the sums of permutations, are taken as the states. These are two realizations of the same states.

Problem VIII.5.a.i–1: Using the rule (sec. V.5.c.ii, p. 166) for the product of two monomials, and the orthogonality relations for the matrices (sec. VII.5.a, p. 193), prove that these states are orthonormal (and therefore linearly independent). How this is related to the ξ's being idempotent?

Problem VIII.5.a.i–2: Prove that these states can be written as linear combinations of those of sec. VIII.3.g, p. 225, and vice-versa, that this transformation is nonsingular, and so there is a one-to-one correspondence between the two sets of states.

Problem VIII.5.a.i–3: The ξ's are sums of permutations σ, and each of these can be written as a product of (neighboring) transpositions, τ_n. Check that

$$\mu_{ij}^{\alpha}(\sigma) = \prod \sum_{lk} \mu_{lk}^{\alpha}(\tau_n), \qquad \text{(VIII.5.a.i–1)}$$

relating their matrix elements. From this show that

$$\xi_{ij}^{\alpha} = \frac{1}{\theta^{\alpha}} \sum_{\sigma} \prod \sum_{lk} \mu_{lk}^{\alpha}(\tau_n)\tau_n, \qquad \text{(VIII.5.a.i–2)}$$

with the product over the (neighboring) transpositions (some may appear several times) giving each σ. It then follows that

$$\xi_{ij}^{\alpha} = \frac{1}{\theta^{\alpha}} \prod \{\varepsilon + \nu_{ij}^{\alpha}(\tau_n)\tau_n\}, \qquad \text{(VIII.5.a.i–3)}$$

where ε is the identity (should it have indices?), and the product is over the (neighboring) transpositions. Evaluate the coefficients ν. Each transposition appears in each pair of braces at most once. If it were more, each pair would give a similarity transformation on the permutation it enclosed, giving another permutation, which is then decomposed into a product, not containing multiples of this transposition. In sec. VIII.3.f, p. 224, we found states by a series of symmetrizations and antisymmetrizations, which can be written as a product of sums of the identity plus (neighboring) transpositions. This is the analog, differing in that there the coefficients are ± 1, here they are (related to) matrix elements.

VIII.5.a.ii *States and standard tableaux*

What do these states have to do with the standard tableaux? Clearly the maximum number of linearly independent states of each representation equals the number of its standard tableaux. Thus we can use these as labels, perhaps in arbitrary ways. However these states do have symmetry, as shown in sec. VIII.7, p. 245, and it is related to that of the tableaux, so these labels are related in a natural way to their representations and states. (However the visual properties of the tableaux are so useful and impressive that we must remember these are labels of states and not states themselves.)

The orthogonal-category states labeled by standard tableaux provide a complete basis for each representation, and are linearly independent, so they form a complete orthonormal basis, as required. These states are thus acceptable, but we must still explain why we have chosen this form for them, and show that our choice has the required properties.

Problem VIII.5.a.ii-1: What are the required properties?

Problem VIII.5.a.ii-2: That these states form a basis for each representation is presumably obvious from the properties of the matrices. And of course they form bases for all representations, so for the regular representation.

Problem VIII.5.a.ii-3: It should be clear (why?) that the basis state operators are linearly independent because the permutations are. However why are the permutations linearly independent? Does this follow from the group axioms, is it an extra postulate (or is the word "assumption"?), or is it necessary for the definition of symmetric groups? Could there be realizations of these groups for which the permutations are not independent? If so, would these basis state operators still form the states of representations of S_n? Might they form states of another symmetric group? How would they be related to the basis state operators of such a group?

VIII.5.a.iii *Why are these the basis state operators?*

The basis state operators are sums of the permutations. They contain all permutations, but are the coefficients correct? The requirements are that these operators be transformed into themselves by the permutations, and that they be orthonormal. We might suspect that the latter follows from the matrices being orthogonal. However this expansion holds for the semi-normal and natural representation categories also, and their matrices are not orthogonal. And we would like the states to have a reasonable relationship to their labels.

To find the coefficients let

$$\xi^{\alpha}_{ij} = \frac{1}{\theta^{\alpha}} \sum_{\tau} \rho^{\alpha}_{ij}(\tau)\tau; \qquad \text{(VIII.5.a.iii-1)}$$

the ρ's are coefficients to be determined. For the identity we take

$$\rho^{\alpha}_{ij}(\varepsilon) \sim \delta_{ij}. \qquad \text{(VIII.5.a.iii-2)}$$

Now

$$\sigma\xi^{\alpha}_{ij} = \frac{1}{\theta^{\alpha}} \sum_{\tau} \rho^{\alpha}_{ij}(\tau)\sigma\tau = \sum_{k} \mu^{\alpha}_{ki}(\sigma)\xi^{\alpha}_{kj} = \sum_{k} \mu^{\alpha}_{ki}(\sigma)\frac{1}{\theta^{\alpha}} \sum_{\zeta} \rho^{\alpha}_{kj}(\zeta)\zeta, \qquad \text{(VIII.5.a.iii-3)}$$

and similarly for action from the right. Since the permutations are independent we equate coefficients and get

$$\sum_{k} \mu^{\alpha}_{ki}(\sigma)\rho^{\alpha}_{kj}(\zeta) = \rho^{\alpha}_{ij}(\sigma^{-1}\zeta). \qquad \text{(VIII.5.a.iii-4)}$$

For orthogonal μ's this equation gives $\rho^{\alpha}_{ij} = \mu^{\alpha}_{ij}$, as we can see by taking $\zeta = \varepsilon$. Thus the coefficients are the matrix elements.

The states are given by the action of operators on a fixed monomial, the operators being sums of the permutations. The only question we had to consider is their coefficients. It is comforting that these are (about) the most intuitive numbers, the matrix elements.

Notice that (essentially) all these results are not for the symmetric groups, but hold for all finite groups. Perhaps this is fortunate as all finite groups are subgroups of symmetric groups.

Problem VIII.5.a.iii-1: The normalization is chosen so that the states are orthonormal. For the product of states we require

$$(T_{\sigma(12\ldots)}, T_{\sigma'(12\ldots)}) = \delta_{\sigma\sigma'}, \qquad \text{(VIII.5.a.iii-5)}$$

that is the product of two monomials is 0 if they have different permutations of their indices, 1 for the same permutation. So

$$(\xi^{\alpha}_{ij}T_{12\ldots}, \xi^{\alpha'}_{i'j'}T_{12\ldots}) = (\frac{1}{\theta^{\alpha}} \sum_{\tau} \mu^{\alpha}_{ij}(\tau)\tau T_{12\ldots}, \frac{1}{\theta^{\alpha'}} \sum_{\tau} \mu^{\alpha'}_{i'j'}(\tau)\tau T_{12\ldots})$$

$$= \delta^{\alpha\alpha'}\delta_{ii'}\delta_{jj'} = \frac{1}{\theta^{\alpha}}\frac{1}{\theta^{\alpha'}} \sum_{\tau} \mu^{\alpha}_{ij}(\tau^{-1})\mu^{\alpha'}_{i'j'}(\tau). \qquad \text{(VIII.5.a.iii-6)}$$

Why? Verify that this does satisfy and that the normalization is correct. Do the matrices have to be orthogonal? Are there conditions on them? Does application of the permutation on the right give further information?

Problem VIII.5.a.iii-2: Show that eq. VIII.5.a-3, p. 233, the idempotency condition, follows from the states being orthonormal. However it also follows from the definition of the ξ 's, before a product was introduced. Does this mean that we are limited in how we can define the product of two monomials? In what way? Why?

Problem VIII.5.a.iii-3: Why are the ξ's idempotent? It is useful to check that if α is the identity representation the normalization is chosen correctly. From the orthogonality conditions, for any representation, except the completely symmetric one, the representation matrices $M(t_i)$ obey

$$\sum M(t_i) = 0, \qquad \text{(VIII.5.a.iii-7)}$$

where the sum is over all group elements (pb. VII.5.c-2, p. 196). Now

$$\xi^{\alpha}_{ij}\xi^{\alpha'}_{i'j'} = \frac{1}{\theta^{\alpha}}\frac{1}{\theta^{\alpha'}}\sum_{\tau}\sum_{\tau'}\mu^{\alpha}_{ij}(\tau)\tau\mu^{\alpha'}_{i'j'}(\tau')\tau'$$

$$= \frac{1}{\theta^{\alpha}}\frac{1}{\theta^{\alpha'}}\{\sum_{\tau}\mu^{\alpha}_{ij}(\tau^{-1})\tau^{-1}\mu^{\alpha'}_{i'j'}(\tau)\tau+\cdots\} = \delta^{\alpha\alpha'}\delta_{ji'}\xi^{\alpha}_{ij'}. \qquad \text{(VIII.5.a.iii-8)}$$

Check that the other terms contain all permutations, so sum to 0. The result should then follow from the orthogonality conditions on the representation matrices.

VIII.5.b Expansion of a monomial in terms of states

The full set of states of all irreducible representations of S_n is complete (as with any finite group, among others) — any monomial with n indices can be written as a sum over them (over the same type of object, say the same function of coordinates, but with different permutations of indices). And any sum of monomials can be expanded in terms of these. However such a sum may not have all states in its expansion; if it were symmetric in two indices, say, it would not have antisymmetric states. Does a single monomial have all states? And what are the coefficients in the expansion?

Problem VIII.5.b-1: Let the monomial be M, the natural monomial N, so

$$M = \sigma N, \qquad \text{(VIII.5.b-1)}$$

where σ is a permutation. Then

$$M = \sigma N = \sum C^{\alpha}_{ij}\xi^{\alpha}_{ij}N, \qquad \text{(VIII.5.b-2)}$$

with the sum over all indices, and the C's the unknowns. The states are orthonormal, and the ξ's idempotent, so show that [Schindler and Mirman (1977a), eq. II-11], where the sum is over?,

$$\sigma = \sum \mu_{ij}^{\alpha}(\sigma)\xi_{ij}^{\alpha}. \qquad \text{(VIII.5.b-3)}$$

This gives the coefficients, the matrix elements of σ. The states that appear are those with a nonzero matrix element for σ, for one state of one representation, at least. Not every state of every representation appears for every σ. For the identity only diagonal elements occur. However every state appears for at least one σ. Why?

VIII.6 THE ORTHOGONAL-REPRESENTATION MATRICES

Now after proving the expansion for the states, all we need are matrix elements. Some symmetric-group representation matrices can easily be found [Elliott and Dawber (1987), p. 445; Hamermesh (1962), p. 214; Schensted (1976), p. 145]; others, unfortunately, cannot. For neighboring transpositions, those interchanging two adjacent numerals, (12), (23), (107 108), say, there is a simple rule for reading the matrix elements directly from the tableaux. For other permutations there is no simple rule, nor any complicated one either. (There are, however, some rules on which matrix elements must be 0 [Mirman (1991e), sec. II].) However any permutation can be written as a product of neighboring transpositions; knowing the matrices of these, we can then, with enough computer time, find the representation matrices of any permutation [Soto and Mirman (1981c)].

Here we give the matrix elements of the neighboring transpositions, for the orthogonal category of representations. Those for the seminormal class can be found by multiplying these by tableau functions [Rutherford (1948), p. 47; Schindler and Mirman (1977a), eq. III-14-16; (1977b), table II]; the ones for the natural representation [Boerner (1963), p. 114; Rutherford (1948), p. 53] are not considered.

Problem VIII.6-1: What would be nice is a formula for the matrix elements given the permutation, frame and indices of an element. We leave it to the reader to find such a formula.

VIII.6.a Matrices of the neighboring transpositions

The matrix for neighboring transposition $(i\ i+1)$ has one or two nonzero elements in each row, and in each column. For any tableau there is another (which may be the same) identical except that i and

$i + 1$ are interchanged. The two tableaux are paired by $(i\ i + 1)$, unless these are in the same row or column. Thus $\begin{array}{|c|c|}\hline 1 & 2 \\ \hline 3 \\ \cline{1-1}\end{array}$ and $\begin{array}{|c|c|}\hline 1 & 3 \\ \hline 2 \\ \cline{1-1}\end{array}$ are interchanged by transposition (23), and left unchanged by (12). The representation-matrix nonzero elements are those four labeled by the two tableaux of the pair (at the intersection of the two rows and two columns labeled by these two tableaux), or the (diagonal) one labeled by the single tableau invariant (up to a sign) under the transposition.

Problem VIII.6.a-1: For the tableaux of S_4, find those taken into themselves by (12), (23) and (34), and the ones paired by these transpositions. Notice that the pairs are different for the different transpositions.

Problem VIII.6.a-2: Can two tableaux be mixed by (12)?

VIII.6.a.i *The axial distance*

Next we need the axial distance, η, between i and $i + 1$; it is found using the first tableaux of the pair (sec. VIII.3.e.ii, p. 224). Then η equals the number of boxes, including that containing i but not $i + 1$, between these two numerals in the tableau, moving from i to the left toward the column in which $i + 1$ is, and then down. The axial distance between 2 and 3 in $\begin{array}{|c|c|}\hline 1 & 2 \\ \hline 3 \\ \cline{1-1}\end{array}$ is 2. For $\begin{array}{|c|c|}\hline 1 & 3 \\ \hline 2 \\ \cline{1-1} 4 \\ \cline{1-1}\end{array}$, it is 1 between 1 and 2, 2 between 2 and 3 and 3 between 3 and 4.

Problem VIII.6.a.i-1: Some readers may feel more comfortable with a formula so we take i in row r_i and column c_i, and $i + 1$ with corresponding coordinates r_{i+1} and c_{i+1}; then

$$\eta = (r_{i+1} - c_{i+1}) - (r_i - c_i). \qquad \text{(VIII.6.a.i-1)}$$

Show that the two definitions are identical.

Problem VIII.6.a.i-2: For S_4, find the axial distances for all pairs in all tableaux.

Problem VIII.6.a.i-3: Why does moving to the left and down go from i to $i + 1$, and not the reverse?

VIII.6.a.ii *The nonzero elements for the transposition*

For transposition $(i\ i + 1)$ the nonzero representation-matrix elements in the rows and columns labeled by tableaux i and $i + 1$ are given by the following rule. If i and $i + 1$ are in the same row, the matrix element labeled by this tableau is 1; if they lie in the same column it is -1. (in this sense, for this transposition, numerals in rows are symmetric, those in columns are antisymmetric.)

If they are not in the same row or column the two nonzero elements in the row labeled by tableau i is $-\frac{1}{\eta}$ and $\sqrt{1-\frac{1}{\eta^2}}$, and in the row labeled by $i+1$, they are $\sqrt{1-\frac{1}{\eta^2}}$ and $\frac{1}{\eta}$, so the 2×2 submatrix (whose rows and columns are usually not adjacent) is

$$M = \begin{pmatrix} -\frac{1}{\eta} & \sqrt{1-\frac{1}{\eta^2}} \\ \sqrt{1-\frac{1}{\eta^2}} & \frac{1}{\eta} \end{pmatrix}. \qquad \text{(VIII.6.a.ii-1)}$$

Notice that it is symmetric ($M_{ij} = M_{ji}$). For the above S_3 tableaux,

$$M_{(2,1)}(23) = \begin{pmatrix} -\frac{1}{2} & \sqrt{\frac{3}{4}} \\ \sqrt{\frac{3}{4}} & \frac{1}{2} \end{pmatrix}, \qquad \text{(VIII.6.a.ii-2)}$$

with the frame and transposition indicated.

The matrices of the other permutations are found from those of the neighboring transpositions by writing the permutation as a product of these transpositions and using matrix multiplication.

Problem VIII.6.a.ii-1: For i, $i+1$ in the same row or column, why is there only a diagonal matrix element?

Problem VIII.6.a.ii-2: Find the representation matrices for all neighboring transpositions, then for all transpositions, for every representation of S_2, S_3 and S_4. Note that some rows and columns have only a single nonzero element, ± 1, others have two nonzero elements. Find the matrices for all permutations for all representations of S_2, S_3.

Problem VIII.6.a.ii-3: Is the product of two symmetric matrices symmetric? Are the representation matrices for all permutations symmetric? Show that the representation matrices of the neighboring transpositions are orthogonal. Are products of orthogonal matrices orthogonal? Are the matrices of all permutations orthogonal because those of the transpositions are? Would there be anything wrong if they were not?

VIII.6.a.iii *The matrices for the semi-normal representations*

For the semi-normal representations the matrix elements are the same except for the 2×2 submatrix. It is

$$m = \begin{pmatrix} -\frac{1}{\eta} & 1-\frac{1}{\eta^2} \\ 1 & \frac{1}{\eta} \end{pmatrix}, \qquad \text{(VIII.6.a.iii-1)}$$

and is not symmetric.

Problem VIII.6.a.iii-1: Is this matrix orthogonal?

Problem VIII.6.a.iii-2: The states are required to be orthonormal. Is this necessary for the orthogonality of the representation matrices?

VIII.6.b Derivation of the transposition matrix elements

These are the neighboring-transposition matrices. How do we know that they are correct [Elliott and Dawber (1987), p. 445; Rutherford (1948), p. 38]? What do we mean by correct? The matrices must obey the same product rules as the permutations — and these we know. But this is for the representation matrices of all permutations, and we have only those for a small subset. To show correctness we need some results which the reader will have no difficulty proving.

Problem VIII.6.b.-1: These matrix elements are correct for S_2. Why? How do we get a proof for S_3? One way is straightforward. Show that the given matrix elements are correct for S_3. That is, for each representation, calculate from these the representation matrices of all permutations, and show that the resultant sets are correct. This implies that such a direct proof is not generally useful. Might induction help?

Problem VIII.6.b.-2: There are properties of transpositions that must hold and these we check next. First, the square of these neighboring-transposition matrices equals the unit matrix E. Also $M(13)M(13)$ equals E. Prove that the square of the matrix of any transposition, using its expansion in terms of the neighboring transpositions and their matrices, equals the unit matrix. Do this for both orthogonal and seminormal representation categories.

Problem VIII.6.b.-3: Check that

$$M(123)M(213) = M(213)M(123) = E. \qquad \text{(VIII.6.b-1)}$$

Find the matrices for (12)(123) and (23)(123) and verify that these are the correct ones for the permutations given by these products. Also check that this is independent of the value chosen for the axial distance. What is the underlying reason that these results are independent of S_n?

VIII.6.b.i *Finding the matrices for S_n by induction*

Now that we are satisfied that these are correct for S_2 and S_3 we consider them for arbitrary n, using induction, assuming them for $n-1$.

Problem VIII.6.b.i-1: Show that $n-1$ and n, unless adjacent, are at the right end of the rows they lie in and the bottom of their columns, for any standard tableau. Removing the box holding n from a standard tableau gives a standard tableau of S_{n-1}. Why? Check that the matrix for a permutation not involving n (in which n is in a single cycle) is block diagonal. What are these blocks? And for such permutations, matrix elements labeled by tableaux in which the positions of n are different, are 0. Thus there is only one permutation we need consider, transposition $(n-1\ n)$. Why?

Problem VIII.6.b.i-2: A permutation not involving $n-1$ or n commutes with $(n-1\ n)$. Acting on a state, $(n-1\ n)$ cannot change the positions in the tableau of $i<n-1$. So if $n, n-1$ are in the same positions in two tableaux $(n-1\ n)$ acts as the identity. Therefore the off-diagonal matrix elements are labeled by these two are 0. For $(n-1\ n)$, the only rows and columns containing nonzero off-diagonal elements are thus those labeled by tableaux in which the numerals $1\ldots n-2$ are in the same positions, in which only the positions of $n-1$ and n differ. The only way they can differ is if n and $n-1$ are interchanged. So for any row r of the matrix of this transposition, there can be only two columns with nonzero entries, the diagonal labeled by tableau r, and that labeled by the single interchanged tableau, and only one if the interchange gives the same tableau (if $n-1$ and n are in the same row or column). Check this for S_3 and S_4. Explain why this holds also for S_m, where $m>n$.

Problem VIII.6.b.i-3: The question is then the values of these matrix elements. Prove that if we require the matrix to be symmetric, the 2×2 submatrix must be that given in eq. VIII.6.a.ii-1, p. 240, since this is a submatrix of a transposition. Show that if symmetry is not required the semi-normal form also satisfies (what?). Why?

VIII.6.b.ii *Computing the axial distance*

This leaves only the the value of η.

Problem VIII.6.b.ii-1: Show that

$$(n-2\ \ n-1)(n-1\ \ n)(n-2\ \ n-1)$$

$$=(n-1\ \ n)(n-2\ \ n-1)(n-1\ \ n)=(n-2\ \ n). \qquad \text{(VIII.6.b.ii-1)}$$

Let us evaluate a 2×2 submatrix of $(n-1\ \ n)$, say of rows and columns r and s of the matrix. This relationship between transpositions gives an equation for the $(n-1\ \ n)$ representation matrix. To find the required submatrix we need only consider the corresponding r,s submatrix of $(n-2\ \ n-1)$ and $(n-1\ \ n)$ to find the r,s matrix elements of these two products. Why? For $(n-2\ \ n-1)$, the diagonal r,r element is $1/\sigma$ and the s,s element is $1/\tau$, where σ is the axial distance from $n-2$ to $n-1$ in tableau r, and τ that in tableau s. Show that τ is the axial distance from $n-2$ to n in tableau r. Remember the definitions of tableaux r and s. By considering the s,r element of the preceding equation, check that

$$\eta = \tau - \sigma. \qquad \text{(VIII.6.b.ii-2)}$$

Show that this η is the axial distance between $n-1$ and n in tableau r. Carry out this induction from S_2 to S_3, and then from S_3 to S_4. Thus the expressions for the matrix elements of the neighboring transpositions has been demonstrated (by the reader) to be correct.

Problem VIII.6.b.ii-2: Any reader who can find an expression of a similar kind for an arbitrary permutation (or even an arbitrary transposition) — say a formula for the matrix elements given S_n, the permutation and the labels of the matrix elements — will reveal great understanding of the subject (probably greater than that of anyone else who has worked on it).

VIII.6.b.iii *The reasons for the form of the matrices*

Why do the matrices for the neighboring transpositions have this form? It is not surprising that the nonzero matrix elements of $(n-1\ n)$ are labeled by tableaux differing only in the positions of these two numerals, or in which they are in the same row and column. But it is true because this transposition commutes with all other neighboring ones — it is not true for other transpositions — so permutations of numerals less than $n-1$ do not affect it. And this links no more than two tableaux. That the matrix element is ± 1 if they are in the same row and column is implied by the symmetry of the tableau. Otherwise the form of the 2×2 submatrix is determined, up to one value, by the condition of orthogonality (for this representation category). Also the form, for all representation categories, must give that the square of the transposition is the identity. The value that appears, the axial distance, for (23), is the most intuitive; what else might we have guessed? And this is why it is picked. (That it is arbitrary, for S_3, should be expected. For S_n we can pick any S_3 subgroup, so any corresponding set of neighboring transpositions, thus their representation matrices must obey the S_3 multiplication rules, whatever the axial distance.) The result for other transpositions then follows by induction, and once we know that for (23), it (perhaps) seems obvious.

Problem VIII.6.b.iii-1: Standard tableaux are used merely as labels for the states. Yet here we find properties of the permutations using them. Why should they give these properties? The essential thing is that S_n standard tableaux are constructed so that they give those of S_{n-1} by removal of a box. This means that states and matrices of the group and its subgroup are closely related. This leads, as should be clear from these problems, to sets of representation matrices being block-diagonal and matrix elements being 0. Show that these results can be obtained from the converse view. There are representation matrices with the properties stated in the previous two problems, and we can consistently label the states, so the matrix elements, by tableaux that have the properties required for them to be standard. This view is perhaps more fundamental.

VIII.6.c Equivalent representations of S_3

As examples of the basis states we use the two two-dimensional representations of S_3 [Schindler and Mirman (1977a), sec. X]. For the representation labeled by tableaux 1,

$$\xi_{11} = 3\{123 + 213 - \frac{1}{2}(132 + 321 + 231 + 312)\}, \qquad \text{(VIII.6.c-1)}$$

$$\xi_{21} = 3\sqrt{\frac{3}{2}}\{132 - 321 - 231 + 312\}, \qquad \text{(VIII.6.c-2)}$$

and for that labeled by tableaux 2,

$$\xi_{22} = 3\{123 - 213 + \frac{1}{2}(132 + 321 - 231 - 312)\}, \qquad \text{(VIII.6.c-3)}$$

$$\xi_{12} = 3\sqrt{\frac{3}{2}}\{132 - 321 + 231 - 312\}. \qquad \text{(VIII.6.c-4)}$$

Problem VIII.6.c–1: Check that these are orthonormal. Calculate all matrix elements for these two representations. For which states are the matrix elements of (12) equal to 1, and for which -1? Find the similarity transformation that transforms these two representations (of the equivalence class) into each other. Show that these two two-dimensional representations are exactly the same — the representation matrices are identical. The two representations are equivalent but their basis vectors are different. The states are actually two distinct realizations of the basis states of the same representation (the representation is realized on two different two-dimensional spaces).

Problem VIII.6.c–2: The orthogonal and semi-normal representations are equivalent — in a sense somewhat different than used here — so there is a similarity transformation linking them. Find it for the two-dimensional representation of S_3. Can it be found in general? Is it unitary?

Problem VIII.6.c–3: Two representations with the same dimension need not be equivalent, e. g., the two one-dimensional representations, the symmetric and antisymmetric ones. For S_3 find, using the above formulas, the representation matrices for these two. For S_4 there is a nontrivial case of two nonequivalent representations of the same dimension. Draw their frames. Prove them inequivalent. Find the matrices.

VIII.7 SYMMETRY OF STATES AND MATRICES

Young tableaux specify how a state changes when numerals in a column are interchanged (the symmetry in rows is removed by antisymmetrization of the columns). But the states considered here are not those given directly by the tableaux, instead are orthogonal states generated using them. What can we say about their symmetry, and that of the matrices?

VIII.7.a Symmetry properties of the states

Permutations can be applied either from the right or the left. That is, a state, a multinomial, is a sum of terms, each a coefficient times a permutation acting on a fixed monomial. A permutation acting on the state acts on each of these permutations of the monomial; action from the right and from the left give different products acting on the monomial. So the resulting sums differ. Tableaux give information about the symmetry properties of the states; if neighboring objects are in the same row (column) we say the tableau is symmetric (antisymmetric) in these objects. But there are two different actions; what symmetry properties are being considered and how do they differ? Of course we might expect two different kinds of symmetry properties since a state is labeled by two tableaux. For the state labeled by $\begin{array}{|c|c|}\hline 1 & 2 \\ \hline 3 \\ \cline{1-1}\end{array}$; $\begin{array}{|c|c|}\hline 1 & 3 \\ \hline 2 \\ \cline{1-1}\end{array}$, does transposition (12) give the state or its negative, or a sum of states?

For states the symmetry properties are given by [Mirman (1991e), appendix A]:

LEMMA: (Lemma on state symmetry) If the left (state-labeling) tableau has neighboring objects i, j in the same row (column) transposition (ij) acting from the left gives the state (minus the state). Acting from the left means that it acts on the tensor given by a sum of permutations acting on the monomial. The tensor ξ that realizes the state is symmetric (antisymmetric) under the interchange of indices i, j (the indices in positions i, j on the monomial). If the right (representation-labeling) tableau has neighboring objects i, j in the same row (column) then (ij) applied from the right (it acts on the monomial and then ξ acts on the result) gives the state (minus the state). A tensor is symmetric (antisymmetric) under the interchange of its indices in positions i, j on each of its terms simultaneously.

Proof: The statements about the action of the transpositions follow from the definition of their effect. The sum of permutations giving state ξ can be written, symbolically,

$$\xi = (\varepsilon + \kappa(ij))P = P'(\varepsilon + \zeta(ij)); \qquad \text{(VIII.7.a-1)}$$

ε is the identity, P, P' permutations, κ, ζ coefficients of (ij), and this is a sum over the permutations.

Problem VIII.7.a-1: We first note that

$$\xi^{\alpha}_{rs}\sigma = \sum_{t} \mu^{\alpha}_{st}(\sigma)\xi^{\alpha}_{rt}, \qquad \text{(VIII.7.a-2)}$$

which is the analogue of the equation in which σ is applied from the left [Schindler and Mirman (1977a), eq. II-12], and is proven the same way (noting eq. VIII.5.b-3, p. 238). Check eq. VIII.7.a-1 by expanding ξ as a sum of permutations, written as products of transpositions, and move the (ij)'s to the right (or left) giving ξ expanded in powers of (ij). Why do only the zeroth and first power appear? We are interested in the cases

$$(ij)\xi = \pm\xi, \quad \xi(ij) = \pm\xi, \qquad \text{(VIII.7.a-3)}$$

for which

$$\kappa = \pm 1, \quad \xi = \pm 1, \qquad \text{(VIII.7.a-4)}$$

respectively. From this schematic form for the states we see that when

$$\xi = (\varepsilon \pm (ij))P \qquad \text{(VIII.7.a-5)}$$

the terms in the tensor come in pairs, identical except that in the second term the indices in positions i, j are interchanged, so the whole sum is symmetric (antisymmetric) in i, j. Similarly if

$$\xi = P(\varepsilon \pm (ij)), \qquad \text{(VIII.7.a-6)}$$

then acting on a monomial it gives a sum of two terms with indices in positions i, j interchanged, and P acts on this sum, permuting the indices giving a sum of two sums. The permutations in P take the numeral in position i and move it elsewhere putting another numeral in this position, and similarly for the one in the j'th position. In the corresponding term in the other subsum, the numerals placed at i, j are interchanged; in one term s, t are at i, j, in the other t, s are. Thus the tensor giving the state consists of pairs of terms, identical (or with a minus sign) except for the interchange of numerals in positions i, j. So the state is symmetric (antisymmetric) under simultaneous interchange of numerals in these positions in each term, giving the result. •

Problem VIII.7.a-2: There is another way of describing this. A monomial can be written $x_1^k x_2^l \ldots$. Permutations on it from the left are defined to act on subscripts while those from the right permute superscripts (the carrier vectors). Action from the left is as described. Action from the right on tensors followed by shuffling of the vectors to return the superscripts to their original order (to compare simply the original and permuted terms) permutes subscripts, now according to their positions on each term, rather than the monomial as in action from the left. Explain why, and show that, these two interpretations of action from the right give the same result.

Problem VIII.7.a-3: Why, and to what extent, does the result and proof fail if the objects are not neighboring?

VIII.7.b Symmetry properties of representation matrices

The symmetry of the states gives information about the matrix elements they label. If in a tableau neighboring objects are in the same row (column), the representation matrix for the transposition of these objects has a 1 (-1) on the diagonal labeled by the tableau and 0's elsewhere in that row and column.

A state, realized by a symmetric-group basis state operator, a ξ, acting on indices of a monomial to give a sum of monomials, a tensor, is labeled by two tableaux whose symmetries (if any) for a pair of objects can be different.

For states the symmetry properties are given by the lemma on state symmetry, sec. VIII.7.a, p. 245, to which we add one point. If the left (state-labeling) tableau has neighboring objects i, j in the same row (column) the (ij) representation matrix has 1 (-1) on the diagonal labeled by that tableau, and 0's elsewhere in that row and column. If the right (representation-labeling) tableau has neighboring objects i, j in the same row (column) the matrix elements of transposition (ij) are the same as for action from the left. The statements about the action of the transpositions follow from the previous result [Schindler and Mirman (1977a), eq. II-12], and eq. VIII.7.a-2, p. 246, for the left and right tableaux respectively, using the matrix elements (± 1) for indices in the same row and column.

Problem VIII.7.b-1: It should be clear that these matrix elements follow from the symmetry of the labeling state (no "s", why?).

Chapter IX

Properties and Applications of Symmetric Groups

IX.1 OBJECTS FOR THE SYMMETRIC GROUP

Finding basis states and representation matrices is the foundation of the study of group representation theory. Having done this (in principle) for the symmetric groups we now need further properties of these representations, for their own intrinsic interest, for the examples they provide for other groups, and for applications. We start the study of these here, and to show that these groups have physical interest, we mention a few ways of utilizing them.

Among the things we need to know about a representation is its dimension. For the symmetric groups this can be found by drawing all standard tableaux and counting. At best, awkward and for large n (essentially) impossible. Fortunately there is a formula, which we study next. Also we need subgroups of groups. Here we only briefly look at one example, alternating groups.

Representations can be found by taking products of representations. How are these defined, how many are there, what is their relevance (and to what?), what representations do they form, and if these are not irreducible, how can they be written as sums of irreducible ones? These questions arise once we have representations, and are considered for the symmetric groups, but this also introduces the concepts, which appear again in the study of other groups.

The states of the representations include two familiar ones, the completely symmetric and completely antisymmetric, describing, among other things, bosons and fermions. Except for these, states have more complicated symmetry. Do other states have physical interpretation, do they appear in physics? If so, why?

IX.2 REPRESENTATION DIMENSIONS

To find the dimension of a symmetric-group representation we do not have to draw all Young tableaux; for S_n there is a formula, which we state and then prove (this is part of the problem of finding characters, which we do not consider [Boerner (1963), p. 188; Elliott and Dawber (1987), p. 431; Hamermesh (1962), p. 182; Littlewood (1958), p. 61, 137, 265 (including tables); Miller (1972), p. 147; Murnaghan (1963), p. 132, 168 (including the alternating group and tables); Robinson (1961), p. 74; Rutherford (1948), p. 62; Schensted (1976), p. 128, 142, 329; Weyl (1931), p. 319; Weyl (1946), p. 212]).

IX.2.a The formula for the dimension

Consider a frame with r rows of length $m_1, m_2, \ldots, m_r$. Define

$$\begin{aligned} l_1 &= m_1 + r - 1, \\ l_2 &= m_2 + r - 2, \\ &\ldots, \\ l_r &= m_r. \end{aligned} \tag{IX.2.a-1}$$

Pictorially,

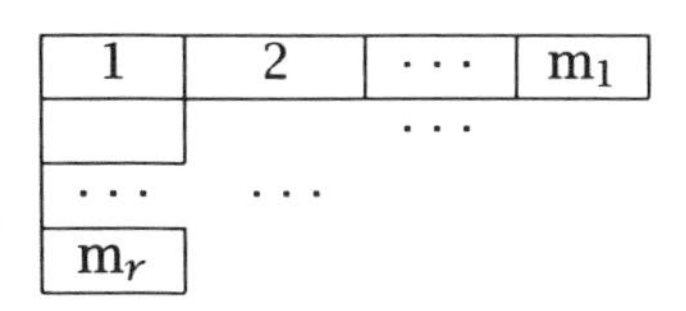

Then for frame α, the dimension of each of its representations (and also their number) is

$$f^\alpha = n!\frac{\prod_{i<k}(l_i - l_k)}{(l_1!l_2!\ldots l_r!)}. \tag{IX.2.a-2}$$

The number of pairs of standard tableaux is the square of this.

Another formula for the dimension is

$$f^\alpha = n!/H^\alpha, \tag{IX.2.a-3}$$

where H is the product of the hooklengths of all boxes in the frame [Patterson and Harter (1976), fig. 1; Robinson (1961), p. 44; Sagan (1991), p.

91]. The hooklength of a box is the sum of the number of boxes below it, to the right of it, and itself. That of box i in row k is

$$H_{ki} = m_k - i + 1 + l, \qquad \text{(IX.2.a-4)}$$

where l is the number of rows below k of length i or greater. Some examples are

$$\boxed{1}\ H = 1; \quad \boxed{1\,|\,2}\ H_1 = 2,\ H_2 = 1;$$

$$\begin{array}{|c|}\hline 1 \\ \hline 2 \\ \hline 3 \\ \hline \end{array}\ H_1 = 3,\ H_2 = 2,\ H_3 = 1,$$

$$\begin{array}{|c|c|}\hline 1 & 2 \\ \hline 3 & 4 \\ \hline \end{array}\ H_1 = 3,\ H_2 = 2,\ H_3 = 2,\ H_4 = 1. \qquad \text{(IX.2.a-5)}$$

Problem IX.2.a-1: What is the sum of the l's?

Problem IX.2.a-2: Check both formulas for all frames of S_1, S_2, S_3, S_4 and S_5. The dimensions should equal the number of standard tableaux, which can be determined by explicit construction. And the two formulas for the dimensions should give the same results.

Problem IX.2.a-3: That the two formulas give the same results is not proven here; prove it. It might help to start with a formula for the hooklength of any box, finding in particular an expression for l in terms of the m's.

Problem IX.2.a-4: It is useful to check that the set of standard frames and tableaux gives all basis states by showing that the total number of mutually orthogonal standard states is $n!$ — the number of states in the regular representation (the number of group elements). So this provides a complete basis. Using the formula(s), show that the sum of the squares of the dimensions equals the order of the group,

$$\sum (f^{\alpha})^2 = n!, \qquad \text{(IX.2.a-6)}$$

where the sum is over all frames α [Boerner (1963), p. 111, 120]. Since f^{α} is the number of standard tableaux, this proves that (all) these (only) label the states; there is a slight hole in this — does it matter?

Problem IX.2.a-5: That the standard tableaux give a complete set of states can be regarded as a combinatorial problem. The number of ways of inserting n numerals into standard frames (remember a state is labeled by two tableaux) should equal n^2. Proving that they are equal shows that the states are complete.

IX.2.b Proof of the dimension formula

The formula, eq. IX.2.a-2, p. 249, has been verified for a few small n, so the proof is by induction. Each S_n tableau is found by adjoining, at the end of a row and column, a box with an n to an S_{n-1} tableau. Thus the number of these tableaux is equal to the number with n at the end of the first row, that is the number of S_{n-1} tableaux obtained from the S_n tableau by removing n from the first row, plus the number with n at the end of the second row, and so on. The exception is that if an S_n tableau has rows equal, a box with n can be added only to the highest of these. Temporarily ignoring this point, and denoting the dimension of the S_{n-1} representation obtained from frame α of S_n by removing a box at the end of row i, by f^{α_i}, the formula gives

$$f^{\alpha_i} = (n-1)!\{(l_1-1-l_2)(l_1-1-l_3)(l_1-1-l_4)\dots\frac{\prod'_{i<k}(l_i-l_k)}{(l_1-1)!l_2!\dots l_r!}$$
$$+ (l_2-1-l_3)(l_2-1-l_4)\dots\frac{\prod'_{i<k}(l_i-l_k)}{l_1!(l_2-1)!\dots l_r!}+\cdots\}, \qquad \text{(IX.2.b-1)}$$

where the prime indicates the product is over all rows except that displayed. If two rows are equal in length then

$$l_i - 1 - l_{i+1} = 0, \qquad \text{(IX.2.b-2)}$$

and the term drops out; thus we can ignore this problem. This now has to be summed over i for all r rows.

Problem IX.2.b-1: This gives [Boerner (1963), p. 123]

$$f^{\alpha} = (n-1)!\frac{\prod_{j<k}(l_j-l_k)}{l_1!l_2!\dots l_r!}\{l_1(l_1-1-l_2)(l_1-1-l_3)(l_1-1-l_4)\dots+\cdots\}$$
$$= (n-1)!\frac{\prod_{j<k}(l_j-l_k)}{l_1!l_2!\dots l_r!}\times\sum_{\rho=1}^{r} l_\rho\frac{\prod_{j\neq\rho}(l_\rho-1-l_j)}{\prod_{j\neq\rho}(l_\rho-l_j)}. \qquad \text{(IX.2.b-3)}$$

The term multiplying the sum should look familiar. Check that if the lengths of the two rows are equal the corresponding term is 0, so this does not cause problems. Also this is correct for $m_r = l_r = 1$ (why does this have to be examined?). We can therefore assume that all l's are different. So all that is necessary is to show that the sum, S, equals n. With

$$f(x) = (x-l_1)(x-l_2)\dots(x-l_2), \quad F(x) = \frac{xf(x-1)}{f'(x)}, \qquad \text{(IX.2.b-4)}$$

show that we get for the sum,

$$S = -\sum_{k=1}^{r} F(l_k). \qquad \text{(IX.2.b-5)}$$

By Taylor's theorem

$$f(x-1)=f(x)-f'(x)+\frac{1}{2}f''(x). \tag{IX.2.b-6}$$

Check that higher-order terms can be ignored — multiplied by x they give a polynomial, $\phi(x)$, of degree $\leq r-2$, and every such ϕ satisfies

$$\sum_{k=1}^{r}\frac{\phi(l_k)}{f'(l_k)}=0. \tag{IX.2.b-7}$$

To show this, label the difference product of the l's by Δ_r, so

$$\Delta_r=\prod_{p<s}(l_p-l_s)=\begin{vmatrix} l_1^{r-1} & l_1^{r-2} & \dots & l_1 & 1\\ l_2^{r-1} & l_2^{r-2} & \dots & l_2 & 1\\ \dots & & \dots & & \\ l_r^{r-1} & l_r^{r-2} & \dots & l_r & 1\end{vmatrix}. \tag{IX.2.b-8}$$

Prove that Δ_r equals the determinant for $r=1,2,3$, and then in general. With $\Delta_{r-1}^{(k)}$ the difference product of the $r-1$ numbers $l_1,\dots,l_{k-1},l_{k+1},\dots,l_r$, show that

$$\sum_{k=1}^{r}\frac{\phi(l_k)}{f'(l_k)}=\frac{\{\phi(l_1)\Delta_{r-1}^{(1)}-\phi(l_2)\Delta_{r-1}^{(2)}+\cdots\}}{\Delta_r}$$

$$=\frac{1}{\Delta_r}\begin{vmatrix} \phi(l_1)\; l_1^{r-2} & \dots & l_1 & 1\\ \phi(l_2)\; l_2^{r-2} & \dots & l_2 & 1\\ \dots & \dots & & \\ \phi(l_r)\; l_r^{r-2} & \dots & l_r & 1\end{vmatrix}=0, \tag{IX.2.b-9}$$

since the first column is a linear combination of the others. Hence, remembering that $f(l_k)=0$, we have for the sum

$$S=\sum_{k=1}^{r}l_k\frac{f'(l_k)-\frac{1}{2}f''(l_k)}{f'(l_k)}=\sum_{k=1}^{r}l_k-\sum_{k=1}^{r}l_k\frac{f''(l_k)}{2f'(l_k)}, \tag{IX.2.b-10}$$

by substitution in eq. IX.2.b-5. The first sum is

$$S_1=\sum_{k=1}^{r}l_k=\sum_{k=1}^{r}m_k+\sum_{k=1}^{r}(r-k)=n+\frac{r(r-1)}{2}. \tag{IX.2.b-11}$$

For the second

$$S_2=-\sum_{k=1}^{r}\frac{l_kf''(l_k)}{2f'(l_k)}=-\sum_{k=1}^{r}l_k\sum_{p\neq k}\frac{1}{(l_k-l_p)}=-\frac{r(r-1)}{2}, \tag{IX.2.b-12}$$

since the $r(r-1)$ summands add in pairs to give 1. Hence

$$S = S_1 + S_2 = n, \qquad \text{(IX.2.b-13)}$$

as required, proving the formula for the representation dimensions.

Problem IX.2.b-2: Can this give a corresponding formula for A_n?

Problem IX.2.b-3: Another problem is finding the number of frames of S_n (pb.VIII.3.a-1, p. 221). What is the maximum dimension of the representation of a frame? There are tables of these values and bounds (which might be improved) [Soto and Mirman (1982)]. An exact formula, in terms of row lengths or frame ordinal, would be interesting.

IX.2.c The dimensions and the regular representation

The number of states equals the number of standard tableaux. The number of states in the regular representation equals the number of permutations. Each permutation is given by a set of cycle lengths, so by a frame, and by the numerals in each cycle and their order, with orderings the same if they differ only in which numeral in a cycle is first (leftmost) — rotating, but not permuting, the numerals does not matter. Thus the states of the regular representation can be assigned to frames, with the states for each frame given by the different arrangement of the numerals. However, unlike frames and equivalence classes, there is no correspondence between partitions and states. Thus for S_3 there is one assignment of numerals to the set of cycles of lengths (1,1,1), three to (2,1) and two to (3), but the number of states given by the frames with these row lengths is $1, 2 \times 2$ and 1. Fortunately however $1^2 + 2^2 + 1^2 = 1 + 3 + 2 = 3! = 6$.

Problem IX.2.c-1: The sum of the dimensions squared of S_n equals its order, $n!$, which however is the number of permutations — the number of ways of breaking n numerals into cycles, and inserting the numerals into them, with the choice of the first in each irrelevant. Hence the two numbers are equal, for all n. It may not be impressive that for S_2, $1^2 + 1^2 = 2 = 2$, but for S_3, $1^2 + 1^2 + 2^2 = 1 + 2 + 3 = 6$, is somewhat less trivial. Check this for S_4 and S_5. This raises interesting questions. Is it possible to use this to find the number of frames of S_n? How about the maximum dimension of a representation? Moreover it is an intriguing number-theoretical result. Can this combinatorial result be proven without group theory; why? Write formulas for these sums, which are equal (and equal to $n!$), for all n. This gives three equations for each n. Can two be found from the third, using number (but not group) theory (or is it just accidental that all three are satisfied, for all n)? It is good that these equalities hold, else symmetric groups would have no representations, which would be unfortunate (that would mean also that if numbers were not nice enough to obey all these relations — and for each n — no group could have representations, causing all sorts of difficulties for physics and mathematics, among other problems). But why should they hold? Having these results, the question arises whether they can be generalized in any way, especially in any interesting way (or might the second power be, for some reason, unusual — a point which arises elsewhere — and if so, does this have anything to do with group theory?). Might such generalizations, or lack thereof, throw light on the theory of symmetric groups, or any other groups? If so, what would it say about number theory and about combinatorics? This also emphasizes something, perhaps more interesting. That physics, the universe, is possible results from, among many other things, the existence of representations of the symmetric groups, and this is possible because all these restrictions on numbers (including small integers) are accidently (?) satisfied (exactly). There are other, additional, restrictions and surprisingly, they also are accidently (?) fulfilled as they must be for there to be a universe [Mirman (1988a,b, 1995a)]. And for electromagnetism and gravitation there are further requirements, and these too just happen to be satisfied [Mirman (1995b)]. There are a large number of different requirements; very few have been indicated here. It seems that at least one should fail, thus so should mathematics and physics, yet strangely, none do.

Problem IX.2.c-2: This can be extended in what are perhaps useful ways. Prove that the order of alternating group A_n is $n!/2$; thus the sum of the squares of its dimensions equals this. Likewise the number of its equivalence classes similarly equals the number of ways of inserting numerals in a subset of cycles. Why? This gives another sum which equals $n!/2$, giving three equations equivalent to those for S_n. Thus we have the analog of the previous problem. Show that these equations are satisfied for all n. Does this follow from the results for S_n, or perhaps conversely? Why? This suggests other questions. Any finite group is a subgroup of a symmetric group, so its order is $d = n!/k$, for some integers n and k. Why? The sum of the squares of its representations equals this, and again the number of its equivalence classes is determined in a corresponding manner by the number of ways of inserting a set of numerals in cycles, a subset of those for S_n, giving again three equations. Is it possible to find expressions for these three (which involves a combinatorial problem of finding the number of ways of inserting the numerals)? This would relate a subset of cycles, determined by their number and lengths, to k. Can these be satisfied for every k (ignoring the trivial case of Abelian groups)? If not this would give the set of allowed orders for non-Abelian finite groups, and also raise the question whether there are finite groups for every allowed order, and whether this gives the number of groups for each allowed order. In any case it is interesting that they can be satisfied — simultaneously — for, at least, some k (for purely combinatorial reasons, having nothing, perhaps, to do with group theory).

Problem IX.2.c–3: Nonsimple group S_3 has two subgroups, A_3, which is invariant, and S_2 for which there are three copies. How can we tell that S_3 is not simple? Because there are three copies of subgroup S_2, it is not invariant. Why? But A_3 being of order 3 must be invariant, so S_3 is not simple. Why? If there were two copies there would be no transformation left to take them into each other. Can this be extended to larger symmetric groups? One problem is that because two subgroups have the same dimension does not mean that they are conjugates, as they must be for S_3. Moreover not all invariant subgroups are of index 2 (the order of the group divided by that of the subgroup equals 2). We leave it to the reader to play with larger groups, to see to what extent this can be generalized. Must all subgroups of a symmetric group with the same dimension be conjugates? Are there counter-examples? If a subgroup appears but once, it does not follow that the group is simple. However the order of any of its subgroups must divide its own order (why?), thus $(n!/k)/k'$ must be an integer, for some k, k'. How can we tell when we have reached a simple subgroup? It is not if it has a subgroup of index 2. Does this determine all the nonsimple groups? Are there other numerical conditions showing that a group is not simple? If it is simple then for every subgroup, whose possible orders are limited, there must be at least one isomorphic subgroup, which presents a possible further restriction on the orders. Does this give any information, or is it trivial? Again the representations give three equations for any subgroup. Can these be found in general? Would doing so be useful? Why can they (fortunately) be satisfied for every subgroup? There are limitations on subgroups, so on all non-Abelian groups, and in particular on simple groups. Also these give information about the set of cycles making up a subgroup, so the inclusion rules, so the (possible) subgroups. To what extent do these restrictions provide information about the possible non-Abelian groups (and do all occur?), especially what simple, finite groups are allowed (and do all occur?)? These are interesting combinatoric and number-theoretical problems, certainly not trivial, and probably not answerable in general. But trying (a large number of) special cases may prove enlightening.

Problem IX.2.c–4: If the inclusion rules for a finite group in a symmetric group are known, the group's representations can be found (in principle) for those of the symmetric groups are known — there are computer programs for them [Soto and Mirman (1981a,b,c)]. Find the representations for a few finite groups, not Abelian and not symmetric, from those of S_n.

IX.3 REPRESENTATIONS OF ALTERNATING GROUPS

Knowing the representations of a group we know those of all its subgroups. But representations irreducible for the group can be reducible for subgroups. How are irreducible representations related? There is one example of, perhaps, special interest. S_n has a subgroup, alternating group A_n. How are their representations related?

For A_3, with three operations, ε, (123), (132), there are three representations, all of one dimension. The basis state is

$$T = 123 + 231 + 312. \qquad \text{(IX.3-1)}$$

Note that

$$T' = 213 + 321 + 132, \qquad \text{(IX.3-2)}$$

is not a state of a different representation, but the same state of a different realization; there are no sums of A_3 permutations taking 123 into T' — it is the same sum of permutations acting on a different monomial, this obtained from 123 with a transposition, and that is not in A_3. The S_3 representation space is a sum of two copies of the A_3 representation space.

Problem IX.3.-1: Find the two separate sums into which each of the four S_3 sums, for both types of states (sec. VIII.3.g, p. 225; sec. VIII.6.c, p. 244), breaks up under A_3. Find the basis vectors of the representations obtained using these sums, and the matrix elements. Are these different for T and T'? Show that A_3 does not have a completely antisymmetric representation. Why? How does the antisymmetric state of S_3 behave under A_3? Are there reducible A_3 representations that are irreducible under S_3?

Problem IX.3-2: Show that the representation matrices are ($\omega = e^{\frac{2\pi i}{3}}$)

E	(123)	(132)
1	1	1
1	ω	ω^2
1	ω^2	ω .

Table IX.3-1: Representation matrices of A_3

Why are all representations one-dimensional? If there is only one basis state what are the three representations? How are the last two related?

How are they related to the representations of S_3; which are subrepresentations of which S_3 representations? What are their basis states?

Problem IX.3-3: Find the representations of A_4 and A_5. Find any relations between them, also relations between these and those of S_4 and S_5. Can the representations of A_n be labeled by (a subset of) frames and tableaux? Are there rules to choose these? How do the properties of these labels relate to those of the representation matrices and states? Are there differences from such relationships for S_n? Why? Do these results suggest any general rules or procedures relating (irreducible) representations of groups and their subgroups? While A_n (for almost all n) is simple (sec. IV.9, p. 143), S_n is not. Can we tell this by looking at basis states?

Problem IX.3.-4: Find a formula for the dimensions of representations of A_n. Relate it to that for S_n. Is simplicity relevant? Why?

IX.4 PHYSICAL REALIZATIONS OF TWO-DIMENSIONAL REPRESENTATIONS

The symmetric and antisymmetric representations are quite familiar, perhaps more than most readers realize, as these are not often discussed in terms of the symmetric group. But do the two-dimensional representations have any physical relevance (sec. V.5.d, p. 167) [Elliott and Dawber (1987), p. 177, 251]?

Consider three (distinguished) particles, named neutron, proton and electron, with orbital angular momenta l_1, l_2, l_3. The product of their statefunctions belongs to a rotation-group reducible representation, as is known from the quantum theory of angular momentum (chap. XI, p. 304). Let the total angular momentum, J, be given. The state with this J can be written in various ways: l_1, l_2, can be combined, then the result added to l_3, and so on, or writing this generally, l_i, l_j, are added, then the result added to l_k, $1 \leq i, j, k \leq 3$, for any choices of i, j, k. There are different ways to couple the angular momenta to get J, so different statefunctions, each a product of statefunctions for the three particles, these indexed by their angular momenta. Any particle can be in any of the (fixed) individual angular momentum states, so the states are obtained from one by permutations. For more particles larger symmetric groups are needed, and representations of larger dimensions appear. But these examples illustrate the meaning of these representations, and their physical significance.

Problem IX.4-1: Show that the different ways of coupling the particles give (four, five, or six?) independent states that form the basis states of a symmetric group representation, and that two (all?) of the representations are equivalent. Find the similarity transformations connecting the different coupling schemes. Repeat for S_n.

Problem IX.4-2: Check that a state of angular momentum J is here an S_3-representation basis state. There are six states. Show that they belong to the symmetric, antisymmetric, and two two-dimensional representations. What is the physical distinction between states of different equivalence classes, of the different representations, and of the states of the same representation? Do all give the same value for J? This gives a systematic way of organizing states and shows why there are two-dimensional representations and what the states mean.

Problem IX.4-3: There should be four ways of combining these states. So l_1 and l_2 can give j_1, or we can use l_1 and l_3, or l_2 and l_3. This gives three states. But also j_1 can have different values; j_1 can be 0, and $l_3 = 1$, to give $J = 1$, and so on. Thus there is another degree of freedom. To get a complete set of states the four states of the two two-dimensional representations are needed. Prove that four states are necessary and sufficient.

Problem IX.4-4: The states giving the same total angular momentum are usually, but not necessarily, degenerate. It is possible there is a Hamiltonian for which the energy of states having l_1 and l_2 coupled is different from that for which l_1 and l_3 are coupled, or for which the energy depends on j_1. One possibility would be inclusion of strong interactions acting on n and p, but not e. Write Hamiltonians (say with strong interactions) having states with different intermediate coupling that are not degenerate. Find one giving the two equivalent representations different energies, and then the two states of each also with different energies. Can you give a physical explanation of the (different sets of) results?

Problem IX.4-5: Another possible way of getting equivalence without degeneracy is with a molecule having three atoms at the vertices of a equilateral triangle and another atom inside the triangle. Take the three atoms identical. The statefunction is symmetric under interchange of the atoms. If the atoms are different, say C, O, and N, different arrangements of them give different states with different energies. Or if another atom were outside the triangle near a vertex (say if the molecule were in a crystal) the energy would be different if, say, C were at that vertex than if N were. Work this out explicitly. Write a Hamiltonian and compute the energies for the different arrangements. Explain physically what is happening. Here there are six arrangements of the atoms — the statefunctions form the regular representation of S_3. The statefunction of this system is $\psi_1(C)\psi_2(O)\psi_3(N)+\psi_1(O)\psi_2(C)\psi_3(N)+\cdots$. Reduce this representation to get the basis states of the irreducible representations. Find their energies for Hamiltonians in which (sets of) the six states have different energies. Interpret physically the two two-dimensional representations, and their states. Explain why physically these representations, and each of their states, are different and what the physical distinction is. Would there be any distinction if they were degenerate? The states with definite energies are eigenstates of the Hamiltonian but not necessarily symmetric-group representation basis states. Find, and explain, these. Assume that the triangle is isoceles. How would the arguments change?

Problem IX.4-6: For a set of spin-$\frac{1}{2}$ particles whose wavefunction is antisymmetric, a symmetric space part means that the spin part is antisymmetric (antiparallel), and an antisymmetric space part requires the spins symmetric (parallel). Thus both symmetric and antisymmetric representations have to be used. Consider states of three nucleons, which have space, spin and isotopic-spin parts, and for which the total statefunction is antisymmetric. Do the two-dimensional representations of S_3 appear in the intermediate stage of coupling two of these parts to find a statefunction, which is then combined with the third part to find the total statefunction?

IX.5 PRODUCTS OF REPRESENTATIONS OF SYMMETRIC GROUPS

A physical system might be labeled by several quantum numbers, say spin and orbital angular momentum, or consist of several parts, such as a set of particles. The statefunctions for each are symmetric-group basis states. If we know the representations of each part, how do we find that of the total system? This symmetric-group representation is reducible; into which irreducible representations does it decompose? It is a sum of irreducible representations; which are in the sum, and how many times each? A state of the system is a sum of irreducible-representation basis states; which are these and what are their coefficients? This, finding the representations in the decomposition of products of representations, and similarly for the states, also is interesting mathematically, for one thing in finding representations of other groups, like unitary ones. So we have to study the various ways of combining representations of groups (taking products), which we do here for the symmetric groups.

For a molecule, the electrons of each atom are (of course) in a completely antisymmetric state. What is the state of all the electrons of the molecule? Less obviously, consider n particles in a state of an S_n representation, and a system of m particles in a state of an S_m representation. The $n + m$ particles are in a state of a reducible S_{n+m}

representation. It can be decomposed into a sum of representations of S_{n+m}. What are these? This is the outer product of a representation of S_n and one of S_m; it is a reducible representation of S_{n+m}, so can be decomposed into a sum of S_{n+m} irreducible representations.

There is another product, the inner product, whose decomposition is the Clebsch-Gordan decomposition. Consider n particles each with two quantum numbers, say spin and isospin. The spin state belongs to a representation of S_n (which need not be symmetric or antisymmetric; for physical particles, only the total statefunction is); the isospin is also described by a state of a representation of S_n. What are the S_n representations and states describing the system, including spin and isospin? This product, giving a state of the same symmetric group, is the inner product. Since this has been discussed in detail elsewhere [Chen (1989), p. 152; Hamermesh (1962), p. 254; Schensted (1976), p. 157; Schindler and Mirman (1977a)], we just summarize the results.

IX.5.a The outer product

A system consists of two subsystems with the state of one labeled by a frame of n boxes with numerals inserted to give a standard tableau, that of the other similarly with m. States of the total system are given by tableaux of $n + m$ boxes; each is a sum of states, each of these labeled by a tableau of $n + m$ boxes [Mirman (1987a)]. How do we get the frames and tableaux in this decomposition of the product? This is the problem of the decomposition of the outer product [Chen (1989), p. 166; Elliott and Dawber (1987), p. 441; Hamermesh (1962), p. 249; Murnaghan (1963), p. 155; Schensted (1976), p. 159].

Each subsystem is given by a multinomial statefunction, one with n, the other with m, indices; the entire system is described by a product of these. We wish to rewrite this product of sums as a sum of sums of products, schematically,

$$(C_{12...n}U_{12...n} + C_{21...n}U_{21...n} + \cdots)(D_{12...m}V_{12...m} + D_{21...m}V_{21...m} + \cdots)$$

$$= C(F_{12...n+m}U_{12...n}V_{12...m} + F'_{12...n+m}U_{12...n}V_{21...m} + \cdots)$$

$$+ C'(F''_{12...n+m}U_{21...n}V_{12...m} + \cdots) + \cdots. \qquad \text{(IX.5.a-1)}$$

Each sum of $n + m$ products, in parentheses, is symmetrized to give a basis state of an irreducible representation of S_{n+m}. The sum over these is a sum over the irreducible-representation states appearing in the decomposition of the reducible-representation basis state (the product on the left). Which representations and states appear in this decomposition (this sum of sums), how many times does each representation appear, and what are the coefficients of the states?

For example the product of two S_1 states $\boxed{\mathrm{a}}$ and $\boxed{1}$ is a state of a reducible representation of S_2,

$$U_a V_1 = \frac{1}{2}(U_a V_1 + U_1 V_a) + \frac{1}{2}(U_a V_1 - U_1 V_a), \qquad \text{(IX.5.a-2)}$$

reduced into a sum of irreducible representation states. Notice that to do this we consider permutations of the n and m numerals, here a and 1. Thus, say, we consider the state of one atom, with the state of its electrons antisymmetrized, and similarly that of the other atom, and then the state of the molecule is found by taking (in addition) the electrons of different atoms antisymmetric.

Formally the decomposition of the direct product is

$$\xi_{pq}^{n\alpha} \times \xi_{rs}^{m\beta} = \sum_{\gamma fg} \Phi_{pqrsfg}^{\alpha\beta\gamma} \xi_{fg}^{(n+m)\gamma}; \qquad \text{(IX.5.a-3)}$$

the problem is to determine the γ, f, g giving nonzero Φ's and what the nonzero values are [Mirman (1987a)]. The superscripts on Φ denote, from left to right, frames of S_n, S_m and S_{n+m}, the subscripts are indices of tableaux of these frames in the same order. On the left is a product of ξ's labeled by standard tableaux of the subgroups, not those of the full group, while the ξ 's on the right are labeled by S_{n+m} standard tableaux. Since a representation of S_{n+m} can appear several times the coefficients bear a multiplicity index, which is included in the γ.

In writing these expansions we assume a choice of subsets of n and m numerals. However given $n + m$ numerals there are many choices [Mirman (1987a)], something we might have to be careful of in a particular application, but which we ignore here.

Problem IX.5.a-1: For the product of the ξ_{11} state of S_2 with the S_1 state, giving an S_3 state (sec. VIII.6.c, p. 244) we have schematically (why?),

$$\begin{array}{|c|}\hline 1\\ \hline 2\\ \hline\end{array} \times \begin{array}{|c|}\hline \mathrm{a}\\ \hline\end{array} = \alpha \begin{array}{|c|}\hline 1\\ \hline 2\\ \hline \mathrm{a}\\ \hline\end{array} + \beta \begin{array}{|c|c|}\hline 1 & \mathrm{a}\\ \hline 2 \\ \cline{1-1}\end{array} = \alpha\xi^1 + \beta\xi^2_{22} + \dots$$

$$= (12 - 21)a = \alpha(12a - 21a - 1a2 + 2a1 + a12 - a21)$$

$$+ \beta\{(12a - 21a) + \frac{1}{2}(1a2 + a21 - a12 - 2a1)\} + \dots. \qquad \text{(IX.5.a-4)}$$

The second state is state 2 of representation 2, this being antisymmetric under (12) applied from either side. Both sides go into their negatives under (12). Find α and β. The other states of frame 2 are found using the S_3 permutations, $(1a)$ and so on, not in S_2, applied from both sides. This should be verified. Try various $S_2 \times S_2$ products.

IX.5.a.i *The frames in the outer products*

Here we give the rules for the frames in the expansion [Hamermesh (1962), p. 249; Littlewood (1958), p. 67, 94], for the product of two; for the general case the decomposition is done in steps.

We start with either frame, called the trunk, and add to it boxes from the other. Label the boxes of the second frame by putting a's in all of the first row, b's in all of the second, and so on. Attach the boxes from the first row of the second frame to the trunk in all possible ways to give standard frames, then from the second row and so on, subject to two exceptions. First, obviously, in attaching boxes from the second frame to the first do not put boxes with the same letter in the same column.

Now read the letters in the frame from right to left (as Hebrew is read), along the first row, then along the second and so on. The second condition is that the only frames allowed are those for which the letters form a lattice permutation of $a, b, c, \ldots$ [Hamermesh (1962), p. 198; Robinson (1961), p. 45]. This means that there should be no fewer a's at any point than any other letter, no fewer b's than c's, d's, and so on, no fewer c's in the same way, ... Thus

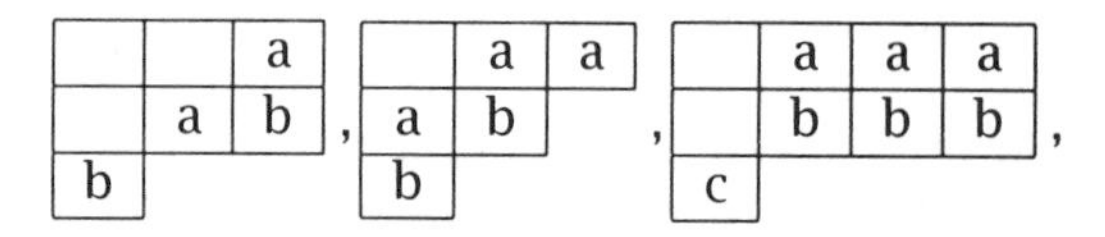

form lattice permutations, but not

Each allowed distribution of the letters gives a frame in the decomposition, the restriction to lattice permutations prevents multiple counting, and this procedure gives all frames, including their multiplicity, in the sum of representations equivalent to the outer product of two representations.

For example, suppose we take the product of frames and . Can appear? It would have

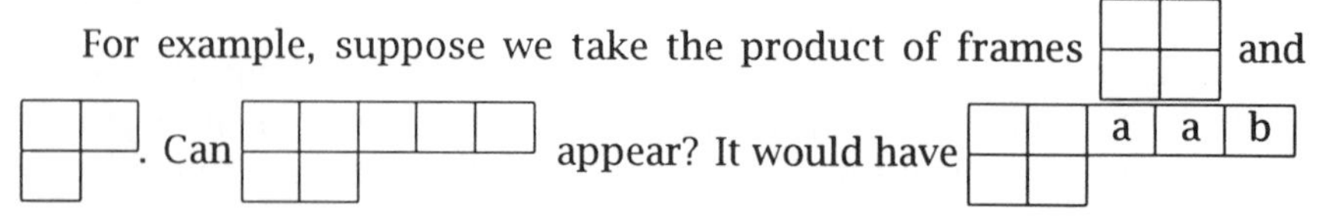

which is not possible; this is not a lattice permutation — two indices antisymmetric in the S_n frame are symmetric in it. Another example is

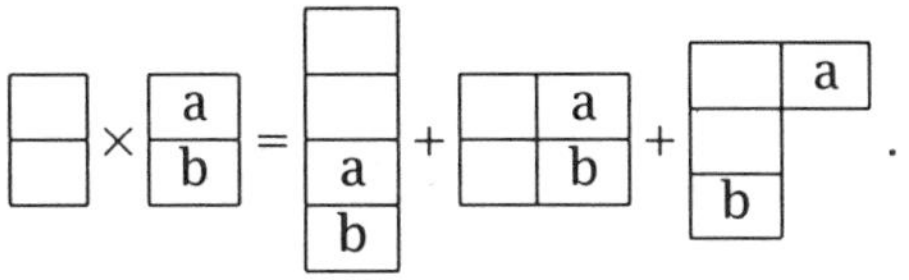

$$\begin{array}{|c|}\hline \ \\ \hline \ \\ \hline\end{array} \times \begin{array}{|c|}\hline a \\ \hline b \\ \hline\end{array} = \begin{array}{|c|}\hline \ \\ \hline \ \\ \hline a \\ \hline b \\ \hline\end{array} + \begin{array}{|c|c|}\hline \ & a \\ \hline \ & b \\ \hline\end{array} + \begin{array}{|c|c|}\hline \ & a \\ \hline \ \\ \cline{1-1} b \\ \cline{1-1}\end{array} . \qquad \text{(IX.5.a.i-1)}$$

Problem IX.5.a.i-1: Show that the outer product is independent of the choice of the trunk, and that it is commutative. Does it depend on n?

Problem IX.5.a.i-2: Find the outer products for: $(2) \times (2)$; $(1,1) \times (2)$; $(2) \times (3)$; $(2) \times (2,1)$; $(2,1) \times (2)$; $(2,1) \times (2,1)$; $(1,1,1) \times (3)$; $(3,1) \times (2,2)$; $(3,1) \times (3,1)$. Calculate the dimensions as a check.

IX.5.a.ii *Frames are independent of choice of subsets*

Given $n + m$ numerals, they can be divided into sets of n, and m, in various ways. Does the choice affect the frames in the product? If there is one product not having a frame in its expansion, then no product from the pair of subgroup frames considered does, for any choice of subsets. The other products are found by having subgroup permutations act on any one (the natural monomial), and these cannot change one S_{n+m} frame to another. The frames do turn out to be independent of how the indices are broken into subsets of n and m numerals [Mirman (1987a), lemma 3.1]. All frames allowed by the rules in sec. IX.5.a.i, p. 262, appear in the expansion of every product of a pair of subgroup states of frames for which these S_{n+m} frames are allowed, for every choice of subsets.

IX.5.a.iii *The states in the expansion*

We know which S_{n+m} frames appear in the expansion of two frames, and what sums of S_{n+m} states appear. Still open is the question of the S_{n+m} standard states in these sums — given a product of an S_n and an S_m state, which S_{n+m} states, of an allowed frame, appear? The expansion of an outer product is given by the states in it and their coefficients. Information about these can be found from the labeling frames and tableaux. Here we start considering how the tableaux determine the states in the expansion, finding rules excluding frames and tableaux, and so states. These are useful for the insight they provide.

We start by applying to eq. IX.5.a-3, p. 261, giving the definition of the expansion of the outer product, permutation σ of S_n, using eq. II-12 of Schindler and Mirman (1977a), to get (suppressing the monomial),

$$\sigma\xi_{pq}^{n\alpha} \times \xi_{rs}^{m\beta} = \sigma \sum_{\gamma fg} \Phi_{pqrsfg}^{\alpha\beta\gamma} \xi_{fg}^{(n+m)\gamma} = \sum_{z} u_{zp}^{n\alpha}(\sigma)\xi_{zq}^{n\alpha} \times \xi_{rs}^{\beta m}$$

$$= \sum_{\gamma'' fgh} u_{hf}^{(n+m)\gamma''}(\sigma)\Phi_{pqrsfg}^{\alpha\beta\gamma''}\xi_{hg}^{(n+m)\gamma''} = \sum_{\gamma' zcd} u_{zp}^{n\alpha}(\sigma)\Phi_{zqrscd}^{\alpha\beta\gamma'}\xi_{cd}^{(n+m)\gamma'}. \tag{IX.5.a.iii-1}$$

Multiplying on the left by $\xi_{ab}^{(n+m)\gamma''}$, on the right by $\xi_{ij}^{(n+m)\gamma}$, and using eq. II-7 of Schindler and Mirman (1977a), we find

$$\sum_{f} u_{bf}^{(n+m)\gamma}(\sigma)\Phi_{pqrsfi}^{\alpha\beta\gamma} = \sum_{z} u_{zp}^{n\alpha}(\sigma)\Phi_{zqrsbi}^{\alpha\beta\gamma}, \tag{IX.5.a.iii-2}$$

for every σ of S_n, where $u_{zp}^{n\alpha}(\sigma)$ is the matrix element.

The multiplicity index has been absorbed into the γ so the Φ's are sums of the Φ's for each occurrence. A similar set of equations is obtained if σ acts on the right, and there is a pair of equivalent sets for each permutation of S_m. We consider how these equations tell which subscripts give nonzero Φ's. Conditions on the states in the expansion can be obtained from the definition of the outer product. Coefficients that do not satisfy are 0. There are such equations for every permutation of S_n, and S_m, but not all are needed as we see from

LEMMA: The necessary and sufficient condition that coefficients satisfy the defining equation, eq. IX.5.a-3, p. 261, for all S_n permutations is that they satisfy for any $n-1$ S_n transpositions (such as neighboring) obeying the requirements of pb. VIII.2.c-2, p. 218 [Mirman (1991e), Lemma II-1].

PROOF: Necessity is obvious. Assume that there are $n-1$ transpositions for which they do satisfy. Then we show they satisfy for all permutations. An S_n permutation can be written as a product of at most $n-1$ of these transpositions. Suppose a set of coefficients satisfies the equations for permutations σ_1 and σ_2. Multiplying the equation for σ_1 by $u_{tb}^{(n+m)\gamma}(\sigma_2)$ and summing, we get

$$\sum_{bf} u_{tb}^{(n+m)\gamma}(\sigma_2)u_{bf}^{(n+m)\gamma}(\sigma_1)\Phi_{pqrsfi}^{\alpha\beta\gamma} = \sum_{bz} u_{tb}^{(n+m)\gamma}(\sigma_2)u_{zp}^{n\alpha}(\sigma_1)\Phi_{zqrsbi}^{\alpha\beta\gamma}$$

$$= \sum_{f} u_{tf}^{(n+m)\gamma}(\sigma_2\sigma_1)\Phi_{pqrsfi}^{\alpha\beta\gamma} = \sum_{z} u_{zp}^{n\alpha}(\sigma_1)\sum_{b} u_{tb}^{(n+m)\gamma}(\sigma_2)\Phi_{zqrsbi}^{\alpha\beta\gamma}$$

$$= \sum_{z} u_{zp}^{n\alpha}(\sigma_1)\sum_{w} u_{wz}^{n\alpha}(\sigma_2)\Phi_{wqrsti}^{\alpha\beta\gamma} = \sum_{w} u_{wp}^{n\alpha}(\sigma_2\sigma_1)\Phi_{wqrsti}^{\alpha\beta\gamma}, \tag{IX.5.a.iii-3}$$

iterating eq. IX.5.a.iii-2. Thus this equation holds for $\sigma_1\sigma_2$. So if the equations hold for $n-1$ transpositions then by iteration they hold for all permutations.•

Any of these $n-1$ transpositions can be used, but the $n-1$ neighboring ones are convenient, since their matrix elements are simple, and simply known. The equations given by $n-1$ transpositions of S_n, and $m-1$ of S_m, provide the conditions. These are not all the $n+m-1$ transpositions of S_{n+m} needed for all its permutations. The other one does not give a condition of the type studied here and is not considered. We can now state:

LEMMA: If a (right or left) subscript on a ξ in the product is an ordinal of a tableau having two numerals, i, $i+1$, neighboring in both the subgroup and the full group, in the same row (column), then in the expansion no ξ appears whose corresponding (right or left) subscript is the ordinal of a tableau with these two numerals in the same column (row).

PROOF: In the defining equation, σ is $(i\ i+1)$, whose matrix elements are 1 (-1) and -1 (1), so we get

$$\Phi^{\alpha\beta\gamma}_{pqrsbj} = -\Phi^{\alpha\beta\gamma}_{pqrsbj} = 0. \quad \bullet \qquad \text{(IX.5.a.iii-4)}$$

These Φ 's are sums over the occurrences, but it is the sums that are the coefficients of the ξ's.

For states that do appear in the expansion we get:

LEMMA: If a subscript on a ξ in the product is the ordinal of a tableau in which two subgroup numerals are in the same row (column), then each ξ in the expansion must belong to a sum, which goes into itself (minus itself) under the transposition of these numerals. If they are neighboring in the full group then the sum consists either of the state or it plus one other standard state.

PROOF: The product, and so the expansion, goes into itself (minus itself) under the transposition, which gives the first statement, with the second true because a neighboring transposition matrix has one, or two, elements in a row and column, and the ξ 's are independent. •

There is a converse problem: given an S_{n+m} state, if we divide the set of numerals into two sets, of n and m numerals, then the state can be written as a sum over outer products of S_n and S_m states. This has been discussed elsewhere [Mirman (1987a), sec. 4]; we do not consider it here.

Problem IX.5.a.iii-1: The question of the coefficients of the states in the expansion is still open. Find a formula (or algorithm) for these.

IX.5.b Inner products

Consider a set of particles i having, say, angular momentum states ω_i and isospin states χ_i. The statefunction of the system is

$$\psi = \omega_1\chi_1\omega_2\chi_2\omega_3\chi_3\cdots, \qquad \text{(IX.5.b-1)}$$

or correctly, this symmetrized over the particles. This symmetrization is performed by interchanging particles, that is interchanging both subscripts 1 with both subscripts 2 and so on. Now suppose we know that the system is in some angular momentum state. Then the angular momentum states alone have a definite symmetry. For two spin-$\frac{1}{2}$ particles in state $J = 1$, the angular momentum statefunction is symmetric under particle interchange. Likewise the total isospin state can be given, so having determined symmetry. If we know the symmetry of these two parts, what are the possible symmetric-group representations and states for the complete statefunction?

This is the problem of finding the decomposition of the inner product of symmetric-group representations and states, the Clebsch-Gordan decomposition. The product of two representations (or states) of the same symmetric group (notice, in the above, permutations act on the indices of both states simultaneously), is a reducible representation of that group, written as a sum of irreducible representations (or states). What frames, representations (or states) appear in this sum? How many times does each frame and representation appear? And what are the coefficients of the states (the tensor-coupling coefficients)?

The representation matrix acting on the product is the product of the representation matrices acting on the individual terms, and is reducible. When reduced, the matrices from which representations appear, and how many times (the same question as for frames)? What is the matrix carrying out this reduction by a similarity transformation — the matrix of the Clebsch-Gordan coefficients (these related to the tensor-coupling coefficients)?

This is analogous to the rotation-group Clebsch-Gordan decomposition. So given a particle with spin and orbital angular momentum, both of which are affected by the transformations of the same group — by the same rotation — what are the possible total angular momentum states, and what are their coefficients? Here there is one aspect not found in the rotation group. A product of two rotation-group states equals a fully determined sum; each state in the sum appears once. However this is not true for more than two states. To find the total angular momentum we combine any two, then combine the result with the third (and so on if there are more states). Thus each total angular momentum state can appear several times in the sum, differing in the intermediate state. For the symmetric group, the product of two states contains in the sum, in general, states appearing more than once; the multiplicity of the states can be greater than one. However here there is no such way, as with the intermediate states of the rotation group, to distinguish them (at least mathematically). Thus the decomposition is partially arbitrary, and conventional.

Unlike the outer product, there are (apparently) no simple (visual) general explanations for which representations and states appear in the decomposition, and their coefficients. We therefore refer to treatments elsewhere [Hamermesh (1962), p. 254; Schindler and Mirman (1977a,b)].

IX.6 AUTOMORPHISMS OF SYMMETRIC GROUPS

The concept of an automorphism (sec. III.4.e, p. 91) may seem somewhat trivial. Why should there be any limit on them; we can always relabel elements? Is anything more involved? To illustrate this we consider the automorphisms of the symmetric groups [Mirman (1991e), appendix C], which should also increase our understanding of these groups, and of how automorphisms of a group are related to its structure. This also emphasizes that the objects permutations act on, being determined up to automorphisms, are not relevant to the properties of permutations, or the groups.

Automorphisms preserve multiplication (pb. IV.2-5, p. 108), so elements in the same class go into elements in the same class (classes go into themselves). The only exception is S_6 (pb. III.4.e.i-6, p. 93; pb. VIII.2.a- 7, p. 216) [Miller (1958); Segal (1940)]. A permutation raised to a power equals the identity and this (invariant) power is determined by the class (the cycle lengths). However it does not determine the class — products of commuting cycles of equal length have the same power as a single cycle; if they have different lengths the power of their class, which depends on both cycle lengths, can equal that of a different class. The number of elements in a class, and the number and lengths of cycles, are invariant, for a permutation of several cycles is a product of single-cycle permutations whose lengths (classes) are invariant, as is the expression of a permutation of the class as this product (pb. III.4.e-3, p. 91). Permutations can be written as products of transpositions, which go into transpositions under automorphisms. An automorphism simply relabels the transpositions; to show this we need:

LEMMA: Any set of transpositions can be relabeled to give any other transpositions which need not have the same neighboring relationships among the numerals, but which contains the same number of appearances of each numeral. This is a similarity transformation on the transpositions (and representation matrices) preserving the product, so commutation relations. The only transformations preserving multiplication are those taking transpositions into transpositions, keeping distinct numerals distinct, and replacing a numeral by a fixed different one in every transposition it appears in.

PROOF: The transformation taking (ij) into (pq) is

$$(ip)(jq)(ij)(jq)(ip) = (pq); \tag{IX.6-1}$$

similarly

$$(jq)(ij)(jq) = (iq). \tag{IX.6-2}$$

In each transposition that i appears in, it is replaced by p, and likewise for the other numerals, so multiplication rules remain the same. The product would be different if a numeral appeared in two transpositions but in only one after the transformation. Thus i is replaced by p in every transposition in which it appears, and p can appear in no transposition except those in which it replaces an i. Only transformations that take distinct numerals into distinct numerals preserve products. •

Obviously, a permutation can be given as a product of transpositions (sec. VIII.2.c, p. 217), and those in the product giving the transformed permutation are transformed transpositions, multiplication being invariant. The most general automorphism of the symmetric group replaces each numeral by another object (different numerals going into different objects) in every permutation in which it appears. If these are the same objects, then the automorphism is an arbitrary permutation of them.

This permutation is a element of the group. Thus all automorphisms are inner (except for a change of the symbols), except for S_6.

Chapter X

The Rotation Groups and their Relatives

X.1 LIE GROUPS

A group is a set of operations, subject to restrictions, and its elements can be finite in number, infinite but discrete (labeled by — put in one-to-one correspondence with — integers; sec. II.3.j, p. 48), or continuously infinite (labeled by continuous parameters — sets of real numbers), or combinations. The elements of continuous groups are (perhaps multidimensional) functions of sets of real (or complex) numbers [Arfken (1970), p. 213; Cornwell (1984), p. 44; Falicov (1966), p. 200; Hamermesh (1962), p. 279, 283; Schensted (1976), p. 186; Wigner (1959), p. 88]. These functions can (within strong limits) be arbitrary. But in physics, and in large areas of mathematics, we expect them to be continuous, for complex parameters hopefully analytic. Thus we come to groups whose transformations are continuous functions of (sets of) real numbers (so, for reasons we do not consider [Varadarajan (1984), p. 41; Wybourne (1974), p. 19], infinitely differentiable, with products which are analytic). These are Lie groups.

Such groups are of fundamental importance in both physics and mathematics, as we see by considering the most familiar of them.

The parameters on which group transformations depend can have ranges finite or infinite. If the parameters all have finite ranges (like angles, going from 0 to 2π, say), the Lie group is compact. The Lorentz group (sec. II.3.e, p. 44), with some parameter ranges infinite (some transformations can be taken as hyperbolic functions), is a noncompact group. Here we discuss only compact groups. The types of Lie groups that we consider, with the exception of a single Abelian one, are all simple groups (semisimple ones are a trivial generalization); but because the group transformations can include both finite and infinite ones, the definition of simplicity is more subtle. Thus the topic here is compact, simple Lie groups (which we take to include ones with invariant — but finite — subgroups). To emphasize this, we accept the redundancy and refer to them as compact, semisimple groups, although compact implies semisimple (though not conversely).

What do we want to know about these? For any group we need the list of transformations and their products, the group multiplication table. But Lie group elements depend on continuous parameters. How then do we list them, and how do transformations of various groups differ — they are all infinite in number? The list of transformations of a Lie group is given by specifying the number and type (real or complex) of parameters and their ranges. Products are given by formulas providing the parameters of the product transformations in terms of those in the product. Knowing these we have a (continuously-infinite) group table; the possible ones are limited, and they fall into sets. We would like to know why, and what.

Then we want representations: the labels of the representations and states, how these are determined by group operators, the basis states (as functions of the parameters) and the matrices carrying out the transformations, and how all these are given by the labels. Also we can take products of basis states, quite familiar in quantum mechanics, and these, basis states of reducible representations, reduce into sums of those of irreducible representations; we need this decomposition. We have to physically interpret the mathematical objects, transformations, basis vectors, labels, and understand what they tell us about the physical world they describe, how, and why.

First we introduce the concepts by considering the rotation groups (sec. II.3.a, p. 41) SO(2), SO(3), these with inversions (sec. II.3.i, p. 46) O(2) and O(3), and SU(2) (sec. II.3.b, p. 43).

X.2 THE ROTATION GROUPS

Among the most intuitive of transformations are rotations, but more than transformations, they are symmetries — the fundamental laws of physics are unchanged by them. What does the nature of these transformations, and that they are symmetries, prescribe for the laws, and for objects whose properties they govern? So we come to the study of their groups, the rotation groups, primarily SO(3) (although much of this holds for SO(n)), and SO(2), and briefly O(2) and O(3) [Bhagavantam and Venkatarayudu (1951), p. 161; Burns (1977), p. 341; Chen (1989), p. 281; Cornwell (1984), p. 433; Elliott and Dawber (1987), p. 130; Falicov (1966), p. 200; Gel'fand, Minlos and Shapiro (1963); Hamermesh (1962), p. 322; Heine (1993), p. 52; Inui, Tanabe, and Onodera (1990), p. 115; Jones (1990), p. 96; Joshi (1982), p. 117; Lyubarskii (1960), p. 192; Miller (1972), p. 222; Murnaghan (1962); Murnaghan (1963), p. 226; Naimark (1964), p. 2; Schensted (1976), p. 258; Tinkham (1964), p. 94; Tung (1985), p. 80, 94, 125; Weyl (1931), p. 140, 185; Wherrett (1986), p. 128]; Hargittai and Hargittai (1987) have many examples of rotationally symmetric objects. There is also a group, SU(2), to which SO(3) is homomorphic, called its covering group [Cornwell (1984), p. 424; Hamermesh (1962), p. 319; Sattinger and Weaver (1986), p. 3, 10], and this has important (physical) consequences. Here we investigate these, their relationships, and consequences for physics.

SO(2) and SO(3) are Lie groups whose transformations are analytic functions of continuous parameters, here regarded as angles (Lie groups that depend on both continuous and discrete parameters are considered later), and are compact. They are the simplest and most familiar of (simple) Lie groups so serve to introduce the concepts. The rotation group, of great physical importance in itself, also serves as a foundation for the study of Lie groups in general.

The analysis of the rotation group has a pseudonym in physics, the quantum theory of angular momentum. (Varshalovich, Moskalev and Khersonskii (1988) has perhaps the most complete discussion of the theory, with many formulas and tables.) One aim here is to see how group theory helps us understand and use angular momentum (not only in quantum mechanics). This is discussed to tie the subjects together and show the group-theoretical meaning of the concepts, and to introduce ideas needed later. Little emphasis is placed on computational aspects, tables or formulas. These are widely available so there is no reason to repeat them.

X.2.a What are the rotation groups?

Physically given a piece of a line, there are various things we can do to it, rotate, displace, shrink or extend, and even bend, it. (The groups we consider are given by a finite number of parameters which — thus — do not depend on space; bending is excluded.) Of these transformations, a rotation is one that leaves fixed one point, and the length of a line, and also the angle between two lines. In particular for three (or in n-dimensional space, n) mutually orthogonal unit vectors (given by real numbers) — a coordinate system — it gives another set of three mutually orthogonal unit vectors, with the same origin, and is the only type of transformation that does. How are these properties of rotations expressed mathematically?

X.2.b Mathematical definitions of rotations

The scalar product of real vectors (all components are real numbers) V and W is

$$V \cdot W = \sum V_i W_i, \qquad \text{(X.2.b-1)}$$

summed over all components. If this is invariant for all vectors, then so are the magnitudes of all, and the angles between them. Thus rotations are those transformations leaving this product fixed, in particular that leave orthogonal vectors orthogonal, so they form orthogonal group O(n), for n dimensions. Here we restrict vectors, so matrix elements, to be real; but it is important that they can be complex [Mirman (1995a)].

Problem X.2.b-1: Show that the necessary and sufficient condition for matrix M to keep scalar products of real vectors unchanged is that it be orthogonal — its inverse equals its transpose $\tilde{M}$,

$$\tilde{M}_{ij} = M_{ji}; \quad \tilde{M} = M^{-1}. \qquad \text{(X.2.b-2)}$$

Check that the determinant of M is ± 1. Also show that an orthogonal matrix with real entries is unitary (sec. II.3.d, p. 43). The group of orthogonal matrices, those leaving orthogonal vectors orthogonal, is the orthogonal group, O(n), in n dimensions — in n dimensions a vector has n components, thus is transformed by $n \times n$ matrices. For SO(n) — "S" means special, the matrices are unimodular — their determinants are 1. If inversions are included the group is O(n). For these, what are the determinants of the transformation matrices? Why? Now we just consider SO(n), and only $n = 2$ and 3. O(n) is an example of a group specified by both continuous parameters, angles, and discrete ones, here only one, the inversion which takes two values, ± 1 — the sign of the determinant. Improper rotations have determinant -1. These do not form a group (the proper rotations, with determinant 1, do). Why? Thus SO(n) consists of all $n \times n$ orthogonal matrices with unit determinant — or perhaps better, these are matrices of one (the defining) representation. Why are these statements equivalent? We can regard the group as those transformations with this representation, though this is somewhat backwards. To give a Lie group we specify a set of parameters, their ranges, and product rules, here ones giving the group whose representation matrices are unimodular, orthogonal matrices. This then is the group, with representations, that describes rotations. To what extent have we provided these?

X.3 THE TWO-DIMENSIONAL ROTATION GROUP

We start with the simplest rotation group SO(2), the transformations, not including inversion, of two-dimensional real vectors — two being the smallest dimension in which rotations are possible; its elements are functions of but one parameter [Hamermesh (1962), p. 322; Tung (1985), p. 81; Wigner (1959), p. 143]. In a one-dimensional space of complex vectors there are equivalent transformations giving the one-parameter group of transformations, unitary group U(1), which leaves fixed absolute values of complex numbers, so the magnitude of a one-dimensional unit complex vector; it is the same group as SO(2).

Problem X.3.-1: Why?

X.3.a Defining SO(2)

A rotation in two dimensions is given by one parameter, but not every one-parameter group is a rotation group — there are also translations along a line. The difference is that for rotations the parameter, angle θ, is limited to a range of 2π (say, $0 \leq \theta < 2\pi$, though any 2π range is equivalent), for translations, the range of parameter x is $-\infty \leq x \leq \infty$. The rotation group is compact, the translation group noncompact. There is another condition: a rotation through a set of angles whose sum equals ϕ (mod 2π) is the same as a rotation through ϕ — the rotation is single-valued. In particular a rotation through 2π is equivalent to the identity.

The product rule is that a rotation of θ_1 followed by one of θ_2 is equivalent to that through $\theta_1 + \theta_2$,

$$R(\theta_1)R(\theta_2) = R(\theta_1 + \theta_2). \tag{X.3.a-1}$$

This, plus the statement of the range of θ (giving the automatic single-valuedness condition), define the group.

Problem X.3.a-1: Why is this condition automatic?

Problem X.3.a-2: The most general vector in two dimensions has two components, thus the most general transformation on it is given by a two-dimensional matrix. Show that

$$M(\theta) = \begin{pmatrix} cos\theta & sin\theta \\ -sin\theta & cos\theta \end{pmatrix}, \tag{X.3.a-2}$$

is orthogonal and does leave the scalar product of two vectors fixed. What is its determinant? Further

$$0 \leq \theta < 2\pi, \tag{X.3.a-3}$$

and M is single valued; for θ outside this range,

$$M(\theta + 2\pi n) = M(\theta), \qquad \text{(X.3.a-4)}$$

where n is an integer. Also show that requiring a 2×2 matrix,

$$T = \begin{pmatrix} a & b \\ c & d \end{pmatrix}, \qquad \text{(X.3.a-5)}$$

to be orthogonal (with unit determinant?) gives relationships among these parameters, reducing their number to one, and that T can be parametrized in terms of θ, thus is equivalent to $M(\theta)$ [Cornwell (1984), p. 49], so that all such matrices give rotations [Tung (1985), p. 93; Aivazis (1991), p. 43]. Further show that $M(\theta)$ satisfies the product rule, eq.X.3.a-1. Verify also that the $M(\theta)$, for θ in the given range, satisfy the group axioms — they form a group. Would this still be so if the condition on the range were changed, say to a different range of 2π, or to one different than 2π? Is this representation irreducible? For an Abelian group, which SO(2) is (why?), irreducible representations are equivalent to one-dimensional ones. Is this true here? With the determinant of $M(\theta)$ equal to 1 (what else could it be?), $M(\theta)$ is a realization of SO(2). For O(2), only the requirement on the determinant is removed. How is it parametrized [Wigner (1959), p. 144]?

Problem X.3.a-3: There is another parametrization of SO(2),

$$R(\theta) = Aexp(in\theta). \qquad \text{(X.3.a-6)}$$

Show that A must be 1, and n an integer. Is R orthogonal? Need it be? What is the equivalent parametrization for O(2)? Repeat the previous problem. How are the two parametrizations related? $R(\theta)$ depends on a complex number, i; could the parametrizations be related if only real numbers are allowed? Can a two-dimensional vector be written so that it can be transformed by $R(\theta)$? What would happen for A, n or θ complex? Suppose n were not integral? Would R be a realization of a (Lie) group and with what product rule? Why does R give a transformation of SO(2) rather than the group of translations? Can $M(\theta)$ be written so that it represents the translation group? The range of the parameter is given by $0 \leq \theta < 2\pi$. Show, for both M and R, that if the range were less than this, there would be group transformations whose product is outside the range — the group axioms do not allow a smaller range — and if it were greater there would be transformations given more than once.

Problem X.3.a-4: Prove that the set of transformations that leave invariant the absolute value (the "unit") of a complex number, $|z|$, is a one-parameter group, with the same product law as SO(2), and the same range, thus is (isomorphic to) SO(2). It is U(1).

Problem X.3.a-5: That SO(2) can be written in two-dimensional form requires, and is allowed by, its parameter range being bounded. The products of SO(2) and of the translation group are given by

$$R(\theta_1)R(\theta_2) = R(\theta_1 + \theta_2), \quad T(x_1)T(x_2) = T(x_1 + x_2). \qquad \text{(X.3.a-7)}$$

Let a group element be

$$U = \begin{pmatrix} f(y) & g(y) \\ h(y) & k(y) \end{pmatrix}, \qquad \text{(X.3.a-8)}$$

where y is the parameter, say θ or x, and f, g, h, k are arbitrary functions. Then show that the group requirement gives conditions on these functions requiring sines and cosines (uniquely?), so that U is of the form $M(\theta)$, and as the functions are sinusoidal the range of the parameter is limited to 2π. Thus the realization — here by 2×2 matrices — and the range, are not independent.

Problem X.3.a-6: The group considered here is SO(2); if we adjoin a reflection (or inversion, which is the same, why?) the group is O(2). By checking their actions on a coordinate system, show that they are not the same: the latter actually has one more transformation. It is an example of a mixed finite-continuous group, having two subgroups, which are invariant (why?), SO(2), and the two-element group consisting of the identity and the inversion. While SO(2) is clearly Abelian, O(2) is not [Jones (1990), p. 119, 269; Wigner (1959), p. 144]. Verify this, and for both groups, find the maximum set of diagonal elements, and their eigenvalues. Interpret these, and explain how they are related to O(2) being non-Abelian.

X.3.b SO(2) is infinitely connected

For SO(2) we can let θ run from $-\infty$ to ∞; we can rotate many times through 2π. However there is an essential difference between this group and that of translations, though we might take the parameter ranges the same. A translation through 2π takes the object to a point other than the origin, a rotation through 2π returns the object to the origin. Thus for the rotation group, there are an infinite number of operators equivalent to (having the same effect as) the identity (and similarly for every other transformation), these being given by $2\pi n$, for all positive and negative integer n's.

The translation group is singly connected, the two-dimensional rotation group SO(2) is multiply-connected, specifically infinitely-connected

[Cornwell (1984), p. 427]. There are an infinite number of operators giving the identity (or any other transformation);

$$R(2\pi n) = exp(2\pi in) = 1, \qquad \text{(X.3.b-1)}$$

for all integer n. One way of visualizing this multiply-connected group is to take a cord and tie its ends together. Marking the transformations on the cord gives a single one for each angle. However a little loop can be made in it, one end put through this loop, and then the ends tied together, so that it forms a double circle. It is then double valued — for each angle there are two pieces of cord, so two transformations, though they have the same effect. This then can be continued, giving the group infinitely-connected.

Why does this matter? The connectedness of a group determines important aspects of it, including the representations, and the physics described. We return to this when we consider SO(3), and its covering group SU(2), and discuss them together to emphasize the effect of their being multiply-connected.

For group G, covering group C is the (unique) singly-connected group to which it is homomorphic. For SO(2) the covering group is that of translations along a line, this being singly-connected. To each transformation of SO(2) given by $2\pi\theta$, there are infinitely many translations, given by x, where $x = 2\pi n\theta$, for all integer n. The multiplication rules are the same, so the groups are homomorphic. We start below to consider the implications of this.

There are only two one-parameter groups, the infinitely-connected SO(2), which is U(1), and its singly-connected cover, the translation group whose range is infinite. SO(2) and U(1) are the same because the space of one complex number is that of two real numbers. The only further possibility is another infinitely-connected group with range different than 2π; as we saw, this does not give a different group.

Problem X.3.b-1: Check that this is actually a homomorphism.

Problem X.3.b-2: A group is connected k times (only $k = 1$, $k = \infty$, for SO(2), and $k = 2$ for SO(n), $n > 2$, are relevant), if there are k operators equivalent to (equal in value to; giving the same transformation as) the identity. Show that k is the same for all transformations; if there are k for the identity, there are k for transformation T.

Problem X.3.b-3: The homomorphism linking SO(2) and the translation group can be stated in another way (why?): the homomorphic mapping of the multiplicative group of positive real numbers onto the group SO(2) [Cornwell (1984), p. 403].

Problem X.3.b-4: Does O(2) also have a covering group?

X.4 DESCRIPTION OF SO(3)

The first step in studying a group is to describe it: list the transformations and the multiplication table. As these are Lie groups, this means giving parametrizations — the numbers, and ranges, of the parameters — and the product of each pair of transformations, the function giving the parameters of the product in terms of those of the terms in the product. To do this, we realize transformations by matrix functions of parameters; products are found by matrix multiplication.

A realization should be justified, mathematically to show that it gives a group, and one with the required properties, and physically to show that it describes the transformations experiment requires. This description, and justification, for the three-dimensional rotation and SU(2) groups, and for the homomorphism relating them — the explanation of how their transformations are related and why there is this relationship — constitute our next task.

X.4.a Parametrizations of the groups

To start we state the parameters, their ranges, and the transformation functions of the groups and then show — really let the reader show — that these do form groups and do have the properties required. After that we explain how to find these (allowing the reader to generalize to n dimensions).

X.4.a.i *The number of parameters*

First it is necessary to know how many parameters are needed. This can be done in two ways.

Problem X.4.a.i-1: We have to give a set of unitary, and a set of orthogonal, matrices. An $n \times n$ matrix with complex entries is specified by $2n^2$ parameters. If matrix A is unitary — its inverse equals its conjugate transpose (sec. II.3.d, p. 43; sec. VII.2, p. 181),

$$A^{-1} = A^{+}, \tag{X.4.a.i-1}$$

$$\sum A^{+}_{ij} A_{kj} = \sum A^{+}_{kj} A_{ij} = \delta_{ik}, \tag{X.4.a.i-2}$$

then for a U(n) transformation it is given by n^2 parameters, and with unit determinant, for SU(n), by $n^2 - 1$. If it is orthogonal, $\frac{n(n-1)}{2}$ parameters are required. This agrees with the results for two and three dimensions. SO(n) and O(n) have the same number of (continuous) parameters; for O(n) the determinant can be only ± 1.

Problem X.4.a.i-2: Besides finding the number of parameters needed to determine the matrices, we can also consider this physically (perhaps geometrically would be better). For a rigid body in three-space, with one point fixed, two angles are needed for the axis of rotation, and one for the angle rotated about this axis, a total of three. Or for a unit vector along one axis, the transformed vector and a rotation about it, given by three numbers, completely specifies the rotation. In n-space, $\frac{n(n-1)}{2}$ angles are needed. Why? And n complex vectors are specified, up to an overall phase, by $n^2 - 1$ parameters (a -1 because that phase is free).

Problem X.4.a.i-3: For two dimensions, denote the variables on which the group acts by η and ζ; they are transformed as

$$\eta' = a\eta + b\zeta, \tag{X.4.a.i-3}$$

$$\zeta' = c\eta + d\zeta; \tag{X.4.a.i-4}$$

a transformation of L(2), the general linear group on two symbols (the transformed variables are linear functions of η, ζ); thus a, b, c, d are unrestricted and complex, giving eight parameters. SL(2) is a special (unimodular) group — the determinant is 1. Show that this requires

$$ad - bc = 1, \tag{X.4.a.i-5}$$

reducing the number of parameters from four complex ones (that is 8) to 7. For SU(2), for which the products $(\eta\eta)$, $(\zeta\zeta)$ and $(\zeta\eta)$ are required to be invariant, check that

$$c = -b^*, d = a^*. \tag{X.4.a.i-6}$$

This reduces the number of parameters to three. Generalize to L(n), the linear group on n variables (also called GL(n), the general linear group) and SU(n). The numbers of parameters of SU(2) and SO(3) are the same. Is this true in general? Explain how this leads to the dimension of space being 3+1 [Mirman (1986, 1988a,b, 1995a)].

X.4.a.ii *Parametrization of SO(3)*

Coordinates x_i are transformed by matrix R,

$$x_i' = R_{ij}x_j; \tag{X.4.a.ii-1}$$

there are three coordinates so R is a 3×3 matrix ($n \times n$ in n-space). However transformations in three-space are given by three angles, so the nine entries of R are not independent — they are functions of three parameters — thus are not proper parameters for labeling group transformations. R acts on three unit vectors, so has nine parameters, but to give it only one unit vector, and a rotation around it, are needed.

What conditions lead to an acceptable set of three, independent, parameters needed to determine a rotation? Rotations around the x, y, z axes can be used but a rotation around one changes the others. The Euler angles do this, but in a more organized way. There are many conventions; we use a standard one [Goldstein (1953) p. 109, eq. 4-46] and refer to that source for explanations of it and how it relates to others.

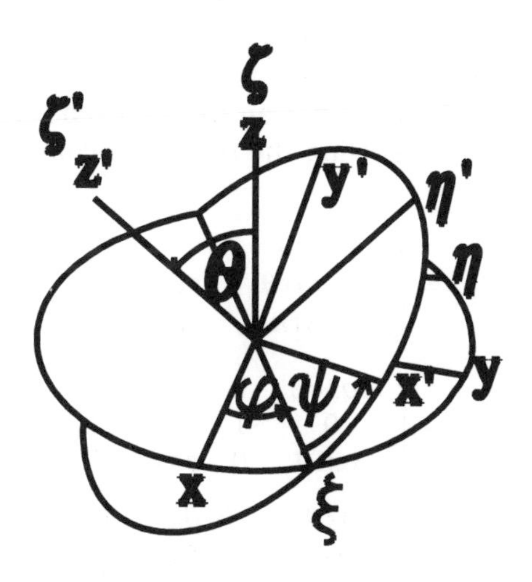

Figure X.4.a.ii-1: EULER ANGLES

We start with a rotation about one axis (here z) through φ, in the direction shown by the arrow, moving the x axis to ξ, and the y axis to η. Then the circle in the xy plane is rotated through θ about ξ, moving η to η', and ζ (which started along z) to ζ'. The final rotation is about ζ' through ψ, giving the new axes, x', y', z'.

The transformation (rotation) matrix, using Euler angles, is

$$R(\theta,\varphi,\psi)=\begin{pmatrix} \cos\psi\cos\varphi-\cos\theta\sin\varphi\sin\psi & \cos\psi\sin\varphi+\cos\theta\cos\varphi\sin\psi & \sin\psi\sin\theta \\ -\sin\psi\cos\varphi-\cos\theta\sin\varphi\cos\psi & -\sin\psi\sin\varphi+\cos\theta\cos\varphi\cos\psi & \cos\psi\sin\theta \\ \sin\theta\sin\varphi & -\sin\theta\cos\varphi & \cos\theta \end{pmatrix}; \tag{X.4.a.ii-2}$$

$$0 \leq \theta \leq \pi, \quad 0 \leq \psi,\ \varphi < 2\pi. \tag{X.4.a.ii-3}$$

The product of two rotations is given by the matrix product using this realization.

Problem X.4.a.ii-1: Any rotation can be written as a product of one about the x axis (through angle θ_x), one around y (through θ_y), and one around z (through θ_z). Why?

Problem X.4.a.ii-2: Check that matrix R is correct (perhaps by writing it as a product of three), and is orthogonal. What is the inverse of R? What parameters, and how many sets, give the identity (transformation)? Find the product law (giving the angles of the product as functions of the angles of the multiplied rotations) [Elliott and Dawber (1987), p. 482]. Write (the simplest form of) the matrix for the product of two of these, and show it is of this form. Why? Verify that these matrices form a group. Also show that every 3×3 unimodular orthogonal matrix can be written in this form, with the stated range — these matrices form the SO(3) group. Can the angles be complex?

Problem X.4.a.ii-3: Labeling the rows and columns of the matrices by x, y, z, let

$$R_x(\theta_x) = \begin{pmatrix} 1 & 0 & 0 \\ 0 & cos\theta_x & sin\theta_x \\ 0 & -sin\theta_x & cos\theta_x \end{pmatrix}, \qquad (X.4.a.ii-4)$$

$$R_y(\theta_y) = \begin{pmatrix} cos\theta_y & 0 & sin\theta_y \\ 0 & 1 & 0 \\ -sin\theta_y & 0 & cos\theta_y \end{pmatrix}, \qquad (X.4.a.ii-5)$$

$$R_z(\theta_z) = \begin{pmatrix} cos\theta_z & sin\theta_z & 0 \\ -sin\theta_z & cos\theta_z & 0 \\ 0 & 0 & 1 \end{pmatrix}; \qquad (X.4.a.ii-6)$$

$0 \le \theta_x, \theta_y, \theta_z < 2\pi$. Write the matrix for the general transformation

$$R = R_x(\theta_x)R_y(\theta_y)R_z(\theta_z). \qquad (X.4.a.ii-7)$$

What axes are these rotations about [Jones (1990), p. 119, 269; Tung (1985), p. 123, Aivazis (1991), p. 45]? Write the expression for a rotation in terms of rotations about the coordinate axes. Is the range correct? Show that R is orthogonal. What is its inverse? Write the matrix for the product transformation and give the product law. Put this matrix into the form of a rotation about a single axis; find the axis and angle [Goldstein (1953), p. 142]. Any rotation is the same as that about a single line since every real orthogonal matrix has an eigenvector with eigenvalue +1 — Euler's theorem [Goldstein (1953), p. 118; Hamermesh (1962), p. 325; Miller (1972), p. 18]. Repeat this for the rotation matrix expressed in Euler angles. Give the relationship between these two parametrizations, relating the parameters [Jones (1990), p. 119, 269; Tung (1985), p. 99, 123; Aivazis (1991), p. 45; Varshalovich, Moskalev and Khersonskii (1988), p. 26]. Are there reasons to prefer one of them?

Problem X.4.a.ii-4: Prove that a rotation — a proper, orthogonal transformation — has an axis (an invariant line), whether the transformation parameters are real or complex, for any odd-dimensional space. What changes if the transformation is not proper — it includes an inversion? The space must be odd; the two-dimensional space is a counter-example. Show that no even-dimensional space has an axis. Why?

Problem X.4.a.ii-5: Why must the range of the angles be the stated one? Given an arbitrary vector, show that R acting on it gives all vectors that start at the same point and have the same length, for this range, but not for a smaller range, and gives some vectors more than once for a larger one. Under the set of all rotations, a unit vector traces out the surface of a sphere. If it did not there would be directions to which it was not rotated. Show, that with this range, the entire spherical surface is covered, once, but is not for a smaller range, and is covered more than once with a larger one. Also more fundamentally, if the range is smaller, there are rotations, no matter how small the range of angles, whose product is outside that range. Prove this and explain why it is fundamental. But, using the properties of sines and cosines, all products are inside the stated range. Thus the group axioms require this range, once the product rule is given. The parameter space is the inside of a sphere. The latitude and longitude are given by θ (that $0 \le \theta \le \pi$ instead of $-\frac{\pi}{2} \le \theta \le \frac{\pi}{2}$ does not matter) and φ. The radius, going from 0 to 2π, is given by ψ. Check that the space over which the transformations are defined is that of the surface of a sphere whose points are labeled by θ and φ.

Problem X.4.a.ii-6: Interpret the product law physically. Does the law found from these matrices (both parametrizations) agree with experiment? Also it should show that these matrices do not commute. This can be checked mathematically, and using a few (simple) sets of angles, physically. Do the results agree?

X.4.b Why rotations do not commute

The order in which rotations are performed on an object is important. Different orders give different final orientations [Tinkham (1964), p. 96]. Why? What properties of rotations require that they not commute? One answer is that if they did commute they would be translations; there are three of these which do commute. And this offers some clues.

Rotations change the axis of rotation so they do affect each other, as translations do not. The rotation group is simple, otherwise a subgroup of rotations would be invariant, thus distinguished from the others, implying all directions of space are not the same. But there is only a single simple, three-parameter, compact group (up to homomorphism) and this determines that rotations do not commute, and their product rule. (Care is needed with these arguments; in four dimensions, only, rotations break up into two independent sets.)

Problem X.4.b–1: In fact, for a simple group, not all pairs of transformations commute. Why? Can any?

X.4.b.i *Noncommutivity and the bounded parameter range*

There is a fundamental difference between rotations and translations, which we used even before finding whether they commute and what their product rules are. For rotations the parameter space is bounded, for translations unbounded. Is this relevant to translations commuting, rotations not, and to their product rules?

The product rule for the rotation group, and that its operators do not commute, follows from its matrix realization. This in turn follows from an arbitrary rotation being the product of three rotations around three different axes, and the form of these. But that, in a way similar to SO(2), follows from, and is allowed by, the parameter range being finite (pb. X.3.a–5, p. 275). Thus the rotation product rule is closely related to the ranges of its parameters.

Problem X.4.b.i–1: The rotation group is also the group of (here three-dimensional) orthogonal matrices, with real matrix elements. Take a general 3×3 matrix, so with nine parameters, all real, and restrict it to be orthogonal. Show that it must have parameters expressed in terms of only three numbers, with the functional form sinusoidal; the numbers must be interpreted as angles, in that they have bounded range appropriate for angles. Show that the expression for R in terms of Euler angles (eq.X.4.a.ii–2, p. 279) is the most general one for an orthogonal matrix, with real entries (any other is a disguised form of it). The orthogonality condition then gives the rotation group, the product rule, the noncommutivity of rotations, and the range of angles. The number, type and range of the parameters are not arbitrary, but are determined by, and determine, the group. (It is outside the scope of this discussion, though not necessarily of the reader's imagination, to allow parameters to be complex, impose orthogonality, which is more restrictive than unitarity, find the general form of the matrices, thus the product rule, and figure out how the resultant group is related to SO(3) [Mirman (1995a)].)

Problem X.4.b.i–2: The geometrical requirement that there be a (sub) set of transformations that leave products of real vectors invariant — physics gives a stronger result, its laws are invariant under these — has major consequences for physics, for example in the role angular momentum plays, although the consequences go well beyond this. The reader might wish to explore whether there are geometries, besides the analogous one with complex vectors, for which this does not hold, and what physics might be like, if it is possible, in them.

X.4.b.ii *Commutators of rotations*

In studying the rotation group we need the rule for the product of two transformations and also for the commutator $R_1^{-1}R_2^{-1}R_1R_2$ — if R_1 and R_2 commute this equals 1, so it gives the deviation from the identity due to noncommutivity.

Problem X.4.b.ii-1: If R_1 commutes with R_2, does R_1^{-1}? Obviously $R_i(\theta)$ and $R_i(\theta')$, rotations about the same axis, commute. So we consider $R_x(\theta_x)$ and $R_y(\theta_y)$. Write the matrix of the commutator of these two transformations when both angles are $\frac{\pi}{2}$. The commutator is a rotation. About what axis is this rotation, and what is its angle? Try this when the angles are $\frac{\pi}{4}$. Also do the cases of one angle $\frac{\pi}{2}$ and the other $\frac{\pi}{4}$, and one $\frac{\pi}{2}$ and the other arbitrary. Repeat for arbitrary angles; find the axis and angle of the rotation. Do the same for the other commutators. Verify that these agree with experiment. Is it possible to write a general formula for the commutator of two rotations giving the matrix and the axis and angle of rotation? Is it possible to find these experimentally? Do theory and experiment agree?

X.4.c The classes of the rotation groups

A class is a set of elements transformed into each other by all similarity transformations using group elements. For the rotation groups, what sets are invariant under such transformations (sec. IV.5.d, p. 125)? To find the class having as a member rotation $R_s(\theta)$ about axis s, we perform a similarity transformation with arbitrary rotation $R_t(\varphi)$, about axis t,

$$R_w(\psi) = R_t(\varphi)^{-1}R_s(\theta)R_t(\varphi). \tag{X.4.c-1}$$

Now $R_t(\varphi)$ changes s to axis s'. $R_s(\theta)$ is then a rotation through some θ about s'. Then $R_t(\varphi)^{-1}$ brings s' to s, and cancels the change of angle. Thus $R_w(\psi)$ and $R_s(\theta)$ are rotations through angle θ but about different axes. The class to which $R_s(\theta)$ belongs consists of rotations through θ about every direction in space. The classes are labeled by θ; thus there are an infinite number of classes, each containing an infinite number of elements [Cornwell (1984), p. 434].

Problem X.4.c-1: If $t = s$ then

$$R_w(\psi) = R_s(\varphi)^{-1}R_s(\theta)R_s(\varphi) = R_s(\theta). \tag{X.4.c-2}$$

Prove this using all expressions for SO(2) rotations. Check it experimentally.

Problem X.4.c-2: Try the general case experimentally and verify that the original and transformed rotations are through the same angle, but about different axes. How are these axes related to the transformation axis t and angle φ? Given the original and transformed axes, are t and φ determined, or only a relationship between them?

Problem X.4.c-3: Axes of rotations are given by two angles so the classes are labeled by the angle of rotation, an element of a class by its axis, so by two angles. How is this shown? A rotation through θ about z is given by

$$R_z(\theta) = \begin{pmatrix} cos\theta & sin\theta & 0 \\ -sin\theta & cos\theta & 0 \\ 0 & 0 & 1 \end{pmatrix}, \quad \text{(X.4.c-3)}$$

when operating on vector (x, y, z). Find the matrix for a rotation through θ about z', where z' and z are at an angle ω. Does the ambiguity in this last sentence matter? Show that a similarity transformation by a group operator transforms the first matrix into the second, and there is no such transformation if the values of θ differ.

Problem X.4.c-4: From the expression for the rotation matrix (use both the simple one and the general one) find the character of each class. Use this to show that transformations of different angles are in different classes. Can this be used to show that all transformations with the same angle are in the same class?

Problem X.4.c-5: Find the classes of SO(n).

Problem X.4.c-6: What are the classes of O(2), O(3), O(n) [Tung (1985), p. 243; Aivazis (1991), p. 97; Wigner (1959), p. 145]?

X.5 THE UNITARY, UNIMODULAR GROUP SU(2)

Besides the rotation group there is another three-parameter, simple, compact group, SU(2), that of 2×2 unitary matrices, the group of transformations on two-dimensional complex vectors leaving fixed their product $|(\omega, \psi)|$, thus the norm of a complex vector $|(\psi, \psi)|$. Because both SO(3) and SU(2) are three-parameter, compact, Lie groups we expect them to be related. They are different, so they are not isomorphic, but homomorphic. First we describe SU(2) then study this relationship.

X.5.a Parametrization of SU(2)

The group of all 2×2 unitary matrices is U(2); if the matrices are unimodular — their determinant is 1 — it is the special unitary group SU(2). One approach is to parametrize the general 2×2 unitary, unimodular matrix. These matrices, with the parameters varying over their entire range, realize the group. SU(2) and SO(3) are given by their product laws, not by specific realizations. But these are useful and so we tend to forget that they are realizations, not the groups. Although it probably does not matter if they are confused, one never knows, and it might in more sophisticated cases.

Problem X.5.a–1: Show that the general 2×2 unitary matrix is

$$U = \begin{pmatrix} a & b \\ -b* & a* \end{pmatrix}. \quad \text{(X.5.a–1)}$$

If it is unimodular

$$|a|^2 + |b|^2 = 1. \quad \text{(X.5.a–2)}$$

Since a and b are complex numbers, with one condition on them, this is a three-parameter group. Derive from this the general transformation matrix [Goldstein (1953) p. 116, Eq. 4-67],

$$Q(\omega, \varphi, \theta) = \begin{pmatrix} exp\frac{i(\omega+\varphi)}{2}cos\frac{\theta}{2} & iexp(\frac{i(\omega-\varphi)}{2})sin\frac{\theta}{2} \\ iexp(\frac{-i(\omega-\varphi)}{2})sin\frac{\theta}{2} & exp(\frac{-i(\omega+\varphi)}{2})cos\frac{\theta}{2} \end{pmatrix}, \quad \text{(X.5.a–3)}$$

$$0 \leq \theta \leq \pi,\ 0 \leq \omega, \varphi < 4\pi, \quad \text{(X.5.a–4)}$$

with parameter range chosen so that the transformation matrix takes all values possible for it. Check that Q has the same form as U. Justify the range. Should it be $0 \leq \omega$, $\varphi < 4\pi$ or $0 \leq \omega$, $\varphi < 2\pi$? Take the angles to run over a smaller range and show that there are products of these matrices with parameters outside that range; for a larger range there are parameter values giving the same matrix more than once. Show directly that Q is unitary and unimodular. What is the inverse of U and of Q? What is the identity? Prove that both sets, U and Q, form a group. Write the matrix (in simplest form) for the product transformation of two (for both U and Q). Show that it is of the same form. Find the parameters of this matrix in terms of those of the matrices being multiplied and so the product law of SU(2). Can the parameters be complex? Explain. While the two-dimensional representation of the U(1) group can be reduced — the matrix can be diagonalized giving two one-dimensional representations — show that the SU(2) matrix cannot be; the basis functions would depend on the transformation parameters. (What is wrong with that?) Thus this realization, which is also a representation, is the smallest SU(2) one. Explain why eq. X.5.a–2, defines a four-dimensional sphere. This is (called) the group manifold of SU(2), and spheres being simply connected, it is a simply connected group.

X.5.b The effect of SU(2) on complex vectors

The group can also be interpreted as the set of transformations leaving the product of two-component complex vectors invariant. Of course it is important to check that both interpretations give the same group.

Problem X.5.b-1: Show that under 2×2 unitary matrices, these products, including norms, are invariant, and that a matrix giving such invariance is unitary (and unimodular?). Check that U and Q satisfy. Why is unimodularity imposed? What would be the effect on the vectors if it were not? Verify that the determinant of a unitary matrix has absolute value 1; it has only a phase. A complex two-dimensional vector has two complex components, it is given by four, and two such vectors by eight, numbers. Unitary transformations do not change their magnitudes (giving two conditions), the angle between them, or their relative phase, giving two more conditions, leaving four parameters. Check that an arbitrary unitary matrix varies both phases by the same amount, the value of its determinant, changing the overall phase. The unimodularity condition keeps this phase unchanged, giving then a simultaneous rotation of both vectors; in addition the group can change the relative phase between two complex coordinates. But it is not merely that we wish to restrict the transformations to only these rotations and changes of relative phase, there is a group-theoretical reason for doing so. Generalize U and Q to unitary matrices with arbitrary determinants, thus matrix realizations of U(2), and check that the set of matrices with the same parameters, but different determinants form a group, thus a subgroup of U(2) — the group of the determinants. Of course the matrices with all values of the parameters, but one value of the determinant, in particular 1, form a group, the SU(2) subgroup of U(2). Both are invariant subgroups; U(2) is not simple, though SU(2) is. (How do we know?) So we can consider each of these subgroups separately. The subgroup of determinants is U(1) for obvious (?) reasons, which is also SO(2); that we have considered. Here we are looking at the other invariant subgroup, SU(2). (The more adventurous reader might like to show this is true for U(n); it has two invariant subgroups U(1) and SU(n), and U(1) changes the overall phase of the n n-component vectors.) There is a difference, in this sense, between U(2) and O(3); both have invariant subgroups, SU(2) and SO(3), but the factor groups are U(1) for U(2), which is a Lie group, but for O(3) it is the two-element finite group.

Problem X.5.b-2: What is the parameter space of SU(2)? What is the space over which the transformations are defined?

Problem X.5.b-3: Find the classes of SU(2) and SU(n) [Cornwell (1984), p. 434].

X.6 THE SU(2), SO(3) HOMOMORPHISM

Both SU(2) and SO(3) are three-parameter, compact, simple groups. With the number of parameters so small, and the simplicity and compactness conditions so strong, we expect that there are not many such groups. There are in fact only these two. But should we expect that they are related [Cornwell (1984), p. 63; Hamermesh (1962), p. 348; Tung (1985), p. 125; Wigner (1959), p. 157]?

Problem X.6-1: One way of approaching this question is by studying the matrices. Given a set of three parameters for the SU(2) matrix we obtain a matrix which we can regard as one for SO(3), since this also depends on three parameters. Try this using simple values, like those giving rotations of 0, or π or 2π, about different axes. It should be clear that for every set of three SU(2) parameters there is a rotation of SO(3), and that for every such rotation there are values of these parameters giving it, but for every rotation there are two sets of SU(2) parameters. Find a general relationship between the SU(2) and SO(3) parameter sets. Also this shows that the multiplication rules of SU(2), as obtained from this matrix realization, and of SO(3), as obtained from its matrix realization, or from experiment, are the same, taking into account that there are two SO(3) matrices for each of SU(2). Thus the groups are related. But this is not a systematic way of seeing it, nor does it explain why they are related — why they are homomorphic, which is what this analysis shows (why?).

X.6.a The relationship of complex vectors to real vectors

SU(2) is the group of transformations in a two-dimensional space of complex vectors, SO(3) that of real three-dimensional vectors. That the groups are related means that a transformation on one space gives transformations on the other, thus relating the vectors. And that the vectors are related means the groups are. What is the connection between these different types of objects?

Problem X.6.a-1: An SU(2) transformation can be written

$$\eta' = a\eta + b\zeta, \qquad \text{(X.6.a-1)}$$

$$\zeta' = -b^*\eta + a^*\zeta, \qquad \text{(X.6.a-2)}$$

with

$$|a|^2 + |b|^2 = 1. \qquad \text{(X.6.a-3)}$$

Define a real three-component vector in terms of (η, ζ) by

$$x = \frac{(\eta^2 - \zeta^2)}{2}, \qquad \text{(X.6.a-4)}$$

$$y = \frac{(\eta^2 + \zeta^2)}{2i}, \quad \text{(X.6.a-5)}$$

$$z = \eta\zeta, \quad \text{(X.6.a-6)}$$

[Hamermesh (1962), p. 348] and show that when (η, ζ) is transformed, it transforms as

$$x' = \frac{1}{2}(a^2 - b^{*2} - b^2 + a^{*2})x + \frac{i}{2}(a^2 - b^{*2} + b^2 - a^{*2})y + (ab + a^*b^*)z, \quad \text{(X.6.a-7)}$$

$$y' = -\frac{i}{2}(a^2 + b^{*2} - b^2 - a^{*2})x + \frac{1}{2}(a^2 + b^{*2} + b^2 + a^{*2})y - i(ab - a^*b^*)z, \quad \text{(X.6.a-8)}$$

$$z' = -(a^*b + ab^*)x + i(a^*b - ab^*)y + (aa^* - bb^*)z. \quad \text{(X.6.a-9)}$$

Check that this leaves $x^2+y^2+z^2$ invariant and that the coefficients are real. Write the matrix acting on (x, y, z), verify that it is orthogonal with unit determinant so of SO(3), and using its parametrization express a and b in terms of angles, and conversely. The x, y, z transformation is a rotation. Verify that every rotation can be so given by writing an SO(3) matrix in terms of a, b. This gives a relationship between SO(3) and SU(2) matrices, so transformations, with two SO(3) matrices for each of SU(2). Note that the SO(3) angles depend quadratically on those of SU(2). The product rules for the groups are the same; if SO(3) matrices $M(\omega_1)$ and $M(\omega_2)$ go with (are mapped to) SU(2) matrices $U(\zeta_1)$ and $U(\zeta_2)$ respectively (ω and ζ are the sets of parameters), then $M(\omega)$ goes with $U(\zeta)$, where

$$M(\omega) = M(\omega_1)M(\omega_2), \quad \text{(X.6.a-10)}$$

$$U(\zeta) = U(\zeta_1)U(\zeta_2). \quad \text{(X.6.a-11)}$$

Given the M's the U's are fixed, but for each U there are two M's. Take a pair of U's, picking reasonable sets of parameters, find the corresponding M's, take the products, giving one product for the U's but four for the M's, and check that there are only two distinct ones, both going with the product of the U's. Explain how the two-to-one relationship between M's and U's is related to (x, y, z) being quadratically dependent on terms in η, ζ. This establishes a — two-to-one — homomorphism between SU(2) and SO(3) (why?); for each SU(2) transformation there is a rotation, the angles of a rotation give an SU(2) transformation, and the product rules are the same, but SO(3) parameters depend on the squares and products of the SU(2) ones, so there are two SU(2) transformations for each rotation.

Problem X.6.a-2: To relate the SU(2) parameters to angles we first let

$$a = exp(\frac{i\varphi}{2}), \;\; b = 0, \quad \text{(X.6.a-12)}$$

which is equivalent to (showing the linear-to-quadratic relationship)

$$a^2 = exp(i\varphi). \qquad \text{(X.6.a-13)}$$

(This linear-to-quadratic relationship is, of course, connected to one set of parameters being expressed in terms of angles, the other in terms of half-angles.) The coordinate transformation reduces to

$$x' = xcos\varphi - ysin\varphi, \quad y' = xsin\varphi + ycos\varphi. \qquad \text{(X.6.a-14)}$$

This is a rotation about z through φ. Next let

$$a = cos\frac{\theta}{2}, \quad b = sin\frac{\theta}{2}. \qquad \text{(X.6.a-15)}$$

Show that this is a rotation about y through θ. Write $sin\theta$ and $cos\theta$ in terms of a and b. Finally let

$$a = exp(\frac{i\omega}{2}), \quad b = 0, \qquad \text{(X.6.a-16)}$$

a rotation again around z, now through ω. Write the matrices giving these three transformations. A general transformation is a product of these three. Show that it is the same as Q (eq. X.5.a-3, p. 285), and further that the three angles can be taken as Euler angles. Find their ranges in terms of those of a and b. This makes explicit the relationship between the SU(2) parameters and the Euler angles of a rotation [Goldstein (1953), p. 109; Varshalovich, Moskalev and Khersonskii (1988), p. 24].

Problem X.6.a-3: Why can an arbitrary SU(2) transformation be written in this form? Were any limitations introduced on these three choices of parameters? Can the ranges for the parameters for one group be found from those of the other?

Problem X.6.a-4: The homomorphism between SO(3) and SU(2) gives a set (why?) of transformations, those between the parameters of the two groups. How much arbitrariness is there? Why? Do these transformations form a group?

X.6.b The rotation group is doubly connected

The translation group is simply connected, there is one value giving the identity (or any other transformation), U(1) is infinitely connected, there are an infinite number of transformations equivalent to the identity (or any other). How multiply connected are the SU(2) and SO(3) groups? As the parameters run through their range each gives a single SU(2), and two SO(3), transformations. SU(2) is singly connected,

SO(3) doubly so. There are two transformations of rotation group SO(3) equivalent to the identity, whereas for SO(2) there are an infinite number; SO(3) is doubly connected, though its SO(2) subgroup is infinitely connected — the connectivity of the subgroup does not give that of the group. Thus the reason that SO(3) is doubly connected is that there are two, and only two, elements (two sets of angles) corresponding to each transformation. Here we consider this.

Using eq. X.4.a.ii-2, p. 279, for rotation $R(\theta, \varphi, \omega)$ we see that for a rotation about z ($cos\theta = 1$) there are two rotations that leave an object unchanged: the identity,

$$cos\chi = 1, \quad cos\varphi = 1, \qquad \text{(X.6.b-1)}$$

and

$$cos\chi = -1, \quad cos\varphi = -1. \qquad \text{(X.6.b-2)}$$

For the latter both rotations are about the z axis, so we rotate first through π and then again through π.

Problem X.6.b-1: Try this experimentally. Prove that there are only two elements that give the identity transformation (not the identity element), and also that there are two corresponding to any transformation.

X.6.b.i *Why the rotation group is not infinitely connected*

The rotation group is doubly connected because there are two values of the parameters giving, say, the identity. However we also get the identity for

$$\chi = 2\pi m, \quad \varphi = 2\pi n, \qquad \text{(X.6.b.i-1)}$$

m, n integers. Why do we not regard it as infinitely connected? To obtain all SO(3) transformations we must take $0 \leq \varphi, \chi < 2\pi$, but the parameters are limited to this; n, $m > 1$ should not be considered — they replicate transformations already given. In this range there are two points that give the identity transformation, and there is no way of reducing this number. It is because the minimum number is 2 that the group is doubly, not singly, not infinitely, connected.

Problem X.6.b.i-1: By the same argument, using eq. X.5.a-3, p. 285 for SU(2) transformations, this group is singly connected. Show this explicitly: there is only one set of parameters, all angles 0, giving the identity transformation (and likewise any another).

X.6.b.ii *Connectivity of groups and subgroups*

SO(2) is infinitely-connected, there are an infinite number of operators corresponding (mapped) to the identity, while SO(3) is merely doubly-connected, only two operators are mapped to the identity. What happened to the other operators? In one dimension the parameter space of the covering group is infinite, for one-parameter subgroups of both SO(3) and its covering group SU(2), the parameter space is finite. How does it being a subgroup result in this limitation on the parameter range?

For the translation group there is no restriction on the parameter range, for SU(2), and therefore any subgroup, there is, because of the unimodularity condition. The translation group, though a one-parameter group, is not a subgroup of SU(2). However U(1) is. While U(1) is infinitely connected, the U(1) subgroup of SU(2) is only singly so, since it cannot have a higher connectivity than its (singly-connected) group. This can be seen from the parametrization of eq. X.5.a–1, p. 285. However in the parametrization using angles, it might seem that there is an infinite number of angles giving the identity transformation, making the group infinitely connected. But if it is singly-connected in one parametrization it must be so in all; the other values of the angles give the same transformation.

The realization of a subgroup is inherited from that of its group, and may thus be different from that of a group to which it is (abstractly) isomorphic. This also emphasizes that the definition of the group elements can be important. For U(1), $\theta = 0$ and $\theta = 2\pi$ label different group elements, though they give the same transformation. But for a U(1) subgroup of SU(2) they label the same group element; for it, the proper parametrization is that of eq. X.5.a–1, p. 285, and these two values of the angle give the same values for a, b, so the same group operator. It is sometimes necessary to make these subtle points about definitions explicit. Because of them there is a distinction between U(1) and the U(1) subgroup of SU(2) — whether this matters has to be seen. The choice of the parametrization is somewhat arbitrary, thus it may seem that whether a group is singly connected or infinitely so, is a matter of the convention we use. That it matters, and that the group really is singly-connected, will be seen when its representations are considered, for it is here that this has meaning. There are restrictions on the type of parameters for a group and these are important in determining the group properties.

The limitations on the ranges can be seen geometrically. If all parameters had infinite range, the group space would be that of a translation group. If some were infinite, but not all, there would be a basic difference between the angles, but then the group could not describe rotations in a space in which all directions are equivalent.

Problem X.6.b.ii–1: Show that if the range of θ were finite, but that of φ infinite, or the reverse, then the space for which this is a rotation group (is it a group?) is not isotropic. Actually the ranges of these angles are different, though both are finite; why can this be?

X.6.c Why only two groups?

The unimodular unitary group is singly connected, there is only one set of parameters that give the identity, the rotation group is doubly connected, there are two sets of parameters that give the identity, so there is a two-to-one homomorphism between the groups. This suggests that there might be three-parameter groups, triply-connected, giving a three-to-one homomorphism, or perhaps one giving a four-to-one homomorphism, and so on. In fact, there are only two three-parameter simple groups, for the same reason that there are two and that they are homomorphic. Every complex number, so every transformation on a complex number, gives a real one, its absolute square, so a transformation on a real number. Every real number, and corresponding transformation, gives complex numbers, and corresponding transformations, though the complex number is not unique. Thus there is a relationship between transformations on complex numbers and on real ones.

While the absolute square of a complex number gives a real one, the cube, or fourth power, ..., does not give any new type of number. There are only two, real and complex. Thus there are two groups which are related — homomorphic — and exactly two. The homomorphism between an orthogonal and a unitary group comes from, and reveals, a relationship between two types of numbers, real and complex.

Problem X.6.c–1: The reader might wish to write down relations, analogous to those for (x, y, z) in terms of (η, ζ), but instead of quadratic expressions, cubic or quartic or ... ones, and see what goes wrong. The three-parameter group of transformations on complex numbers generates a three-parameter group on real ones. Is it possible to get another similar, but distinct, three-parameter group?

X.7 ANGULAR MOMENTUM OPERATORS AND THEIR ALGEBRA

The laws of physics are invariant under the rotation group, and this leads to conservation of angular momentum. But what is angular momentum, and in what way is it related to the rotation group [Varshalovich, Moskalev and Khersonskii (1988), p. 36]? Why is it conserved? Why does it have the form that it does, why is it related to linear momentum in the way that it is? In quantum mechanics we learn that the components of angular momentum do not commute, so cannot be simultaneously diagonalized. And the diagonal operators have a certain, limited, set of eigenvalues. Why do angular momentum components have the commutation relations that they do, the eigenvalues that they do? Why are they related as they are? And what does all this have to do with the rotation group?

The characteristics of angular momentum that we learn in both elementary physics and quantum mechanics are really expressions of properties of SO(3), and the fundamental experimental, and geometrical, fact that the objects of physics are transformed by it, and further that physical laws are invariant under the transformations. This we study next, so introducing a key concept, Lie algebras.

X.7.a The transformation matrices when angles are small

The product of two group operators is a group operator. If we limit the parameters to a small region, then there are products of operators from this region that are outside it. Continuing we get the whole group. This raises the question whether knowing the group in some small region — say around the identity — might give the entire group and (almost) uniquely [Hamermesh (1962), p. 293]. So what is the rotation group like in the neighborhood of the identity [Elliott and Dawber (1987), p. 135; Inui, Tanabe, and Onodera (1990), p. 119; Jones (1990), p. 101; Schensted (1976), p. 258; Tung (1985), p. 99]?

X.7.a.i *The SO(2) group for small angle*

A two-dimensional realization of SO(2), to first order, is

$$M(\theta) = \begin{pmatrix} 1 & 0 \\ 0 & 1 \end{pmatrix} + \theta \begin{pmatrix} 0 & 1 \\ -1 & 0 \end{pmatrix}. \qquad \text{(X.7.a.i-1)}$$

Problem X.7.a.i-1: Why? Multiply $M(\theta)$ by itself several times and check that it leads to eq. X.3.a-2, p. 273, for finite θ. In this case having the expression for the group operator in the neighborhood of the identity does (almost?) lead back to the group. (This can easily be shown more rigorously.) Can the translation group be obtained this way? What is the difference?

Problem X.7.a.i-2: Thus we have an operator, in matrix form, that gives the group near the identity, and from which the entire group can be obtained by taking products — by exponentiation. Justify the last word. How do we know that we get the full group by taking products?

X.7.a.ii *The SO(3) and SU(2) groups when parameters are small*

The SO(2) group, having a single parameter, so a single operator, is not very interesting. Can anything similar be done with more complicated groups? Would it be useful? As the covering group SU(2) has earned the right to be considered first.

Problem X.7.a.ii–1: From the expression (eq. X.5.a–3, p. 285) for the SU(2) matrix, find the realization when the parameters are small and show that, to first order, it can be written as a sum of the 2×2 unit matrix (of course) plus these parameters multiplied by (sums of) Pauli matrices — which any 2×2 matrix can, although the coefficients might be arbitrary functions — but with real coefficients, this being more restrictive. Moreover the Pauli matrices

$$\sigma_z = \begin{pmatrix} 1 & 0 \\ 0 & -1 \end{pmatrix}, \quad \sigma_x = \begin{pmatrix} 0 & 1 \\ 1 & 0 \end{pmatrix}, \quad \sigma_y = \begin{pmatrix} 0 & -i \\ i & 0 \end{pmatrix}, \qquad \text{(X.7.a.ii–1)}$$

can be seen to obey the commutation relations

$$[\sigma_i, \sigma_j] = 2i\varepsilon_{ijk}\sigma_k, \qquad \text{(X.7.a.ii–2)}$$

where the commutator is

$$[a, b] = ab - ba; \qquad \text{(X.7.a.ii–3)}$$

ε_{ijk} is the totally antisymmetric symbol; it is 0 if two indices are equal, +1 if they are in the order 123 (or xyz), and -1 in the order 132 (or xzy),

$$\varepsilon_{iij} = \varepsilon_{iji} = \varepsilon_{ijj} = \cdots = 0, \quad \varepsilon_{231} = \varepsilon_{312} = \cdots = 1, \quad \varepsilon_{321} = \cdots = 1. \qquad \text{(X.7.a.ii–4)}$$

Check that all σ_i are hermitian. Why are there four operators? It is somewhat harder to show that if a set of transformations is written in this form — the sum of coefficients times the 2×2 unit matrix plus the Pauli matrices — then exponentiation gives the SU(2) finite-angle realization. Thus these operators, the Pauli matrices, give the group. What do the Pauli matrices give when exponentiated? Find the determinants of the operators obtained by exponentiating the Pauli matrices. Do these have unit determinants? Or is there a subset that does?

Problem X.7.a.ii–2: What would happen if the coefficients were not real, would a group be obtained upon exponentiation? If so which? How is it realized? What is its relationship to SU(2)? What might its physical (or geometrical) significance be? Unfortunately discussions of these questions are outside the scope of this book, but some readers might wish to think about them.

Problem X.7.a.ii–3: It is reasonable to try the same expansion for SO(3) and not difficult to show, using its realization, eq. X.4.a.ii–2, p. 279, that for small angles any rotation can be written as a sum of the 3×3 unit matrix plus three (hermitian) matrices, these multiplied by sums of the angles, with real coefficients, and these three matrices are [Schiff (1955), p. 146]

$$L_z = \begin{pmatrix} 1 & 0 & 0 \\ 0 & 0 & 0 \\ 0 & 0 & -1 \end{pmatrix}, \; L_x = \begin{pmatrix} 0 & 1 & 0 \\ 1 & 0 & 1 \\ 0 & 1 & 0 \end{pmatrix}, \; L_y = \begin{pmatrix} 0 & -i & 0 \\ i & 0 & -i \\ 0 & i & 0 \end{pmatrix}, \tag{X.7.a.ii-5}$$

which is more interesting, as not all 3×3 matrices can be written as a sum of these four. Moreover show that these obey the commutation relations (for angular momentum, familiar from quantum mechanics),

$$[L_i, L_j] = i\varepsilon_{ijk}L_k, \tag{X.7.a.ii-6}$$

which is the same as those for the 2×2 Pauli matrices. Why?

X.7.b Lie Algebras of Lie groups

A set of (abstract) symbols whose sum is defined — as we can immediately do here using the matrix realizations — with a product taken as the commutator of the symbols, is called a Lie algebra. Every Lie group has a corresponding Lie algebra, and conversely, though the algebra may give more than one group (or perhaps for algebras of the type that we definitely do not consider here, the group might have unusual properties).

Lie groups SU(2) and SO(3) have the same Lie algebra, though the symbols are different, and are here differently realized, as 2×2 and 3×3 matrices. But both have three symbols (more often called elements, generators or operators) and the same commutation relations, thus the algebras, and around the identity — for small transformations — the groups, are the same. For large transformations they are not, but are related. There can be more than one group with the same Lie algebra; these are, as here, homomorphic.

X.7.c A differential realization of the rotation algebra

A Lie algebra, like a (Lie) group, is a set of abstract symbols with rules for combination, addition plus a product, commutation. We have realized the algebra of SU(2) — and of SO(3) — in two different, though related, ways, as 2×2 and as 3×3 matrices. But these are not the only realizations.

Problem X.7.c-1: Show that the three operators

$$L_p = -i(x_q \frac{\partial}{\partial x_r} - x_r \frac{\partial}{\partial x_q}), \tag{X.7.c-1}$$

where $1 \leq p, q, r \leq 3$, and are in the order 1, 2, 3 (or x, y, z), obey the commutation rules as for the L's (eq. X.7.a.ii-6, p. 295). These form a differential realization of the algebra. Introduce spherical coordinates and show that these operators can be written as

$$L_x = i(sin\varphi\frac{\partial}{\partial\theta} + cot\theta cos\varphi\frac{\partial}{\partial\varphi}), \quad \text{(X.7.c-2)}$$

$$L_y = i(-cos\varphi\frac{\partial}{\partial\theta} + cot\theta sin\varphi\frac{\partial}{\partial\varphi}), \quad \text{(X.7.c-3)}$$

$$L_z = -i\frac{\partial}{\partial\varphi}, \quad \text{(X.7.c-4)}$$

and that they obey the same commutation relations [Schiff (1955), p. 75, 142]. Thus we now have two matrix realization and two differential ones, all obeying the same commutation relations.

Problem X.7.c-2: One way to get this realization of the generators is to introduce variable r, with components x, y, z and

$$p_i = -i\frac{\partial}{\partial r_i}. \quad \text{(X.7.c-5)}$$

Show that

$$L = r \times p, \quad \text{(X.7.c-6)}$$

where $\times$ is the usual cross product. There is good reason for the way angular momentum is defined in elementary physics. The operators $p_\mu, \mu = 1, \ldots, 4$, are the generators of the (four-dimensional) translation group, which is (fortunately) Abelian. It is useful to give them a name; the p_μ are called the momentum (operators).

Problem X.7.c-3: The symbol

$$L^2 = L_x^2 + L_y^2 + L_z^2, \quad \text{(X.7.c-7)}$$

commutes with all three L's, using just the abstract symbols and the commutation relations, or any of the realizations, including the Pauli matrices. Check this all ways [Tung (1985), p. 124; Aivazis (1991), p. 48]. Thus its eigenvalues are constant under rotations. To define this operator a new combination rule, besides the algebra product (commutation), has to be defined, that of the product of symbols, which can be realized as the matrix product. How is it realized if the L's are derivatives? Using spherical coordinates [Schiff (1955), p. 75, 142], show that it is

$$L^2 = -\frac{1}{sin\theta}\frac{\partial}{\partial\theta}(sin\theta\frac{\partial}{\partial\theta}) - \frac{1}{sin^2\theta}\frac{\partial^2}{\partial\varphi^2}, \quad \text{(X.7.c-8)}$$

and is the same as the Laplacian operator (giving the angular part of equations like Schrödinger's equation for spherical potentials). Would you expect this? Why? What does it have to do with the rotation group? What is the relationship of this operator to these Schrödinger equations? Why does it not appear for other Schrödinger equations?

Problem X.7.c-4: Besides the L operators it is often convenient, perhaps necessary, to take linear commutation of them. Show that the operators L_z (also called L_o) and

$$L_{\pm} = L_x \pm iL_y, \qquad \text{(X.7.c-9)}$$

obey the commutation relations

$$[L_z, L_{\pm}] = \pm L_{\pm}, \qquad \text{(X.7.c-10)}$$

$$[L_+, L_-] = 2L_z. \qquad \text{(X.7.c-11)}$$

Obtain the expressions for $L_{\pm}$ for all realizations. Find L_x and L_y, and L^2, in terms of $L_{\pm}$. Show that L^2 commutes with $L_{\pm}$. Find the corresponding forms for the Pauli matrices, and their commutation relations (pb. II.4.g-1, p. 56). While a Lie algebra is unique, the form in which it is written is not.

Problem X.7.c-5: Does it mean anything to exponentiate the differential generators? What? Explain.

Problem X.7.c-6: For SU(2), find the (four) differential operators acting on complex variables z_1, z_2, and those obtained by exponentiation, for both ways of writing the algebra (J_z, J_x, J_y, and J_z, $J_{\pm}$). Verify that the norm of products is invariant under differential operators (which?), and corresponding Pauli matrices, and also under their exponentiated forms. Find the matrix, and the differential and exponentiated operators, that commute with all others. Does the exponentiated operator commute with all differential operators? With all exponentiated operators? Are these questions related?

Problem X.7.c-7: We have two realizations of the transformations of the rotation group, one in terms of angles, the other in terms of the transformation parameters of complex numbers, both using sets of three parameters. This might imply, and intuitively we might expect, that since this is the group of transformations in three-dimensional space, it must be parametrized with three numbers, and that these are angles, or closely related to them. However a group is an abstract concept, and it can have many different interpretations, thus realizations. Show that the generators of the algebra can be parametrized using two variables

$$J_z = \frac{\partial}{\partial y}, \quad J_{\pm} = exp(\pm y)\{\pm\frac{\partial}{\partial x} - k(x) + j(x)\}, \quad E = \mu, \qquad \text{(X.7.c-12)}$$

where E commutes with the J's, μ is a constant (which can be taken 0 to give three generators), and k and j are functions, which have to be determined [Miller (1968), p. 45, 49]. This might not be surprising since we can write rotations in terms of two angles. However show that there is also a realization using a single variable

$$J_z = \lambda + z\frac{d}{dz}, \quad J_+ = j_1(z) - j_2(z)\frac{d}{dz}, \quad J_- = k_1(z) - k_2(z)\frac{d}{dz}, \quad E = \mu; \tag{X.7.c-13}$$

λ, μ are complex constants, the j's and k's functions to be determined. Can the group transformations be written in terms of these parameters?

X.7.d What the commutation relations require

The generators of the algebra have been realized in several ways. How are these related? Do the basis states of a matrix realization, column vectors, have anything to do with those of a differential realization, these functions of parameters? Why can we realize the operators as both 2×2 and 3×3 matrices? Might there be others? The commutation relations allow two of the operators to be simultaneously diagonalized, L^2 and (say) L_z. What are their eigenvalues and eigenvectors, with what limitations; what is the significance, group theoretical and physical, of the eigenvalues, eigenvectors, and limitations? How are they related to angular momentum? Are these the same for the various realizations, say the matrix and the differential ones?

Many of these equations involving the L's are the same as ones in quantum mechanics (except, perhaps, for choice of units so of a constant). Why are the quantum-mechanical equations what they are, how are they determined by the rotation-group Lie algebra?

The eigenvalues, and their meaning, are determined by the Lie algebra, not by a realization, as are many of the properties, though not necessarily the specific form, of the operators and eigenfunctions. There are different realizations, and representations, but the underlying properties are the same. And these give the related quantum-mechanical equations — and much of quantum mechanics.

So this we have to study, the forms of the operators and their eigenfunctions, and the eigenvalues, and how these are determined by the Lie algebra, and their significance. We have to find the representations of the algebra, and the different expressions of these representations, and how, and why, they are related.

X.7.e The physical significance of the rotation Lie algebra

Lie groups lead to further operators, suggestive of their derivatives around the identity, and to a new system, an algebra. The group operators have a clear physical interpretation, that of transformations of objects (or conversely, coordinate systems) but what are those of the algebra (besides generators of small transformations)?

It is clear that the L's are components of angular momentum, both from their expressions as vector products of coordinates and momenta, introduced in elementary physics, and from their related expressions as derivatives in both Cartesian and spherical coordinates, well known in quantum mechanics. There is a slight point that angular momentum operators are multiplied by a constant, but that merely reflects the choice of units and has, besides that, no importance. (To avoid confusion Planck's constant and the speed of light should always, except in calculations, be chosen as 1. This will avoid such absurdities as taking the limits $h \Rightarrow 0$, or $c \Rightarrow \infty$. It really does not make sense to take the limits $1 \Rightarrow 0$, or $1 \Rightarrow \infty$.)

The Lie algebra operators, for the rotation Lie group, are angular momentum components. But why? What does the Lie algebra have to do with angular momentum? However that is not the proper question. The question is why a quantity called angular momentum is introduced into physics, and why it has the properties that it does.

That transformations in space (not merely symmetries) are given by the rotation group leads to a Lie algebra and its operators. We should expect (should we not?) that these operators, being related so fundamentally to the properties of space, would have physical significance, and would appear in physics even without knowledge of the group that gives rise to them. But there is another reason. The operator L^2 is invariant under rotations — it commutes with all — so should have physical meaning. And these four operators, L_i and L^2 are conserved, they are constant in time. This comes from the Poincaré group (sec. II.3.h, p. 45) which gives a time-translation operator, the Hamiltonian, and one which is rotationally invariant (since space is invariant under the Poincaré group). Thus the Hamiltonian commutes with all angular momentum operators, so they are the same for all time.

These operators have entered physics because they (more properly their eigenvalues) are constant in time, and the importance of that is clear even in elementary physics, and becomes more so in quantum mechanics. Thus these Lie algebra operators are significant because they express a fundamental property of space — transformations are given by the rotation group — and because they are conserved. Their name, and the constant introduced by the units, are merely matters of convention. Their importance, and meaning, come from geometry and from the structure of the group expressing it.

Problem X.7.e-1: Though we have not considered the Poincaré group the reader should be (at least) able to show that commutation with the time-translation operator means that the eigenvalue of an operator is constant in time.

Problem X.7.e-2: Operator L^2 is related to the Laplacian (sec. X.7.c, p. 295), thus to the angular part of Schrödinger's equation for spherical potentials. From this would you expect that L^2 should commute with all rotations? Explain.

Problem X.7.e-3: Repeat this discussion for linear momenta.

Problem X.7.e-4: The Lie algebra of SO(4), thus SO(4), is not simple, being only semisimple; it splits into two isomorphic simple algebras, both (isomorphic to) that of SO(3) [Cornwell (1984), p. 489, 866; Schweber (1962), p. 38; Talman (1968), p. 169; Wybourne (1974), p. 48]. Write its algebra, and check this. Rotations have an axis only in odd-dimensional spaces (pb. X.4.a.ii-4, p. 280); is this relevant?

X.7.f Why Lie algebras are important

We now have two methods for studying Lie groups, directly, and using their algebras. Why bother with doubling our work using two methods instead of just one?

There are several reasons why Lie algebras are important. First, their generators have physical meaning. For the rotation group they give quantities of great importance physically — it is difficult to imagine physics without the concepts of angular (or linear) momentum. And it is difficult to see how the properties of angular momentum would be found without introducing these operators, whether as generators of a Lie algebra or in disguise, as is done in elementary physics and in quantum mechanics. For the translation group the Lie algebra generators are the momenta (whose fourth component is the energy), among the most fundamental of physical quantities.

Also groups are related, as here with SU(2) and SO(3), and this is a consequence, at least in part, of their having the same Lie algebra. Thus the algebra leads to relations between groups, and other properties, that we might not see, or be able to see, in other ways.

Also algebras are simpler than groups, though this might not be obvious looking at their products; commutation might seem more complicated than ordinary (matrix) multiplication. But groups are functions, and complicated functions, of parameters. Algebras are not. Thus algebras are simpler to study, relate, classify and so understand. And their representations, or at least properties of them, and so the representations of the groups, are in ways easier to obtain.

The meaning and properties of many physical quantities come from their being (eigenvalues of) Lie algebra generators. Many of their properties come from the structure of the algebra itself, others from their realizations as matrices. It is perhaps fortunate that matrix multiplication is noncommutative so that matrices can represent Lie algebras. Here we have found the Lie algebras of Lie groups. But Lie algebras are structures defined by their product — commutation. This suggests a richness that may not even be hinted at here. For the present, all that we have room for is an introduction to the Lie algebras of the few Lie groups that we study. It should be emphasized how limited our scope is.

X.8 CONSERVATION LAWS AND SPACE

The quantum numbers, angular momentum and so on, are important because they are conserved. If they changed in time they would not be good state labels. But why are they conserved? Here we briefly illustrate the relationship between conservation and the nature of space, specifically isotropy and homogeneity.

X.8.a Conservation and commutation

The state of a system is described by a statefunction $\psi(x)$, where x schematically represents the variables describing the system. A transformation is an operator that acts on the statevector to produce a different vector. So a transformation labeled by operator L_z (whose relationship to the rotation group has to be specified) is given by

$$L_z\psi(x) = \psi'(x). \qquad \text{(X.8.a-1)}$$

If the state vector is an eigenfunction of this operator,

$$L_z\psi(x) = l_z\psi(x), \qquad \text{(X.8.a-2)}$$

where l_z is the eigenvalue, then this eigenvalue can be (one of) the label(s) of vector $\psi(x)$. If operator A commutes with L_z, there is a statevector $\varphi(x)$ simultaneously an eigenfunction of A and L_z,

$$L_zA\varphi(x) = AL_z\varphi(x) = l_zA\varphi(x), \qquad \text{(X.8.a-3)}$$

so $A\varphi(x)$ is an eigenfunction of L_z.

If the eigenvalues of operator A are conserved (do not change with time), it commutes with that operator responsible for the time development of the system, the (exponential of the) Hamiltonian H,

$$exp(itH)\psi(x,t=0)=\psi(x,t), \quad \text{(X.8.a-4)}$$

and

$$Aexp(itH)\psi(x,t=0)=exp(itH)A\psi(x,t=0)$$
$$=exp(itH)a\psi(x,t=0)=aexp(itH)\psi(x,t=0)=a\psi(x,t). \quad \text{(X.8.a-5)}$$

So the eigenvalue at time t is the same as the eigenvalue at time 0; if the two operators commute (written

$$[A,H]=0), \quad \text{(X.8.a-6)}$$

it is the same for all time.

Problem X.8.a-1: The time-translation operator is (really) the exponential of H, yet the requirement is that A commute with H. Why?

X.8.b Why momentum must be conserved

The eigenfunction of the momentum z-component,

$$p_z=-i\frac{d}{dz}, \quad \text{(X.8.b-1)}$$

is $exp(ipz)$,

$$p_z exp(ipz)=-i\frac{d}{dz}exp(ipz)=pexp(ipz); \quad \text{(X.8.b-2)}$$

the eigenvalue is p. If this were not constant then at time t the eigenvalue would be $(p+\delta p)$ with eigenfunction $exp(i(p+\delta p)z)$. This has the unphysical result that the eigenfunction is not a function of time at $z=0$ but is at any other z. Hence, the eigenvalues of p_z should be conserved and

$$[p_z,H]=0. \quad \text{(X.8.b-3)}$$

Problem X.8.b-1: What assumptions are used in this argument?

X.8.c Homogeneity, isotropy, conservation

This assumes that space is locally homogeneous ("like generated" — the same at all points) and also isotropic ("equal directions" — all directions are the same). Hence, diagonal operators for rotations and translations (angular momentum and momentum operators) should have conserved eigenvalues and these operators should commute with the Hamiltonian. The transformation properties of physical systems, stemming from the properties of space, lead to conservation of angular momentum and momentum (eigenvalues).

The description of physical systems, say, the hydrogen atom and the linear harmonic oscillator, are mandated, in part, by the nature of space, and the consequent requirement that rotation and translation operators commute with the time-translation operator, the Hamiltonian.

One of the uses of group theory is to obtain mechanisms for translating such general properties into specific functions, so for describing given systems.

Chapter XI

Representations of Groups SO(3) and SU(2)

XI.1 REPRESENTATIONS OF GROUPS AND OF THEIR ALGEBRAS

In the previous chapter we described the SU(2) and rotation groups and their algebras. For physical application we need representations. But beyond applications, to physics, mathematics, or elsewhere, these provide insight about the algebras and groups themselves, about the physics, and their relationships. What such insight is, and why for it representations are needed, or at least helpful, is also something we must determine.

There are two ways of finding representations of a Lie group. One is to find the representations of the algebra, then from these the group representations. The other is to find directly those of the group (which can then be used to find — perhaps only some — of the representations of the algebra). Both are useful, with advantages and disadvantages. We therefore study both. Here we begin with the algebra representations, limited to that of the rotation and SU(2) groups. Representations of groups can be obtained directly, without using their algebras (much). This we consider next.

XI.1.a Representations of linear groups

A group representation is a set of matrices, functions of the group parameters, which have the same product law as the abstract symbols making up the group, plus basis vectors — states — on which the matrices act. These vectors are the physical statefunctions and we are interested in the effect of the group transformations on them. So we have to determine what these states are. Also, having them provides a useful way of finding the matrices. The only problem is that "are" is undefined.

There is an important assumption made throughout, one on which the whole discussion, including that for the states, is based. Here we make it explicit, as it should be, and because we need to use it. The groups we are forced by the geometry (or should it be the physics?) to use are linear. They are realized, and defined, by linear transformations:

$$x_i = a_{ij}x_j', \tag{XI.1.a-1}$$

so x depends linearly on x'. It is possible to have nonlinear transformations, in which x is a nonlinear function of x', and also linear groups realizeable using nonlinear transformations [Mirman (1995b)]. However we consider only linear transformations, linear groups and linear realizations. The term representation is used for those realizations that are linear; thus linear representation is redundant.

The analysis uses as a basis, monomials (single terms, or names), and sums of them, multinomials (having multiple terms). A multinomial is a homogeneous polynomial, a sum of terms of the same degree, say $3z_1^3 + 5z_1^2z_2 + 7z_2^3$, a general polynomial (many names) is a sum of terms of different degrees. The basis states of irreducible representations, as we will see, are terms all of the same degree. This follows from the group being realized linearly, and it results in quantum mechanics — the physical expression of the group representations — being linear, giving the superposition principle, a fundamental attribute of it, which is merely an expression of this property of the groups and the realizations (used) [Mirman (1995a)].

Problem XI.1.a-1: Show that the operators of the groups, both differential ones and those giving finite transformations, leave unchanged the degree of any multinomial on which they act. Why does this mean that the bases of a representation are multinomials, all of the same degree? Need it be irreducible? Is this related to linearity? How? It is on this that the construction of the representations rests. A linear transformation acting on a product of variables leaves the power of the product (the sums of the powers of the variables in the product) invariant, so leaves invariant the power of a homogeneous sum of products (a sum of terms all generated the same — all of the same total power). These sums then are the basis states of the representations. This is not a property of a particular type of group, but of the linearity of the group and realization. Because of linearity, the basis vectors are homogeneous functions of the variables — multinomials, not general polynomials. However this may not always be apparent. There may be terms that transform as scalars, thus are constant under the group, and so suppressed, making a multinomial look like a polynomial.

Problem XI.1.a-2: It is an interesting question, which we must leave open, whether there are geometries in which transformations are (must be) nonlinear, and whether physics is possible in them, and what kind. Also interesting and left open, and apparently not (much) investigated, is whether there is a difference between nonlinear groups, and nonlinear realizations of linear groups; whether there are groups not (also) realizable linearly.

XI.1.b Representations of SO(3) and SU(2)

There are several ways to obtain group representations and we use different ones for SU(2) and for SO(3) (interestingly, the latter does not give all those of the former). These methods are not only valuable in themselves, and provide different insights into the groups, their meaning and applications, but suggest approaches generalizable to other Lie groups. However care is needed. SU(2) and SO(3) are so simple that some aspects, essential in the development of representations of larger groups, do not appear for them.

Problem XI.1-1: Prove, analogously to finite groups, that the rotation group must have an infinite number of representations. Notice however that the classes are labeled by a continuous parameter, an angle, while the representations are labeled by a discrete one, j (the angular-momentum quantum number), an integer or half-integer. However a little thought will show that this is not as strange as it might seem.

XI.1.c Representations of algebras

A representation of a group is given by an assignment of a matrix to each group element, with the matrices obeying the same product rule as the group operators. A representation of a Lie algebra is an assignment of a matrix to each algebra element, with the matrices obeying the same product rule as the algebra operators — the matrices have the same commutation relations as the elements of the algebra. Also for each representation, of a group and of an algebra, there is a set of basis vectors on which the representation matrices operate. However the algebra suggests a more general definition, with the word "matrix" being replaced by "operator", as in differential operator (which is not to imply that this modifier is required). So we get another question, how are the different realizations of the (presumably) same representation related?

Thus given an algebra we have to find a set of matrices (or other operators) with the same commutation relations, where the product of matrices in the commutator is matrix multiplication,

$$[P, Q] = PQ - QP; \tag{XI.1.c-1}$$

PQ is matrix multiplication. For other types of operators, this product must be defined, though there are restrictions on the definitions; for one thing the results obtained must agree with those found using matrices (assuming that this is possible).

XI.2 REPRESENTATIONS OF THE ALGEBRA OF SO(3) AND SU(2)

The representation matrices of the algebra obey the same commutation relations as its symbols, but could there be additional conditions — could there be matrices with these commutation relations that do not give (acceptable) representations of the groups? For the rotation, and unitary, groups, what might these be [Cornwell (1984), p. 438]? And why? And what general guidance can be obtained from these examples?

XI.2.a The reality conditions

Of the objects that we transform some, those of geometry say, like displacement, or of (classical) physics, like force, are real, others, as those of quantum mechanics, statefunctions, are complex. The transformations (that we consider) leave their magnitudes — lengths of vectors, absolute values of statefunctions — unchanged. Thus the transformations are orthogonal or unitary. Indeed the groups, as we have defined and realized them, are picked — restricted — by these criteria. Their transformations are orthogonal or unitary.

For the algebras however this restriction does not limit the abstract structures themselves. SU(2) is the group of 2×2 unitary matrices, SO(3) of 3×3 orthogonal ones. Yet they have the same algebra. In fact there are other (similar) groups (which we have carefully avoided) with this algebra, given by analogous matrices, but without such restrictions on their elements as imposed by unitarity or orthogonality. But if an algebra is realized so as to give unitary or orthogonal groups (perhaps better, realizations of them) there must be restrictions on it. What are these? Are they important? Why?

Problem XI.2.a-1: With M is a matrix, let the operator

$$T(\omega) = exp(i\omega M), \tag{XI.2.a-1}$$

be unitary. Explain what that means. M is the Lie algebra operator for T. Prove that the necessary and sufficient condition for T to be unitary is that M be hermitian; a matrix is hermitian if it is equal to its hermitian conjugate (pb. V.4.d-2, p. 162). Note that every realization given here for Lie algebras is hermitian. What does this mean for differential realizations? There is a slight exception. Generators J_x, J_y, J_z are hermitian. However check that J_+ and J_- (eq. X.7.c-9, p. 297) are hermitian conjugates of each other (they go into each other under hermitian conjugation); but since they are sums of hermitian operators they give unitary group realizations. Why? Verify that if two of J_x, J_y, J_z, are hermitian, the third must be (which is why i is in the commutation relations), and that if J_z is, then J_+ and J_- are hermitian conjugates. As they are encountered below, check that the matrices realizing (here) the SU(2) and SO(3) groups are unitary, but that they can be changed, still giving groups, to nonunitary ones — but then their algebra operators are not hermitian. Change the algebra operators of the different realizations to nonhermitian ones, but ones still obeying the commutation relations. See if it is possible to get realizations of the group from them, and check that these are not unitary. Suppose the group matrices are orthogonal, is there any condition on the algebra generators? Thus for the group operators to be unitary (or orthogonal, a subcase) the algebra elements must be hermitian, and conversely. For physical reasons (at least) we want unitary group operators, so limit consideration to hermitian algebra generators. However the nonhermitian (realized) algebra operators can be replaced with hermitian ones, though these have slightly (?) different commutation relations. The limitation to hermitian matrix representations is the key to the properties of the algebra representations. While fundamental, too often it is not made explicit. It is conceivable that in some cases it does not hold; the results obtained here would be for these at best partial.

Problem XI.2.a-2: There is another point, which is completely obvious therefore never made explicit, but not always true. The operators are taken to have discrete eigenvalues (translations of a free particle do not). Might these points be related?

Problem XI.2.a-3: Although the form of transformation T may look special, being an exponential, it is completely general; all operators of every Lie group can be written in this form, where M is the corresponding generator (realized as a matrix or not) of the Lie algebra of the group. Unfortunately showing this is outside the scope of the present book, but any reader who wants to try is strongly encouraged to do so.

XI.2.b Labeling the representations and states

The first step in finding the representations is to determine the labels. We look for operators that provide these. They are diagonal.

Problem XI.2.b-1: Why?

XI.2.b.i *The diagonal generators*

The states are labeled by the eigenvalues of the diagonal operators. For the rotation group how many operators, and which, are simultaneously diagonal(izable)?

Operator J^2 commutes with the three algebra generators (pb. X.7.c-3, p. 296). Such an invariant operator (multinomial function of the algebra generators) is a Casimir operator. For this algebra it is the only independent Casimir operator; powers also commute, but are not independent. Thus we can take diagonal J^2 and (any) J_z, and no further operators (z is just a label; the three generators are the same, but distinguished).

XI.2.b.ii *The eigenvalues of the diagonal generators*

Using the hermiticity, we can find the eigenvalues of the two commuting generators, J^2 and J_z. We write an eigenstate of these operators as $|j,m)$, with j giving (not equaling) the eigenvalue of J^2 and m the eigenvalue of J_z,

$$J_z|j,m) = m|j,m). \tag{XI.2.b.ii-1}$$

Problem XI.2.b.ii-1: Applying $J_\pm$ we get two states

$$J_+|j,m) = c_m|j,m_+), \tag{XI.2.b.ii-2}$$

$$J_-|j,m) = c_m{}'|j,m_-); \tag{XI.2.b.ii-3}$$

c, c' are coefficients to which we must return. Why is j the same for all the states? From the commutation relations, show that

$$m_+ = m + 1, \tag{XI.2.b.ii-4}$$

$$m_- = m - 1, \tag{XI.2.b.ii-5}$$

for all states (no matter what the values of j and m are). This explains the subscripts on the J's; they increase and decrease the eigenvalue by 1. They are, therefore, step (or ladder) operators (the choice of the name depending on whether we like to go one step at a time, or all at once). We now know the action of all three J's on every state. Using it, we can see that

$$J^2|j,m) = j(j+1)|j,m), \tag{XI.2.b.ii-6}$$

which should look familiar from quantum mechanics. While j gives the eigenvalue, it does not equal it, but is a more convenient label. Why? Explain why the eigenvalue of J^2 is independent of m. Since the J's, the operators of the algebra, do not change the value of j, the states of the same j belong to the same representation — the step operators go from one state of a representation to another; states of different j belong to different representations. Thus j, which gives the total angular momentum (squared), labels the representation, m, which gives the z component, labels the state. States of the same representation have different angular momentum z-component, and states with different (total) angular momentum are in different representations. Explain, physically, the reason for this. Restate this without using any concepts from physics.

XI.2.b.iii *The values of the labels*

This tells how the representation and state labels are given by the eigenvalues of the diagonal operators, and how the state labels are changed by the step generators. But what are the values of these labels, what determines them [Edmonds (1960), p. 13]? The matrix elements of a product of operators can be written

$$(j', m'|J_aJ_b|j, m) = \sum(j', m'|J_a|j'', m'')(j'', m''|J_b|j, m), \tag{XI.2.b.iii–1}$$

with the sum over a complete set of states (the mathematical question of whether such exists can, fortunately, be ignored here); since the operators cannot change the representation $j' = j'' = j$. Moreover since each operator acting on a state gives a unique state, because the eigenvalue m is determined and there is nothing else to distinguish states, in the sum there is only one state. Applying this to the commutator, eq. X.7.c–11, p. 297, for which only diagonal elements are nonzero, we have

$$(j, m|[J_+, J_-]|j, m) = 2m = |c_{m-1}|^2 - |c_m|^2, \tag{XI.2.b.iii–2}$$

where we have used the condition that J_+ and J_- are hermitian conjugates of each other; all results from here on are based on this.

Problem XI.2.b.iii–1: Check that hermiticity results in the c coefficients being related (how?), giving this expression, with absolute values. Also since J_z is hermitian, m must be real. Why? The commutation relations require that $J_{\pm}$ be hermitian conjugates if J_z is hermitian, and conversely.

Problem XI.2.b.iii–2: This expression is a linear difference relation, and it should be clear that its solution is, with K a constant,

$$|c_m|^2 = K - m(m + 1), \tag{XI.2.b.iii–3}$$

and the left side is always positive (or 0). But J_+ increases the value of m, leading eventually to a contradiction. Thus by the hermiticity of the operators, there must be some value of m for which $|c_m|$ is 0. For that largest value, m_h,

$$J_+|j, m_h) = 0. \qquad \text{(XI.2.b.iii-4)}$$

There can be no states with an m value larger than m_h (for the given j). There is also a smallest m, by the same argument, using J_-. Do this explicitly. Show that the smallest value m_s is

$$m_s = -m_h. \qquad \text{(XI.2.b.iii-5)}$$

The states then are labeled by a set of m's (the largest labeled now j — why?), symmetric around zero,

$$m = -j, -j+1, \ldots, j-1, j. \qquad \text{(XI.2.b.iii-6)}$$

Check that this means $2j$ is an integer, so j is an integer or a half (odd) integer — explaining one of the fundamental facts of physics: particles have spin which is only an integer or a half (odd) integer, and only integral orbital angular momentum (we have to see the reason for this extra limitation). Also m changes by integer steps, so must be either an integer or half-integer. To find the largest m show that

$$J^2|j, m) = \{m^2 + \frac{1}{2}(J_+J_- + J_-J_+)\}|j, m). \qquad \text{(XI.2.b.iii-7)}$$

Take the state with the largest (or smallest) m, for which J_+ (or J_-) gives 0, and evaluate the J_+J_- term to get

$$J^2|j, m_h) = (j^2 + \frac{1}{2}|c_{m_h-1}|^2)|j, m_h) = j(j+1)|j, m_h), \qquad \text{(XI.2.b.iii-8)}$$

giving the eigenvalue of J^2. Thus $j(j+1)$ is the square of the angular momentum (up to a constant), the angular momentum z-component has all values ranging from $-j$ to j, in integer steps, and j is either an integer or half-integer.

Problem XI.2.b.iii-3: It is now possible to evaluate constant K and show that

$$|c_m|^2 = j(j+1) - m(m+1) = (j-m)(j+m+1). \qquad \text{(XI.2.b.iii-9)}$$

This gives the action of the step generators on the states, thus their matrix elements. There is one slight problem, which with a little care can turn into quite a mess. This gives the absolute value of the matrix elements allowing a choice of phases. Different choices give matrix elements which look, but are not really, different. And different choices do exist. However we take the phases 0 (which we can do consistently, here). Show that then

$$(j, m+1|J_+|j, m) = \{(j-m)(j+m+1)\}^{\frac{1}{2}}. \quad \text{(XI.2.b.iii-10)}$$

Find the matrix elements for J_-. How are they related? Why? So we now have the labels of the representations and states, and the action of all algebra generators on the states, that is their matrix elements, giving the algebra representations.

XI.2.b.iv *Values of the labels differ for the two groups*

The representations of SU(2) are given by all integer and half-integer values of j, those of the rotation group, for reasons we have to see, by only the integer values. But the algebra is that of both groups, so its representations give (in a way to be determined) the representations of both of them. This raises the question, to which we must return, why a group does not have all the representations of its algebra, and what the other representations are.

This then gives the representations of the Lie algebra of SU(2) and SO(3), the matrix elements of the diagonal operators, so the representation and state labels, and their matrix elements. These results come from the commutation relations, not from the realization. They hold thus for all realizations, though the specific forms of the states may differ. Here we have realized the states only by giving their eigenvalues — the results then are general — but they can also be realized in other ways, say as column vectors, or functions of the parameters. But this, especially the latter, involves the group rather than the algebra (but thus also the algebra), and to the groups we therefore now turn.

XI.3 THE REPRESENTATIONS OF SO(2)

The simplest rotation group is SO(2) (or U(1)); it is very simple being a single parameter, Abelian group (however while it is indeed a simple group it is not semisimple, being the only simple group that is not). So its representations are one-dimensional — the group, having an infinite number of classes, has infinitely many one-dimensional representations [Lyubarskii (1960), p. 192]. They are trivially given by

$$V(\theta) = Cexp(i\theta). \quad \text{(XI.3-1)}$$

Any value of θ gives an acceptable group operator. We cannot arbitrarily limit it to a range of values. But all values of θ that are equal (mod 2π) give the same transformation. So the representations are multi-valued; there is an infinite number of transformations having the same effect. This is the real import of the statement that the group is infinitely connected.

The group elements (which are here somewhat difficult to distinguish from the basis vectors) can be written as

$$R(\phi) = Aexp(il\phi + i\eta); \quad \text{(XI.3-2)}$$

A, l and η are constants. Since

$$R(\phi)R(\phi') = R(\phi + \phi'), \quad \text{(XI.3-3)}$$

$A = 1$, $\eta = 0$. Each value of ϕ gives a class, each of l a representation (since the product does not change l). There are an infinite number of classes, so an infinite number of representations. Since ϕ is a continuous variable it appears that there would be a disagreement with the requirement that the number of representations equals the number of classes if l were not also continuous.

Problem XI.3-1: Discuss whether the form of the representations is so trivial. Under what conditions is the group unitary? Why is the group simple, but not semisimple? Why does it have an infinite number of classes? What are they? Explain why the group is infinitely multi-valued (or connected). The interesting question is how the infinite connectivity of the parameter space affects the representations.

Problem XI.3-2: Each of these R's is a basis vector of the regular representation, and linear combinations are basis vectors of the irreducible representations. It should be clear that sums of these do not give irreducible representations; these are given by each R. If now we identify (regard as identical) transformations differing by 2π then the set of l's, instead of being continuous, becomes discrete. Explain. However this reduces the number of representations, though not the number of classes. For each value of l then there must be an infinite number of representations. The (normalized) basis vectors are thus given by

$$V_{\zeta,n}(\phi) = exp(in\phi + i\zeta), \ \ 0 \leq \phi < 2\pi, \quad \text{(XI.3-4)}$$

with n an integer. Why? A group transformation does not change n or ζ, thus n and ζ label the representations; these are one-dimensional. Justify them. What is the significance of ζ ? Explain why, with these basis vectors, the number of representations equals the number of classes. And what does this have to do with the group being infinitely connected? Find the matrix elements of the transformations. Show that these vectors are orthonormal — with what definition of a product? The angle is defined over a 2π range. But the particular 2π range should be arbitrary. Does this affect the representations?

Problem XI.3-3: Show that the representations with these basis vectors are irreducible, unitary, orthonormal, and complete — any (well-behaved) function of ϕ defined over the (proper) range can be written as a sum over them (Fourier's theorem) [Arfken (1970), p. 643, 673; Courant and Hilbert (1955), p. 69; Margenau and Murphy (1955), p. 241]. What is the proper range for Fourier's theorem? Why? Do the ranges agree? Because they are complete, there are no other irreducible representations of SO(2). Are there reducible representations? Can they be (completely) reduced?

Problem XI.3-4: The (normalized — in what sense?) basis vectors of the representations of the (one-dimensional) translation group, the covering group of $U(1)$, are given by

$$\psi(x) = exp(ikx); \tag{XI.3-5}$$

k is any real number. Why must it be real? Is a phase needed? What labels the representations? Are the numbers of representations and classes equal? Show that these are irreducible, unitary, orthonormal and complete (Fourier's theorem) — any (well-behaved?) function of x with the (proper?) range can be written as a sum over them. Since they are complete there is no other irreducible representation of the translation group. Are there reducible ones? Compare Fourier's theorems for this and the preceding problem. Can one set of representations be written as an expansion over the other? Why? Find the transformation matrix elements. The basis vectors in more than one dimension are the same with kx replaced by the scalar product $k_v x_v$. Why? Using the Lie algebra generators (eq. X.7.c-5, p. 296), check that $\psi(x)$ is an eigenfunction, all are of this form, and the eigenvalue of p_v is k_v, called in physics the momentum (component).

Problem XI.3-5: Find the representations of O(2); how are these related to those of SO(2) [Jones (1990), p. 119, 269; Tung (1985), p. 215; Wigner (1959), p. 146]? Is there anything equivalent for the translation group?

Problem XI.3-6: Which representations of SO(2) are faithful [Tung (1985), p. 85]? How about O(2)?

XI.4 THE REPRESENTATIONS OF SO(3)

A group representation is given by the matrices of the transformations and the states they act on. For applications, especially to physics, explicit basis states, the statefunctions of the system, functions of variables giving the group transformations, are often most useful. These can be helpful (although probably not necessary) for calculations; also their form, say when graphed, gives some feeling for the behavior of the objects they describe. So for the rotation group we must find those functions of the angles that are the states.

Thus, here, to obtain the representations of SO(3), we start by finding the basis states, and then from them, the matrices [Cornwell (1984), p. 444; Lyubarskii (1960), p. 200]. The method is familiar, but in a different context: it is to solve a differential equation whose solutions are the basis states. Historically, the equation arose in many physical situations and in studies of physical systems, with no use of group theory, or even an understanding of its relationship to groups. Thus the solutions are well-known, being certain special functions, though the reasons for the equation or the form of the solutions, is less so. But it should be clear why this equation, and why these solutions, appear for so many different physical systems, and what these have to do with the rotation group, and what this group has to do with these systems. (This hints that the theories of special functions and of groups are closely related, and how they are. We cannot go further into this here, except to note the importance, and correctness of this hint.)

One way of solving the equation is standard, and considered in many places. Here we use a different method, given by group theory, and the reader should compare it with the standard one; it should be clear that this, using a more fundamental understanding of the equation, is simpler, as is the method for determining the equation itself. The method then is a differential one, and if the reader has not managed to forget the process of finding the representations of the algebra, he may notice some similarities.

Problem XI.4-1: Why are explicit functions often not needed?

Problem XI.4-2: Explain why the defining requirement of orthogonal groups — they leave invariant products of real vectors — gives that all representations are (equivalent to) unitary ones, and thus are decomposable.

XI.4.a An equation invariant under rotations

The basis vectors of a representation are a set (here of functions) that are transformed among themselves, but not mixed with others, by the group operators. We want functions, so we look for a differential equation whose solutions are so mixed. Thus we have differential operator D such that

$$Df_{\alpha}(x_i) = 0; \qquad \text{(XI.4.a-1)}$$

the f_α's are the solutions, and the x_i's the variables acted on by the group, with the condition that there be f's that are intermixed by the group. If D is invariant under the group, then for transformation T,

$$TDf_\alpha(x_i) = DTf_\alpha(x_i) = 0, \tag{XI.4.a-2}$$

so both f_α and Tf_α are solutions, and the functions obtained from f_α by using all T's of the group are the basis states of a representation. This does not say that all solutions belong to the same (irreducible) representation, only that if a basis state is a solution, all basis states of an irreducible representation are. On the other hand, if D is not invariant under T, then f_α and Tf_α obey different equations, so the solutions of the equation are not intermixed by the transformations; not all basis states are solutions. If the operator is a group invariant, it cannot change one basis vector into another. Hence we look for an operator invariant under the rotation group.

Problem XI.4.a-1: It should be clear that

$$D^2 = \frac{\partial^2}{\partial x^2} + \frac{\partial^2}{\partial y^2} + \frac{\partial^2}{\partial z^2}, \tag{XI.4.a-3}$$

is invariant under all rotations, and is the only differential operator that is (except for powers of D^2, which are irrelevant — why?). How do we know that this is unique? The phrase "differential operator", as used here, implies a restriction on the types of operators being considered. Why?

XI.4.a.i *Solution by separation of variables*

We want thus the solutions to the equation, Laplace's equation [Hamermesh (1962), p. 333],

$$D^2 f = (\frac{\partial^2}{\partial x^2} + \frac{\partial^2}{\partial y^2} + \frac{\partial^2}{\partial z^2})f = 0. \tag{XI.4.a.i-1}$$

For the rotation group the best variables are not Cartesian ones, but angles. Hence we rewrite this in spherical coordinates [Schiff (1955), p. 70],

$$D^2 f = \{\frac{\partial}{\partial r}(r^2\frac{\partial}{\partial r}) + \frac{1}{sin\theta}\frac{\partial}{\partial\theta}(sin\theta\frac{\partial}{\partial\theta}) + \frac{1}{sin^2\theta}\frac{\partial^2}{\partial\phi^2}\}f = 0, \tag{XI.4.a.i-2}$$

and separate variables (noting that separation of variables and group theory are, perhaps not surprisingly, closely connected). The equation for the radial variable is not relevant — it does not change under rotations — so we get two ordinary differential equations for angles θ

and ϕ. (Though the group transformations depend on three parameters, here we have only two angles; a point in space — the radius is irrelevant — is given by just two angles.)

Problem XI.4.a.i–1: Why is (part of) this the same as eq. X.7.c–8, p. 296? If the solution of the eigenvalue equation is $F(r, \theta, \phi)$, show that writing

$$F(r, \theta, \phi) = f(r)Y(\theta, \phi), \tag{XI.4.a.i–3}$$

where f and Y are functions to be determined, gives two differential equations, one for f, depending only on r, and one for Y depending only on θ and ϕ. Show further that the latter equation can be separated into two equations by writing

$$Y(\theta, \phi) = P(\theta)\Phi(\phi), \tag{XI.4.a.i–4}$$

giving thus a set of three ordinary differential equations. Explain why the method of separation of variables gives all solutions.

XI.4.a.ii *Group theoretical reasons for solution by separation*

Why does this method work? Is it an accident, a nice but unusual property of these particular equations, or is there some general reason for it? The solutions of these equations are all basis vectors of all irreducible representations of the rotation group; the coefficients are arbitrary functions of r (only).

As we will see the representation basis vectors form a complete set [Barut and Raczka (1986), p. 172; Miller (1972), p. 215; Tung (1985), p. 134]; any (well-behaved) function of the group variables, defined over the same domain as these, can be expanded in terms of them, in particular any solution of these equations, and with coefficients functions of r (only) — this is the import of the Peter-Weyl theorem. The solutions are basis vectors times functions of r and the general solution is a sum over these. So rotational invariance requires that a solution be a sum of products of functions of r with ones of the angles. This is the group-theoretical reason the equation can be separated into such a product. (Completeness comes from the ability to produce transformations in space; invariance under these is not needed. Completeness is a weaker condition.)

Problem XI.4.a.ii–1: Why are the solutions all basis vectors, with coefficients functions of r (only)?

Problem XI.4.a.ii–2: What is the group-theoretical reason that the equation for Y can be separated? Thus explain why there is a good group-theoretical reason for the equation being separable, and why any equation like this coming from group theory allows such a separation.

Problem XI.4.a.ii-3: There are constants introduced in the separation procedure, usually called l and m. Show that the solutions with the same l are mixed among themselves by the group transformations but not mixed with those of other l's — l labels irreducible representations and functions with a given l form a basis for it; m labels states. Prove that the representation is irreducible. Show that l and m are integral, $l \geq 0$, and $-l \leq m \leq l$. Check that functions differing in either l or m, are orthogonal. Thus the set of functions for all l and m are the complete set of (orthogonal) representation basis vectors of the rotation group. So this method, of solving the eigenvalue equation, gives (here) the representation and state labels, l and m, and the orthogonal basis states of each representation.

Problem XI.4.a.ii-4: Why are functions with different l not mixed among themselves? The separation process gives a differential equation for $Y(\theta, \phi)$ which is invariant under rotations. Why? This depends on a constant, l, and since it is invariant, rotations cannot change l, but these do change one basis vector, one solution, into another. So functions with different l cannot be mixed, but those with the same l are. Can functions with different m be mixed? By what generators? Why does this argument not hold for m?

Problem XI.4.a.ii-5: What is the group-theoretical reason states are orthogonal? To what extent does it depend on the definition of a scalar product? Is there a (group-theoretical) reason for the scalar product usually used? Would you expect that these states can be normalized?

XI.4.b The representation basis states

To find the explicit functions we start with the ϕ equation,

$$i\frac{\partial^2\Phi}{\partial\phi^2} = \Phi, \tag{XI.4.b-1}$$

which has solutions

$$\Phi = \Phi_o e^{im\phi}, \tag{XI.4.b-2}$$

where m is a constant; we need not consider the (rotational invariant) Φ_o which is determined by the normalization convention. For this to be single valued m must be an integer, a point we note with interest.

Since this is a representation basis state we ask the effect of group operators on it. For transformation $\phi \rightarrow \phi + \omega$ it goes to

$$\Phi_\omega = \Phi_o e^{im(\phi+\omega)}, \tag{XI.4.b-3}$$

thus the same function, but with the angle measured from a different point, really not surprising. While the angle changes, m does not.

XI.4.b.i *The P equation*

The next step is to solve the equation for θ, but this is too much work, and besides the equation is more complicated.

Problem XI.4.b.i-1: However we know that if $\theta \to \theta + \chi$, then

$$cos(\theta) \to cos(\theta + \chi) = cos(\theta)cos(\chi) - sin(\theta)sin(\chi), \quad \text{(XI.4.b.i-1)}$$

that is a function linear in θ goes to a sum of functions both linear in θ, with coefficients functions of transformation parameter χ. This can be generalized. Show that

$$cos^l(\theta + \chi) = \sum a_{l,k}(\chi)cos^k(\theta)sin^{l-k}(\theta), \quad \text{(XI.4.b.i-2)}$$

$$sin^l(\theta + \chi) = \sum b_{l,k}(\chi)sin^k(\theta)cos^{l-k}(\theta), \quad \text{(XI.4.b.i-3)}$$

so that these functions go into sums, each with the same power, which is the power of the original function. But that is what we mean by representation basis states. Thus the rotation-group basis states are sums of terms like $cos^k(\theta)sin^{l-k}(\theta)$, where l labels the representation.

Problem XI.4.b.i-2: This leaves open the question of whether there might be different sets, all with the same l, that go into themselves, but are not mixed with states from different sets with that l. We would suspect not for there is only one label, l, to distinguish representations. In fact, show that starting with

$$|l, 0) = cos^l(\theta), \quad \text{(XI.4.b.i-4)}$$

for any l, the group transformations give every term of the form

$$|l, k) = cos^k(\theta)sin^{l-k}(\theta), \text{ for } 0 \leq k \leq l; \quad \text{(XI.4.b.i-5)}$$

k being every integer in the range. The representations then are given by the power, l, and the states by k. If we had solved the equation explicitly we would have found, of course, that $k = |m|$, as we would suspect from their ranges. So the basis states of the representations are (up to normalization)

$$|l, m) = cos^m(\theta)sin^{l-m}(\theta)exp(im\phi); \quad \text{(XI.4.b.i-6)}$$

l gives the representation, m the state. Note that k, so l, or conversely, must be integral. Why? Orbital angular momentum, because of its explicit dependence on the angles, must be given by integers. Explain the relevance, and effect, of "explicit".

Problem XI.4.b.i-3: One way of finding the meaning of m is to return to the algebra. Write L_+ in spherical coordinates (sec. X.7.c, p. 296) and notice that its action on $|l,k)$ is to decrease the power of $cos(\theta)$, and increase that of $sin(\theta)$, both by the same value, of course 1, while increasing that of m, also by 1. L_- has the opposite effect. Thus from $|l,0)$, these generate the states of the representation. Why is there one state on which L_+ (and another on which L_-) give 0? Which? Find the action of L_x and L_y on these states, and also check eq. X.7.c–4, p. 296. Does it seem reasonable? How about the actions of L_x and L_y? (The action of L_+ and L_- obviously is.) Explain.

Problem XI.4.b.i–4: These solutions of Laplace's equation, the representation basis states, are given by a label l, which here, being a power, must be a positive integer; a point we note. Suppose it were not. Check that if it is negative, or a positive number not an integer, say a half-integer, the sets of functions do not go into themselves under rotations, and do not even form closed sets. Are there other problems? So the representations of the rotation group, being explicit functions of the angles, are given by a positive (or zero) integer l, and by all. Not all representations of the Lie algebra of the rotation group are representations of the group itself. The algebra does not contain the angles, the group does. And this provides the extra condition to limit the representations (so orbital angular momentum). As SO(3) does not have all the representations of its algebra, or of SU(2), these groups cannot be isomorphic, but only homomorphic.

Problem XI.4.b.i–5: There is an alternate way of finding the solutions, using Cartesian coordinates. Under a rotation x, y, z are mixed among themselves, which is one definition of a rotation. These form a representation of the rotation group, the defining representation — their transformations define the group. The variables on which the group acts are the basis vectors of the defining representation. (The defining representation of the rotation group is that for which $l = 1$; this is the group of transformations on a space of three-dimensional real vectors.) It is then easy to show that, with

$$l = p + q + r, \qquad \text{(XI.4.b.i–7)}$$

the states

$$|p, q, r) = x^p y^q z^r, \qquad \text{(XI.4.b.i–8)}$$

go into sums of themselves, for each l. These are monomials. They are also solutions of Laplace's equation written in Cartesian coordinates. Why? Express these in terms of spherical coordinates and show that they are the same as the states of eq. XI.4.b.i–6, p. 319 (multiplied by r^l — would you expect them to all be multiplied by the same power of r?). Find the action of all (five) L's and also L^2 on these (finding their matrix elements). Relate the matrix elements to those of the L's in spherical coordinates. Under what conditions does an L give 0? Explain why. Give a qualitative explanation of why these are the basis states of representations of the rotation group and why they are related as they are to the states expressed in spherical coordinates. Physically these states with definite z component are what we want. A nice accidental result. But of course, we want these states because of their group properties. That we get what we want is not an accident. We want what we get because that is what we get.

XI.4.b.ii *Reducing the representations*

These multinomials are representation basis states, but not of irreducible ones. The distance squared between two points,

$$r^2 = x^2 + y^2 + z^2, \qquad \text{(XI.4.b.ii-1)}$$

and the square of a vector

$$V^2 = V_x^2 + V_y^2 + V_z^2, \qquad \text{(XI.4.b.ii-2)}$$

are rotationally invariant. They are realizations of the basis states of the scalar representation.

Problem XI.4.b.ii-1: Check that the representation matrices (which?), and the various algebra operators, do leave these invariant. Thus a homogeneous polynomial is a basis state of a representation but not of an irreducible one. From the multinomials of second order, x^2, y^2, z^2, xy, xz, yz, we can subtract r^2 to give two sets of states, r^2, which gives one representation (for it l =?) and $x^2 - r^2$, $y^2 - r^2$, $z^2 - r^2$ (only two are independent), xy, xz, yz, which gives a second. Show that both sets go into themselves under rotations. For third-order multinomials show that terms like xr^2 can be subtracted off giving again two sets, these and $x^3 - xr^2, \ldots$. Thus we still have to find the basis states, and matrix elements, of the irreducible representations. These sums giving the (incompletely) reduced states have terms all of the same power — as they must — and since r^2 is a constant the total power of variables x, y, z is either all even or all odd.

Problem XI.4.b.ii-2: This reduces the representations given by multinomials. The resultant representations are still not irreducible. Show that $u_iv_j + u_jv_i$ and $u_iv_j - u_jv_i$, the symmetric and antisymmetric sums, go into themselves (where u, v are each sets of three variables transformed among themselves by rotations) [Tung (1985), p. 124; Aivazis (1991), p. 53]. Also, as $1 \le i, j \le 3$, there is only one antisymmetric term; it is a scalar. So to reduce the representations we symmetrize, and have to show this does give irreducible representations. (The $l = 2$ representations of the last two paragraphs are symmetric because the special case of $u = v$ was taken). For third order, there are three ways of symmetrizing, given by the three S_3 irreducible representations. Should the symmetrization be done before or after the subtraction of lower-order terms, or does it not matter? This procedure has to be systematized. For the rotation group the result is well known. There are three ways of taking products of two vectors, the dot (scalar) product, which gives a scalar, the cross (vector) product — of distinct vectors — which gives a vector, and the direct product, which gives a second-rank tensor, symmetric in the two vectors. The moment-of-inertia tensor is an example of the latter. Explain the relationship of these three products to the reduction of a representation of degree two. Check that these do transform properly.

Problem XI.4.b.ii–3: There are other ways of finding the basis vectors. That described in this problem is perhaps the worst. But it may add insight. The basis state of the $l = 0$ representation is a constant. The $l = 1$ representation, which is irreducible, has basis vectors x_1, x_2, x_3 (or x, y, z). Basis vectors for $l = 2$, are second order multinomials, but these are not irreducible. They are (unnormalized) $a_{ij}x_i x_j - b_{ij}r^2$ (no sum over repeated indices). The coefficients are determined by the requirements that these go into themselves under rotations, and that they are orthogonal (when integrated over a sphere — why?) [Arfken (1970), p. 546, 561]. It is a useful exercise to try this and find the coefficients. Then the basis vectors should be normalized. This can be continued, say for the $l = 3$ representations. Matrix elements for these representations can also be found.

XI.4.c Spherical harmonics

For the rotation group, representation bases are homogeneous functions of the coordinates, which when reduced into irreducible representations, are called spherical harmonics [Inui, Tanabe, and Onodera (1990), p. 132; Tung (1985), p. 143; Varshalovich, Moskalev and Khersonskii (1988), p. 130]. So we next find these explicit functions of the coordinates (here angles), that is of transformation parameters, and see how their properties follow from their being basis states. Later we derive the expression for them. In giving this expression some care is needed because minus signs can be put in different places. We therefore copy accepted formulas [Wigner (1959), p. 154].

The Legendre polynomials are defined as

$$P_l(\theta) = \frac{1}{2^l l!}\frac{d^l(cos^2\theta - 1)^l}{d(cos\theta)^l}. \qquad \text{(XI.4.c–1)}$$

Then the spherical harmonics are

$$Y_{lm}(\theta,\phi) = \Phi_m(\phi)\Theta_{lm}(\theta), \tag{XI.4.c-2}$$

with

$$\Phi_m(\phi) = \frac{1}{\sqrt{2\pi}}e^{im\phi}, \tag{XI.4.c-3}$$

and, $m \geq 0$,

$$\Theta_{lm}(\theta) = (-1)^m\sqrt{\frac{2l+1}{2}\frac{(l-m)!}{(l+m)!}}sin^m\theta\frac{d^mP_l(\theta)}{d(cos\theta)^m}. \tag{XI.4.c-4}$$

Problem XI.4.c-1: It is an interesting exercise to show that these are actually basis vectors of rotation-group irreducible representations and are orthonormal. Check that $Y_{lm} = 0$ if $|m| > l$. Verify that $L_\pm$ acting multiple times on Y_{l0} gives Y_{lm}, including the coefficients (even without this, the exponents of $sin\theta$ and $cos\theta$ should be clear), and eventually 0 — why? Also find the matrix elements. In particular write the spherical harmonics for several values of l and show that they agree with their defining expressions. Check that, except for Y_{l0},

$$Y_{lm}(0,\phi) = 0, \tag{XI.4.c-5}$$

since ϕ is undefined for $\theta = 0$. The representations are given by multinomials of every power, and it is clear that every power appears here. Further notice that both the Legendre polynomials and the spherical harmonics are sums of terms, and while the powers differ, for each the total power, of $cos\theta$, $sin\theta$, or of their product, is even for all terms, or odd for all terms. Explain why. Thus Y_{lm} is a descending series in $cos\theta$, starting with $cos^l\theta$, with the power of the terms decreasing by two, and the last term either a constant or $cos\theta$.

Problem XI.4.c-2: This form of the spherical harmonics should seem reasonable, except for one point. These are series in $cos\theta$, picked to give irreducible representations. The terms in the sums we might guess, but why are the coefficients correct, why does the expression for $P_l(cos\theta)$ give irreducible-representation basis vectors? Show that [Arfken (1970), p. 557]

$$\int_{-1}^{1} x^kP_n(x)dx = 0, \text{ for } k < n. \tag{XI.4.c-6}$$

Why is this relevant? The basis states of an irreducible representation are orthogonal to those of all other representations. The states of a representation for $l < n$ are sums of x^k, for $k \leq l$. Thus P_n, so Y_n, is orthogonal to all representation basis vectors for $l < n$. Thus the basis vectors of different representations are mutually orthogonal. Explain, using integration by parts, why the basis vectors are as stated. This is an explanation of the form of the spherical harmonics, and why these are the correct functions, including coefficients, for the basis states of the irreducible representations.

Problem XI.4.c-3: In what sense are the coordinates transformation parameters?

Problem XI.4.c-4: Extend this to the representations of O(3) [Tung (1985), p. 221].

XI.4.d Spherical harmonics and rotations

The spherical harmonics are rotation-group basis states. Thus we have to study how rotations affect them. The easiest part to study is the ϕ dependence. A rotation about the z axis by β gives $\phi \to \phi + \beta$.

Problem XI.4.d-1: But the spherical harmonics are functions, there is no axis involved — so what is the z axis? Apply the transformation matrix for a rotation about z through β to the definition of these polynomials (eq. XI.4.c-2, p. 323), and check that it behaves correctly,

$$R(0,\beta,0)Y_{lm}(\theta,\phi) = exp(im\beta)Y_{lm}(\theta,\phi). \qquad \text{(XI.4.d-1)}$$

Explain whether, and how, axes are involved in the definition of Y_{lm}. For a general rotation,

$$R(\alpha,\beta,\gamma)Y_{lm}(\theta,\phi) = \sum_{m'=-l}^{l} M(\alpha,\beta,\gamma)^{l}_{m'm}Y_{lm'}(\theta,\phi), \qquad \text{(XI.4.d-2)}$$

where $M^{l}_{m'm}$ is the $(m'm)$ matrix element for the matrix of representation l representing element (α,β,γ). These equations give

$$M(0,\beta,0)^{l}_{m'm} = exp(im\beta)\delta_{m'm}. \qquad \text{(XI.4.d-3)}$$

The representation matrix is diagonal for a rotation about z, as we would expect. Why? Write the matrix for this rotation.

Problem XI.4.d-2: For an arbitrary rotation θ is changed. How? First consider a rotation changing only θ. Substitute the changed value in the equations for the spherical harmonics, and find the expressions for these functions in the rotated coordinate system, in terms of those in the initial system. Is this expected? Compute the matrix elements. What is the effect of the rotation $R(0,0,\gamma)$? What is its significance? Why? Find the expressions for the rotated basis vectors, and the matrix elements, for rotation $R(\alpha,\beta,\gamma)$.

Problem XI.4.d-3: A rotation leaves the product of vectors invariant. Vectors and spherical harmonics are different realizations of the same basis vectors, so it would be expected that a product of the latter is invariant also. Prove that

$$B(\theta,\phi,\theta',\phi') = \sum_{-l}^{l} Y_{lm}^{*}(\theta,\phi)Y_{lm}(\theta',\phi') \qquad \text{(XI.4.d-4)}$$

is invariant and explain why the product has this form.

XI.4.e Properties of spherical harmonics

The spherical harmonics are quite well-known, but it is not often emphasized that they are rotation-group representation basis states. That is why they exist, and why they are important. And their properties are not strange accidents, as one might believe from their usual derivations, but are specific examples of properties of group representations. They, like the sines and cosines appearing in Fourier series and integrals, are specific cases of special functions. These have group-theoretical significance; others do also.

Problem XI.4.e-1: Show that the addition theorem for the spherical harmonics is a relationship between untransformed and transformed basis states [Arfken (1970), p. 222, 581, 761; Elliott and Dawber (1987), p. 532; Inui, Tanabe, and Onodera (1990), p. 133; Tung (1985), p. 145]. How do the transformation angles appear in the formula? Explain why the coefficients are the SO(3) matrix elements. How is the addition theorem related to the scalar product?

Problem XI.4.e-2: Show that the product of two spherical harmonics (eq. XI.4.d-4) is given by the Legendre polynomial (eq. XI.4.c-1, p. 322),

$$P_l(cos\gamma) = \frac{4\pi}{2l+1}\sum_{-l}^{l} Y_{lm}^{*}(\theta_1,\phi_1)Y_{lm}(\theta_2,\phi_2) \qquad \text{(XI.4.e-1)}$$

where

$$cos\gamma = cos\theta_1 cos\theta_2 + sin\theta_1 sin\theta_2 cos(\phi_1-\phi_2). \qquad \text{(XI.4.e-2)}$$

a form of the addition theorem. What is the significance of this?

Problem XI.4.e-3: Find all other properties listed for these functions [Arfken (1970), p. 569; Courant and Hilbert (1955), p. 510; Margenau and Murphy (1955), p. 344; Schiff (1955), p. 69; Talman (1968); Varshalovich, Moskalev and Khersonskii (1988), p. 130; and many other places] and explain their group-theoretical significance.

XI.4.f The addition formula for spherical harmonics

Products of basis states appear in various group-theoretical contexts, and in physical ones, but often in disguise. One such leads to the addition formula for spherical harmonics. First we ask how to form a scalar product of spherical harmonics.

Problem XI.4.f-1: Show that

$$S = \sum_{m=-l}^{l} Y_{lm}^*(\theta_1, \phi_1) Y_{lm}(\theta_2, \phi_2) \tag{XI.4.f-1}$$

is a scalar, just using the transformation properties of the Y's under rotations and the fact that the representation is unitary. Notice that the Y's depend on different angles. However their behavior as basis states depends only on l and m, independent of the angles. And it is the transformation matrix that determines how they transform. Suppose the l's for the Y's were different. How would the product transform? Compare this expression to the dot product of two vectors, which is a sum over components.

Problem XI.4.f-2: Using $cos\gamma$ defined in the previous section check that the addition theorem is

$$Y_{l0}(\gamma) = \frac{4\pi}{2l+1} \sum_{m=-l}^{l} (-1)^m Y_{l(-m)}(\theta_1, \phi_1) Y_{lm}(\theta_2, \phi_2). \tag{XI.4.f-2}$$

Why? The spherical harmonics form a complete set, as do the basis vectors of the representations of (at least) any finite or compact semisimple group. Thus the product must be expandable in terms of them. Since every Y_{lm} is multiplied by a Y_{l-m}, each term in the product transforms as the $m = 0$ component of a spherical harmonic. Why? Is this the complete explanation for the addition formula? What is the geometrical significance of γ?

Problem XI.4.f-3: Consider the $l = 1$ case. Then we have, suppressing numerical coefficients,

$$S \sim cos\theta_1 cos\theta_2 + \frac{1}{2} sin\theta_1 sin\theta_2 \{e^{i(\phi_1-\phi_2)} + e^{-i(\phi_1-\phi_2)}\}$$

$$= cos\theta_1 cos\theta_2 + sin\theta_1 sin\theta_2 cos(\phi_1 - \phi_2), \tag{XI.4.f-3}$$

which agrees with the addition formula since $S = cos\gamma$, by definition of $cos\gamma$. Is $cos\gamma$ a scalar?

Problem XI.4.f-4: Verify the addition formula for $l = 2$, and also that the result is a scalar. Check the correctness of the numerical factor.

Problem XI.4.f-5: To show the addition formula in general we set $\theta_2 = 0$ so $\gamma = \theta_1$ (and for ease in writing, $\phi_2 = 0$). Does this matter? We then get

$$Y_{l0}(\theta_1) = \frac{4\pi}{2l+1} \sum_{m=-l}^{l} Y_{lm}^*(\theta_1, \phi_1) Y_{lm}(0, 0)$$

$$= \frac{4\pi}{2l+1} \sum_{m=-l}^{l} (-1)^m Y_{l(-m)}(\theta_1, \phi_1) Y_{lm}(0,0). \qquad \text{(XI.4.f-4)}$$

Check that the numerical factor is correct. So all that is left is to explain why the left-hand side does not depend on ϕ_1. Why does it not? Thus the scalar product of basis states of the same representation, which are functions of different angles, is a function which is a basis state of the same representation, but a function of an angle that depends on the four angles. In "scalar product of basis states of the same representation" are the last four words necessary? Thus for a system having two angular momenta we know how this product transforms — it is a scalar. But having this expression we are now free to vary only one pair of angles. Might this be useful?

XI.4.g The number of parameters in spherical harmonics

For rotations we need three angles: the rotation, through one, is about an axis (given by two other angles). To locate a point particle, perhaps an electron in an atom (at fixed distance from the center), only two angles, giving the direction of the line to the particle, are needed. A rigid body, say a set of axes, or the "orientation" of an electron, requires three angles to describe its rotation, thus a three-parameter group, SU(2); the position of a particle described by SO(3), requires two angles, and it is two angles that the SO(3) basis vectors, the spherical harmonics, depend on.

It might seem strange (it should) that these groups with the same algebra have representations with different numbers of parameters. SO(3) transformations also depend on three parameters, the angles of rotation about the three axes. Why then do the spherical harmonics depend on only two? To find the direction to a particle we can rotate first about x, then about y, to give the vector pointing to the particle. We can now rotate about z, which we take as the line to the particle. This does not change the direction of the line; what it does change is the phase of the statevector. But the orthogonal group is defined as acting on real vectors, and it takes them to real vectors. Is there a contradiction in saying spherical harmonics are complex? That the group is defined by its action on real vectors does not mean that it cannot act on complex ones. And quantum-mechanical statefunctions are complex. To describe a rigid body, a classical concept, three angles are needed. Thus classically, functions describing rigid bodies depend on three parameters. In quantum mechanics there are no rigid bodies, only point particles. Two parameters are needed to give their angular coordinates. Thus the third parameter gets hidden, but it must still be there. And it is, being now a phase.

Problem XI.4.g–1: Try this experimentally. Take a beam of atoms with orbital angular momentum z-components all the same, and put it in successive magnetic fields to rotate the orbital angular momentum l (using a method to eliminate the effects of spin) first around x, then around y. Then split the beam into two, and put one in a potential for a time to change the energy, and thus the phase of the electrons (in effect rotating them about the direction vector from the nucleus to the electron), with the potential so chosen that the energy change depends on l. Combine the two beams to get an interference pattern, this determined by the phase change, and l dependent. The spherical harmonic statevectors now have to be regarded as complex numbers, with a phase (and phase difference between states of different orbital angular momentum), and this is the third parameter they depend on. Spherical harmonics seem to depend on only two angles because of the way they are usually written. Nevertheless they are actually complex numbers, there is a phase, and this has experimental effects.

Problem XI.4.g–2: Realize the SO(3) states using three angles [Edmonds (1960), p. 53; Margenau and Murphy (1955), p. 352; Talman (1968), p. 143; Tinkham (1964), p. 109; Tung (1985), p. 141; Varshalovich, Moskalev and Khersonskii (1988), p. 72]. Do this by finding the representation matrix elements in terms of the three transformation angles, and have these act on a fixed state. The result is, for arbitrary angles, an arbitrary state. How are these functions related to the transformation matrices? Notice that the states bear three sets of labels (here each set contains one label), as we expect for the equivalence class, representation, and state. Prove that all representations of an equivalence class are equivalent. Are those of different classes equivalent? Why?

XI.4.h Irreducibility of the rotation group representations

Multinomials are basis vectors, but not of irreducible representations. Are the spherical harmonics irreducible? The proof that they are uses Schur's lemma by finding a matrix that commutes with all rotation matrices (every angle, every axis) and showing that it is a multiple of the unit matrix [Wigner (1959), p. 155].

Problem XI.4.h–1: Show that the matrix for rotation α about z can be diagonalized with elements $exp(im\alpha)$, $-l \leq m \leq l$. This can be shown from the expression for an arbitrary rotation. However it should also be expected for other reasons. So the only matrix that commutes with this representation matrix, for all α, is diagonal and thus a matrix commuting with all rotations is diagonal. For rotation β about y, show that no matrix elements of the 0'th row ($m = 0$) are 0 for all β. Prove that the diagonal matrix commuting with all rotation matrices must have all diagonal elements equal, giving the result. Above, heuristic arguments were given for the representations being irreducible (pb. XI.4.a.ii–3, p. 318; pb. XI.4.c–1, p. 323; pb. XI.4.c–2, p. 323). How are they related to this, rigorous, argument?

Problem XI.4.h–2: SO(3) has the unusual property that its defining representation is also its regular (adjoint) representation, the set of transformations of the group on itself. It, like the other representations, is irreducible. Are these points related to this being a simple group?

XI.5 THE REPRESENTATIONS OF SU(2)

SU(2) is defined as the simple group of unimodular transformations on a two-dimensional space of complex vectors keeping fixed absolute values of products of two vectors $|uv|$, thus also the norm of a vector, $|v|$. Both it and SO(3) are three-parameter, simple groups with representations, and similarly states, labeled by a single number, integers — for SU(2) also half-integers. Why only a single number, and why do the conditions on it differ for these groups?

Problem XI.5–1: Why is "simple" included in the definition?

Problem XI.5–2: The representations are defined over a specific range of parameters. However we can take different ranges. Are these different representations so obtained equivalent? How many equivalence classes of representations are there? Each class of elements is given by a set of angles. But these sets are repeated a denumerably-infinite number of times. How many such classes are there? Are the numbers of classes and of equivalence classes equal?

XI.5.a Basis states of the SU(2) defining representation

The defining representation of a group is that used to define it; for SU(2) it is the representation acting on a (the?) two-dimensional complex vector space. The bases of the defining ($j = \frac{1}{2}$) representation are two vectors written z_1 and z_2, or

$$\zeta = \begin{pmatrix} 1 \\ 0 \end{pmatrix}, \quad \eta = \begin{pmatrix} 0 \\ 1 \end{pmatrix}. \qquad \text{(XI.5.a–1)}$$

There are four operators acting on these (leaving norms of products invariant) which we can find by exponentiating $z_1 d/dz_1, z_2 d/dz_1, \ldots$, or the four Pauli spin matrices (the fourth being the identity).

Defining-representation basis states are supposed to be complex vectors. However the entries in these vectors look real, being 1 and

0. Actually they are complex numbers with the particular values 1 and 0 (which are just as complex as i and 0). A general SU(2) transformation acting on these vectors gives ones with complex entries. Since transformations acting on real numbers cannot give complex numbers, we must regard the entries in the vectors as complex.

Vectors, z_1, z_2, and ζ, η, are bases (basis states, basis vectors) of the SU(2) defining representation. But they are different realizations of this representation, as variables, and as column vectors, but the vectors of each are both transformed among themselves the same way. Physically these states can be the spin-up and spin-down statefunctions of a spin-$\frac{1}{2}$ particle. This interpretation is another realization in addition to the purely mathematical symbols; another is isospin up and down states.

Problem XI.5.a.-1: For $\begin{pmatrix} 1 \\ 0 \end{pmatrix}$ and $\begin{pmatrix} 0 \\ 1 \end{pmatrix}$ find the vectors obtained using the SU(2) transformation matrices (eq. X.5.a-3, p. 285) and show that SU(2) gives complex vectors. Note that a rotation around z (these states are taken to be eigenstates of J_z) changes the phase of the state, so that the 1 can be replaced by an arbitrary, thus undeterminable, phase. What are the states with eigenvalue -1? Write the states "up" in an arbitrary direction as a sum of the J_z eigenstates. How much freedom is there in choice of phases? Is there physical significance to this?

XI.5.b The basis vectors of arbitrary SU(2) representations

How do we find the bases (called spinors [Cartan (1981); Gel'fand, Minlos and Shapiro (1963); Lyubarskii (1960), p. 212; Naimark (1964); Zelobenko (1973), p. 335]) of other representations? Can those of the defining representation help? And what do we mean by the basis vectors of other representations? What "are" they?

As is well-known in quantum mechanics every angular momentum state can be written as a product of spin-$\frac{1}{2}$ statefunctions — any SU(2) state of an irreducible representation, however realized, transforms the same as (which does not imply it is the same as) one realized as a product of spin-$\frac{1}{2}$ states. SU(2) operators do not change the degree of a multinomial in these variables; basis states of irreducible representations thus are all of the same degree.

The basis vectors that we want are functions of the two vectors of the defining representation, however we might realize them, specifically multinomials. But does the degree of the multinomials realizing the basis vectors determine the representation uniquely; can there be more than one irreducible representation of a single degree? What are the irreducible-representation basis states — the complete, orthonormal set of multinomials? What are the representation matrix elements? The answer to the first question can easily be guessed. For SU(2), but not larger groups, an irreducible representation is determined by a single number, the angular-momentum quantum number. This is given by the degree and does in fact uniquely determine the representation. This is also shown by the derivation of the matrix elements, as we now see.

XI.5.c Computing the SU(2) matrix elements

The key to finding the irreducible representations is the invariance of the degree of multinomials under group transformations [Hamermesh (1962), p. 353; Wigner (1959), p. 163]. Thus the basis vectors are multinomials with each degree giving, for SU(2) uniquely, a representation. The degree is written $2j$ (being influenced by quantum mechanics) and the multinomial for state μ is

$$f_{\mu}^{j}(\zeta,\eta)=\frac{\zeta^{j+\mu}\eta^{j-\mu}}{\sqrt{(j+\mu)!(j-\mu)!}}. \quad \text{(XI.5.c-1)}$$

The irreducible representation basis vectors must have this form, up to notation, except that we must justify the normalization. With group operator P acting on the multinomial,

$$\begin{aligned}Pf_{\mu}(\zeta,\eta)&=f_{\mu}(a\zeta+b\eta,c\zeta+d\eta)\\&=\frac{(a\zeta+b\eta)^{j+\mu}(c\zeta+d\eta)^{j-\mu}}{\sqrt{(j+\mu)!(j-\mu)!}}.\end{aligned} \quad \text{(XI.5.c-2)}$$

Problem XI.5.c-1: Using the binomial theorem show that,

$$\begin{aligned}Pf_{\mu}(\zeta,\eta)&=\sum\sum(-1)^{\kappa}\frac{\sqrt{(j+\mu)!(j-\mu)!}}{\sqrt{(j+\mu-\kappa)!(j-\mu-\kappa')!}}\\&\quad\times\frac{a^{\kappa'}d^{j+\mu-\kappa}b^{\kappa}(-c)^{j-\mu-\kappa'}\zeta^{2j-\kappa-\kappa'}\zeta^{\kappa+\kappa'}}{\kappa!\kappa'!}\\&=\sum\sum(-1)^{\kappa}\frac{\sqrt{\{(j+\mu)!(j-\mu)!}\sqrt{\{(j+\mu')!(j-\mu')!\}}}{\kappa'(j-\mu'-\kappa)!(j+\mu-\kappa)!(\kappa+\mu'-\mu)!}\\&\quad\times a^{j-\mu'-\kappa}d^{j+\mu-\kappa}b^{\kappa}(-c)^{\kappa+\mu'-\mu}f_{\mu'}(\zeta,\eta),\end{aligned} \quad \text{(XI.5.c-3)}$$

$$\mu'=j-\kappa-\kappa', \quad \text{(XI.5.c-4)}$$

with μ' being integral, and half-integral, when j is. The binomial coefficients determine the limits of the sums, the first being from $\kappa=0$ to $j+\mu$, the second from $\kappa'=0$ to $j-\mu$. For SU(2),

$$c=-b^{*},\quad d=a^{*}. \quad \text{(XI.5.c-5)}$$

From these coefficients show that j is an integer or half-integer and that μ goes in integral steps from $-j$ to j. What goes wrong for μ outside this range? Note that j is not changed and that all these values of μ appear; the representation is determined by j. Why does this show that the representations are irreducible? The coefficients of f are the matrix elements of representation j, with the states labeled by μ. Explain. Thus demonstrate that

$$M^j_{\mu\mu'}(a,b,c,d) = \sum\sum(-1)^\kappa \frac{\sqrt{(j+\mu)!(j-\mu)!}\sqrt{(j+\mu')!(j-\mu')!}}{\kappa!(j-\mu'-\kappa)!(j+\mu-\kappa)!(\kappa+\mu'-\mu)!}$$

$$\times a^{j-\mu'-\kappa} d^{j+\mu-\kappa} b^\kappa (-c)^{\kappa+\mu'-\mu}, \qquad \text{(XI.5.c-6)}$$

the matrix element in representation j between states μ and μ'. Notice that the matrix elements depend on parameters. How are these related to those in the transformation matrix? Show that the powers of ζ and ζ must be positive for the representation to be finite-dimensional. Prove that the dimension is $2j+1$, that $-j \leq \mu \leq j$, and that j and μ are (both) either integers or half-integers.

Problem XI.5.c-2: Compute the matrix elements for $j = \frac{1}{2}$ and show that with this normalization the representation is unitary. Repeat for $j = 1$. Do this for both SU(2) and SL(2) (pb. X.4.a.i-3, p. 278) and see if they are both unitary [Miller (1972), p. 199, 233]. How is the difference between these two put in? Would they be unitary if the normalization was different?

Problem XI.5.c-3: Show that the states are orthonormal (states for either, or both, different j or μ are orthogonal).

Problem XI.5.c-4: What are the basis states and matrix elements for U(2)? How about SL(2) and L(2)? Are these orthonormal?

Problem XI.5.c-5: For SO(3), realizing the states as explicit functions of the variables showed that j must be integral. For SU(2) the states are also realized as explicit functions of the variables. Yet j can also be a half-integer. Why? These states seem to be realized as functions of four variables (two complex ones), which is one too many. Explain.

XI.5.d The unitarity of the representations

Thus we have representations, but are they what we want? Are they all? They must be shown to be unitary, irreducible, and complete (there are no other inequivalent irreducible representations). First we consider unitarity.

Problem XI.5.d-1: Find $\sum f_\mu f^*_\mu$, where the sum goes from ? to ?, and also $\sum P f_\mu P f^*_\mu$, the sum over the transformed f's, which should demonstrate that this product is invariant under unitary transformations. It should then be clear that

$$\sum\sum M^{j}_{\mu\mu'} f_{\mu'} M^{*j}_{\mu\mu'} f^{*}_{\mu'} = ?, \qquad \text{(XI.5.d-1)}$$

and so is invariant. Explain the normalization chosen for the states. Obviously the $(2j+1)^2$ functions, $f_\mu f^*_\mu$ are linearly independent. Why? Is this used in the discussion? How?

Problem XI.5.d-2: State the requirement of unitarity for the matrix elements and demonstrate that they satisfy.

Problem XI.5.d-3: Note the importance of the basis states being linearly independent. Otherwise some couldbe eliminated andtherewould be relations between matrix elements. Verify this. What would happen to unitarity if there were such relations? Demonstrate explicitly for two-dimensional matrices. Discuss this for U(2), SL(2) and L(2).

Problem XI.5.d-4: For SU(2), using the realization of the operators as (exponentials of) derivatives, compute the matrix elements for the defining, adjoint, and general, representations. Show that these are (equivalent to) unitary representations. Are they unitary for SL(2), that is without the condition on the parameters?

XI.5.e Completeness

The (irreducible) representations of SU(2) are complete; there are no other inequivalent ones. Any (well-behaved) function of the variables on which the group transformations are defined, here z_1 and z_2, can be expanded as a sum over all states of all group representations.

Problem XI.5.e-1: Why are these two sentences equivalent?

Problem XI.5.e-2: Now any such function can be expanded as a polynomial in these variables (Taylor's theorem; the expansion is Taylor's series). Why is Taylor's series, rather than the Laurent series, the relevant one? Show that the Taylor expansion of a function of these two variables is equivalent to an expansion over all states of all irreducible representations. Why are only irreducible ones considered? Thus the irreducible representations of SU(2), as realized here, are complete.

Problem XI.5.e-3: Could there be a realization of all irreducible representations that is not complete?

Problem XI.5.e-4: What about the representations of linear group SL(2), with unitarity not imposed; are they complete? Are there two sets of complete representations or are they related? Are the functions expanded the same for the two groups? Why? How about the variables on which the groups act? Do these points affect the answers to the questions? Answer the questions for U(2) and $L(2)$.

XI.5.f Connectivity

The representations of SU(2) are given by every half-integer value of j, one representation for each, thus though infinite in number, they are discrete, unlike the representations of U(1) which are labeled by a continuous parameter. SU(2) is singly connected, U(1) infinitely so, and the type of representation label is determined by, and gives meaning to, this degree of connectivity. The number of representations equals the number of classes; for U(1) these need both be continuously infinite, thus the domain of definition goes to infinity. But it is periodic, so we must regard it as infinite, but wound around itself, like a rope forming a circle, and wound around itself an infinite number of times, thus infinitely connected. For SU(2), and so any of its U(1) subgroups, to do that would give more classes than representations. Hence it is singly connected.

Problem XI.5.f-1: Check that with these views of the connectivity, the number of classes and the number of representations are equal, for both groups. Does this give the correct result for the U(1) subgroups of SU(2)?

XI.5.g Irreducibility

Are these representations irreducible? This is shown by demonstrating that any matrix that commutes with all of the representation is a multiple of the identity.

Problem XI.5.g-1: Why? Take first

$$b = 0, \quad a = exp(-i\alpha/2). \tag{XI.5.g-1}$$

Find the representation matrix T for this transformation and show that it is diagonal. Any matrix M that commutes with this set is thus also diagonal. Next take the j'th row of the commutator of M and T (and explain why no element of this matrix can vanish for all values of the parameters) to show that all diagonal elements of M are equal. M is proportional to the identity — the representation is irreducible. We expect that basis vectors of each degree belong to irreducible representations since these are labeled by only one variable. How is this used in the proof? What would go wrong otherwise? How is this argument related to the one (sec. XI.4.h, p. 328) for SO(3)? To the argument coming from the derivation of the matrix elements (sec. XI.5.c, p. 331)? Should you expect these relations? The representations of SU(2), the basis vectors — explicit multinomials — and matrix elements have now been found and shown to be unitary, complete and irreducible. Is anything left?

XI.5.h How SO(3) and SU(2) representations are related

SO(3) and SU(2) are related; their representations should also be. Those of SO(3) can be derived in a way similar to that for SU(2); find the action of the operators on multinomials. The first difference is the number of variables, two complex ones for the latter, three real ones for the former. Does this affect the values of j? The number of variables does not; we can consider the SO(2) subgroup and identify ζ with x and η with z. But these are now real. To find the representations of SO(3) from the expression for the SU(2) matrix elements, we express the SU(2) transformation parameters in terms of half-angles, giving the transformation parameters of SO(3) (sec. X.6.a, p. 287).

Problem XI.5.h-1: Do this, and find the matrix elements. Define

$$\nu = 2\mu, \quad m = 2\kappa, \quad l = 2j, \tag{XI.5.h-1}$$

rewrite the matrix element (sec. XI.5.c, p. 331) in terms of these variables, and show that the only nonzero matrix elements (and so representations) are those for which l is an integer, and

$$-l \leq \nu, m \leq l. \tag{XI.5.h-2}$$

Problem XI.5.h-2: Check that these representations are unitary, and discuss completeness. In particular show two things, first, any function defined over the group (any well-behaved function in x, y, z) can be expanded in terms of these basis vectors, second, there are no other irreducible representations. (Is this one statement, or two?)

Problem XI.5.h-3: From the SU(2) transformation matrix (sec. X.5.a, p. 284) find the parameters giving the identity of the rotation group. Show that both

$$Q = \begin{pmatrix} 1 & 0 \\ 0 & 1 \end{pmatrix}, \tag{XI.5.h-3}$$

and

$$Q' = \begin{pmatrix} -1 & 0 \\ 0 & -1 \end{pmatrix}, \tag{XI.5.h-4}$$

give the identity, and that only these two do. Take any other rotation and show that there are two SU(2) transformations, only, giving it. This means that the product of two rotations, for half-integral representations, is determined only up to a sign. If we take the identity E, and T a rotation of π, as

$$E = \begin{pmatrix} 1 & 0 \\ 0 & 1 \end{pmatrix} \tag{XI.5.h-5}$$

and

$$T = \begin{pmatrix} i & 0 \\ 0 & -i \end{pmatrix}, \qquad \text{(XI.5.h-6)}$$

and take their product we get T (of course). But, for a 2π rotation, $T^2 = -1$. Thus these half-integral representations of SU(2) are not good rotations, and do not form rotation-group representations. Now the representations of SO(3), with integral j, are complete for any well-behaved functions of x, y, z. This emphasizes that half-integral representations of SU(2) are not SO(3) representations. It would be interesting to solve for ζ and η in terms of x, y, z (sec. X.6.a, p. 287). Can these functions be expanded in terms of the complete set of SO(3) representations? Would that contradict these assertions?

Problem XI.5.h-4: For SU(2) the basis vectors of the irreducible representations are multinomials, while for SO(3) they are polynomials (in different variables). Explain. This is true generally for SU(n); however the set of all multinomials of a given degree is not irreducible as it is for SU(2). Why?

Problem XI.5.h-5: A set of eigenfunctions (in three variables) of J_z and J^2 with both integral and half-integral eigenvalues are the representation matrix elements for finite transformations [Edmonds (1960), p. 53, 64; Elliott and Dawber (1987), p. 519; Inui, Tanabe, and Onodera (1990), p. 125; Tinkham (1964), p. 109; Varshalovich, Moskalev and Khersonskii (1988), p. 72; Wigner (1959), p. 166]. Find these finite transformation matrices and discuss their relationship to the spherical harmonics and SU(2) eigenfunctions. These are the matrix elements of an arbitrary rotation — of a rigid body, which needs three angles (why?); is it reasonable that they give spin states?

XI.5.i The characters of the representations

Characters are class functions, so to compute them we take the simplest member of each class. The classes consist of all transformations of the same angle; only one is needed since the other two label the axes, and these give the members of the class. Thus the simplest SO(3) transformation is a rotation through ϕ about the z axis for which the other two angles are 0.

Problem XI.5.i-1: What is the corresponding simplest SU(2) transformation?

Problem XI.5.i-2: Show that for these transformations, both the SU(2) and SO(3) representation matrices are diagonal, with diagonal elements $exp(im\phi), -j \leq m \leq j$, for representation j, which can be integral or half-integral. Thus the character, for both SU(2) and SO(3), of representation j and class ϕ, is

$$\zeta_j(\phi) = \sum_{m=-j}^{m=j} e^{im\phi}. \tag{XI.5.i-1}$$

Problem XI.5.i-3: Now the characters of any other irreducible representation must be orthogonal to these. (Does the proof of this hold for continuous groups?) Show, by Fourier's theorem, that this cannot be, so there are no other characters. This is another proof that the representations found are complete [Wigner (1959), p. 166, 168]. Why are the integral representations complete for SO(3), but not for SU(2)?

Problem XI.5.i-4: As with finite groups the transformations of Lie groups divide into classes with the characters class functions and orthonormal. Prove this, at least for these groups.

XI.6 WHAT HAVE WE LEARNED FROM THESE GROUPS?

The rotation group and SU(2) are the simplest (nontrivial) simple Lie groups. They provide examples of group transformations with continuous parameters given by functions which are infinitely differentiable and "well-behaved". Their parameter ranges are chosen to give all values of the transformation functions. Representations are found by having the transformations act explicitly on basis functions, these being polynomials in variables transforming in the same way as (covariantly — they vary with) the coordinates of the spaces over which the groups are defined. This gives also the matrix elements. Irreducible-representation basis functions, here multinomials for SU(2) but polynomials for SO(3), form a complete set allowing the expansion of any "well-behaved" functions of these variables. And they are a complete set of irreducible representations. These representations can be made unitary (to have unitary matrix elements) by proper choice of the normalization of the basis vectors.

The classes and characters have been found and shown to behave as required. And, as with finite groups, there are invariant subgroups and factor groups, although here we only considered hints of these. But these two groups are related, they are homomorphic, which was explicitly demonstrated and interpreted.

Thus there are similarities between Lie and finite groups; also clearly, major differences. But these are so simple that many other aspects of Lie group theory do not appear for them.

The representations of the rotation group, and of SU(2), and the relationship of these groups, their labels, that these labels are discrete and integer or half-integer (and that the states are discrete and finite in number for each representation), and the discrete labels for their states, are familiar from quantum mechanics. But for those unfamiliar with quantum mechanics these results, or at least their physical interpretations, are often surprising. But distressing (as it often is when we first learn atomic physics and quantum mechanics) or not, the rotation group (and its covering group SU(2)) determine the structure of matter, and the nature of the world we live in. The application of rotation operators to polynomials and the resultant algebra may seem simple, even trite, but the results and limitations obtained are of the greatest importance.

The physical significance of these results is familiar, so familiar as to often be beyond notice. That rotations are possible, that they are smooth functions of the angles, that these parameters have the ranges that they do, is so central to our nature as to be almost incomprehensible. It would be very difficult to imagine a world for which these were not true.

Problem XI.6-1: But probably not impossible.

Problem XI.6-2: In finding the representations of the groups, to what extent was knowledge of the algebras, and their representations, needed and helpful?

Problem XI.6-3: How have the representations themselves, and to what degree, provided insight into these groups, into group theory, and into the geometry and physics they describe? What insight have they provided?

Problem XI.6-4: We have asked various questions above about these groups and their algebras. To what extent have these been answered (or even considered)?

Chapter XII

Applications of Representations of SO(3) and O(3)

XII.1 PROPERTIES AND APPLICATIONS OF THE REPRESENTATIONS

The rotation groups are important, because of their physical relevance, but also as a model for groups with more structure and greater complexity. Here we develop properties and techniques needed for their application, and then study some uses in physics to understand both the physics and the value and limitations of these groups as models. For these, and for other applications, we often need products of representations — reducible representations — so we need also their reduction. For physical applications we have to study physical operators and their action on physical states; this action given by the matrix elements of the operators. These matrix elements are related to the reduction of product representations and their properties depend on both the operator and the group. Which properties are determined by the operator, and which by the group; how can we furnish those parts of matrix elements that depend only on the group and the state and representation to which the operator belongs, thus leaving the terms for whose determination we need knowledge of the (physical properties of the) operators?

Transformations of geometry (and physics) are given by O(3), under which most physical laws are invariant. Extension of the results for

SO(3) to include inversion is (rather) simple, and previously hinted at; here we make it explicit.

Thus we will have developed the tools to study the effects of rotational symmetry. But some of the most important applications are to physical situations in which there is not complete symmetry, in particular in the presence of fields, here electric and magnetic fields, giving the Stark and Zeeman effects. How does the rotation group help us understand these effects, and in general, symmetry breaking in physical systems? And why is the group relevant if (the part that is considered of) the system is not invariant under it? To examine situations with symmetry breaking, combination of several angular momenta is needed; so we first study products of representations.

XII.2 PRODUCTS OF GROUPS AND REPRESENTATIONS

The state of an electron in an atom is a rotation-group basis state (not necessarily of a decomposed representation). What is the state of two electrons? An electron has, besides orbital angular momentum, spin; what is the complete state of the electron? And what is the state of an atom, with substates including the (not quite obvious) states of the electron(s), and of the nucleus, this determined by the states of the individual nucleons (each with orbital and spin angular momentum)? There are several rotation groups: each object can be rotated separately, also each can be moved in its orbit, or its orbit's direction varied, without changing its spin direction, and conversely. There is a set of rotation operators, each acting on different states (say, functions for orbital angular momentum, or ones giving spin states, or those of different objects), and these commute. Thus we have a group consisting of all products of these operators, the group that is the direct product of these groups (sec. III.5.c, p. 101) [Wigner (1959), p. 171]. What are the representations? What do they tell us?

The Hamiltonian of an atom is not invariant under rotations of electrons individually, or changes of spin without variation of orbital angular momentum; these change it (although they give information, so are of interest). It is invariant under rotations of the atom — rotations of the coordinate system. Product states are bases of a subgroup of the group of all products of operators; this subgroup, the one rotating the entire atom, is (isomorphic to) the rotation group. Such product states describe the atom as unit, or an electron with some total angular momentum.

More generally, we can take functions, rotation basis states, and multiply them. The product is also a basis state — a rotation transforms it. But which state, of what representation, reducible or irreducible, and how is this state related to those of the product? Individual states can be functions of different variables, so of what variables is the product a function of, and what operators act on it?

To find these states, we consider not the representations of the direct product of groups (each a copy of the same group), but the direct product of representations of a single group. These representations arise from the multiplication of representations. The reason that we can attack the problem this way is that all representations of a group are the same no matter how arrived at; they are a property of the group. But they can be realized differently.

What questions about these are interesting physically, and mathematically, and why? The behavior of a system, an atom, or nucleus, say, is dependent on its angular momentum, as well as that of its components. Thus we want not only the total angular momentum, but the spin, orbital, and total angular momentum of each object, and those of various combinations. Moreover we want the statefunction of the system, and wish to understand how it is related to the statefunctions of the individual particles. For, say an atom — helium is an example — with electrons having angular momenta $j_1, \ldots$, what is the probability of finding total angular momentum J? Or given J, what is the probability of finding the electrons with angular momenta $j_1, \ldots$?

This is physics; what are the mathematical questions? While the product of basis states is, must be, a basis state, it need not be one of an irreducible representation. But the set of these representations is complete, so any product can be written as a sum over irreducible-representation basis states. A product of states of representations $j_1, j_2, \ldots$, is a state of a reducible representation, and can be reduced to a sum of states of irreducible representations. Those of an irreducible representation can appear in the decomposition of some products, but generally will not appear in all. In these sums which states appear — what representations contribute states, how many times does each contribute; what are the states in the sum, what are their forms? In the decomposition of the product to a sum, what are the coefficients? For a given product do the answers to these questions vary with the terms (particularly the representations) in the sum? Conversely, if we write a state of angular momentum J as a sum of product states, which appear, and with what coefficients? How are these two sets of appearances, coefficients, and questions, related?

The sum of representations into which a product of representations decomposes is the Clebsch-Gordan decomposition, and the coefficients of the states are the Clebsch-Gordan (or Wigner, or vector-coupling) coefficients (the states in the sum that equals the product of states, are those with nonzero coefficients) [Cornwell (1984), p. 449; Edmonds (1960), p. 31; Hamermesh (1962), p. 147, 367; Inui, Tanabe, and Onodera (1990), p. 137, 212; Jones (1990), p. 109; Schensted (1976), p. 267; Tinkham (1964), p. 115; Tung (1985), p. 117; Varshalovich, Moskalev and Khersonskii (1988), p. 235; Wigner (1959), p. 184, 188]. There being no distinction here between orbital and spin — integral and half-integral — angular momentum, this applies to the direct product of orbital with orbital, spin with spin, and spin with orbital angular momenta; the results hold for both SO(3) and SU(2), and other Lie groups — the theory is general, but these are the only cases considered explicitly here.

Problem XII.2-1: Why must the product of basis states be a basis state?

XII.2.a The vector addition model

It is helpful to look at this semi-classically (especially since the picture, the vector addition model is popular in physics) [Wigner (1959), p. 184]. If we add two angular momentum vectors, labeled j, j', the resultant angular momentum J is maximum if the two vectors are parallel, minimum if they are antiparallel, and varies through this range as the angle between them does. In quantum mechanics (that is group theory) we cannot give precisely the direction of angular momentum, only its projection along one axis, though we might guess that J is restricted to integral steps, and

$$|j - j'| \leq J \leq j + j'. \tag{XII.2.a-1}$$

This is well-known to be correct. Why is it correct?

This gives the limits of the range for J, which it must attain. But need every (integral or half-integral) J in the range occur? Might some appear more than once, might there be more than one sum of states forming the same basis state of the same representation? For the product of two SU(2) representations, each representation in the allowed range appears once, and only once. This is not true for larger groups. Nor is it true for SU(2), if there are more than two terms in the product. To give the orbital state of, say, three electrons, it is necessary to give the state of each of these, and the total orbital angular momentum J, but the state for J is found by first coupling any two, finding its orbital state, and then coupling that to the third, so the intermediate state must be given. There is more than one way to get the same final J. Given the states of the individual objects, and the angular-momentum quantum numbers for the state of the system, there are several different statefunctions, these depending on the pair coupled to give the intermediate state, and that state's quantum numbers.

Problem XII.2.a–1: To show that the product of more than two terms is not simply reducible — a total J can occur more than once — take an electron and a nucleus, the discussion is clearer if these are different. The electron has orbital angular momentum L, and spin $s(=\frac{1}{2})$, the nucleus has spin S. Find all ways of combining these three angular momenta to give angular momentum J. What is the range of J? Is the number of ways dependent on J? For several electrons, we can find the total angular momentum by finding the total for each electron separately, giving the j for that electron, then adding — jj coupling — or adding all orbital angular momenta, giving L, and adding all spins, giving S, then combining — LS (Russell-Saunders) coupling [Elliott and Dawber (1987), p. 173; Hamermesh (1962), p. 417; Heine (1993), p. 84; Wigner (1959), p. 272]. Or if we want, and (usually) no one does, we can use intermediate combinations. The sets obtained by jj coupling and LS coupling are different, so might have different energies and transition probabilities. If they are complete (within the approximations used) any state can be expanded as a sum over either of them.

Problem XII.2.a–2: The decomposition of products has been discussed for symmetric groups (sec. IX.5, p. 259). Compare that discussion with this one. Which type of product is being considered here?

XII.2.b The representations in the decomposition of products

We thus have to find the range of the representations, the number of times each occurs, and then the states and their coefficients.

XII.2.b.i *The range*

First we justify the range. One argument uses the Lie algebra (the angular momentum operators). Since the operators for the representations j, j' in the product commute, the operator for the representation J of the total system is their sum. So for the z component

$$J_z = j_z + j_z'. \tag{XII.2.b.i–1}$$

The state is the product of the states, thus the corresponding eigenvalues, M, m, and m' are related by

$$M = m + m'. \tag{XII.2.b.i–2}$$

So

$$|j - j'| \leq M \leq j + j'. \tag{XII.2.b.i–3}$$

For any J we can pick the axes to give any M in the range

$$-J \leq M \leq J. \tag{XII.2.b.i–4}$$

Thus J cannot lie outside this range, and all M occur within it.

Problem XII.2.b.i-1: There is a simple, but limited (to SU(2)), way of finding the representations in the decomposition of a product of two representations — the Clebsch-Gordan decomposition. If the states of representations j and j' in the product have their maximum m values, so does the resultant state, this being

$$M = J = j + j'. \qquad \text{(XII.2.b.i-5)}$$

This state is unique; there is only one product of states giving it. From it we can obtain the others of representation J by using the step-down operator. For

$$M = j + j' - 1, \qquad \text{(XII.2.b.i-6)}$$

there are two states, the terms in the product having

$$m = j,\ m' = j' - 1, \text{ or } m = j - 1,\ m' = j'. \qquad \text{(XII.2.b.i-7)}$$

From these two states we form two linear combinations; one goes with representation J. Why? The other can only go with $J - 1$, and there is no other state that does. Why? Hence this representation appears in the decomposition, and only once. Applying the step operator gives the other states of representation $J - 1$. Show that each time the value of M is decreased (to 0 or $\frac{1}{2}$, whichever occurs) the number of ways of getting the state increases by 1. Thus for $M = J - 2$, there are three states. For each M, all states belong to representations with J greater than M, except one. That must then belong to the representation K, with $K = M$. Hence K appears in the decomposition, for all K in the stated range, and since there is only one way of getting K, it appears but once. Thus all representations from the largest

$$J_{max} = j + j', \qquad \text{(XII.2.b.i-8)}$$

to the smallest

$$J_{min} = |j - j'|, \qquad \text{(XII.2.b.i-9)}$$

appear, and just once. How does this argument give J_{min}? This is an explanation of why the representations appear and the number of times; it depends on the number of states available, and which must be used in giving the representations in the decomposition. Try this with a few products of representations. Here the number of representation labels, and the number of state labels, are equal; both are 1. It should be clear that the argument does not work for more complicated situations, like products of several terms (why?) and the result does not hold in general. Try it, and check. If there is more than one representation label, the representations cannot be ordered as here. And representations do appear more than once in the decomposition of products of larger groups, and not all representations appear in a range.

Problem XII.2.b.i-2: Characters give another way of reducing [Wigner (1959), p. 186]. For each class the character of the product representation equals the product of the characters of the representations in the product, which should first be proven (for Lie groups). This gives (sec. VII.6.a, p. 199)

$$\eta(\phi) = \eta^j(\phi)\eta^{j'}(\phi) = \sum exp(im\phi) \sum exp(im'\phi) = \sum \eta^J(\phi). \tag{XII.2.b.i-10}$$

The question is what terms appear in the sum over J, and how many times. It should be easy to show that the range is given by eq. XII.2.a-1, p. 342, all J's in the range appear, and each once.

XII.2.b.ii *Finding the linear combinations*

To see how this works it is useful to try an example. We consider the product of two orbital angular momentum representations, both with $l = 1$.

Problem XII.2.b.ii-1: Each of these representations has three states, so the product has nine, and

$$L = 2, 1, 0. \tag{XII.2.b.ii-1}$$

This holds no matter how the representation is interpreted. It is only when we take a specific form for the states — spherical harmonics — that we imply orbital angular momentum. The state is then

$$|L, M) = \sum C(L, M; l, m; l', m')Y_{lm}(\theta, \phi)Y_{l'm'}(\theta', \phi'). \tag{XII.2.b.ii-2}$$

The C's are the relevant coefficients; these are Clebsch-Gordan (Wigner or vector-coupling) coefficients. The state vectors depend on different variables; these might be the coordinates of two electrons. We can also write the states using Cartesian coordinates. Define (how?) x_+, $x_0(= z)$, x_-. For a rotation, of the coordinate system,

$$\theta \Rightarrow \theta + \omega, \quad \theta' \Rightarrow \theta' + \omega, \tag{XII.2.b.ii-3}$$

and similarly for ϕ, ϕ'. Show, using both the angular momentum operators, and by this change of angles, that (up to normalization)

$$|2, 2) \sim sin\theta sin\theta' expi(\phi + \phi') \sim x_+x_{+'}, \tag{XII.2.b.ii-4}$$

is one of five states, and has angular momentum 2. Find the others. Also the states

$$|1, 1) \sim cos\theta sin\theta' exp(i\phi') - cos\theta' sin\theta exp(i\phi) \sim x_0x'_+ - x'_0x_+, \tag{XII.2.b.ii-5}$$

$$|1,0) \sim sin\theta sin\theta' sin(\phi - \phi') \sim x_- x'_+ - x_+ x'_-, \qquad \text{(XII.2.b.ii-6)}$$

$$|1,-1) \sim cos\theta sin\theta' exp(-i\phi') - cos\theta' sin\theta exp(-i\phi) \sim x_0 x'_- - x'_0 x_-, \qquad \text{(XII.2.b.ii-7)}$$

are mixed among themselves, and have $L = 1$. The corresponding states with the plus sign have $L = 2$. There are three states with $M = 0$; the third has $L = 0$. Find it. Normalize all these using both expressions for the states (justifying the normalization). Find the Clebsch-Gordan coefficients. Notice that for $L = 1$, say, the states belong to the same representation as $Y_{1m}(\theta, \phi)$, but are realized differently, in one case using the coordinates of one object, for the other using the coordinates (and states) of two. There are an infinite number of realizations of each basis state of every representation. Another is obtained by taking products of spin-$\frac{1}{2}$ states. Show this for $L = 0, 1$.

Problem XII.2.b.ii-2: Another way of realizing the $l = 1$ states is by column vectors. The matrices acting on these are, say, those given in eq. X.4.a.ii-2, p. 279. Realize the states that are the products of these, and the 9×9 matrices acting on these products. If a computer program can be written to find the direct product of any number of these matrices it would be even more interesting. The product state has nine components. Write it as a five-component object (the other four components are 0), plus a three-component object (six of the nine components are 0) plus a one-component object. Do this by block diagonalizing the 9×9 matrix. Compare the 3×3 matrix obtained in this decomposition with the matrices in the product. Find the Clebsch-Gordan coefficients and check that they are the same as those just found using spherical harmonics. The Clebsch-Gordan coefficients are the matrix elements of the similarity transformation block-diagonalizing the representation matrices. Find them by carrying out this diagonalization, and check that they are the same as found from the states. The coefficients are a property of the group, and the representations, not of the realization.

Problem XII.2.b.ii-3: Suppose that the two vectors (belonging to the two $l = 1$ representations) are interpreted as electric and magnetic fields. There are three different objects obtained by taking their products, going with the three representations. What might physical interpretations of them be?

XII.2.b.iii *Computing the Clebsch-Gordan coefficients*

Two main problems in studying the decomposition of products of representations into sums of irreducible representations are finding the number of times each irreducible representation appears in the sum, which in this simple case has been solved, and finding the Clebsch-Gordan coefficients. Finding these coefficients can be a complicated algebraic problem. There are many ways of doing so [Edmonds (1960), p. 42; Hamermesh (1962), p. 367; Heine (1993), p. 176; Joshi (1982), p. 209; Lyubarskii (1960), p. 227; Miller (1972), p. 256; Tinkham (1964), p. 117; Wigner (1959), p. 184]. Since these are readily available we do not repeat them here. The result is a (complicated) formula; it is generally not used since there are tables available [Biedenharn and Van Dam (1965), p. 317; Cornwell (1984), p. 456; Heine (1993), p. 432-445; Miller (1972), p. 263; Tinkham (1964), p. 123; Varshalovich, Moskalev and Khersonskii (1988), p. 270-288; Wigner (1959), p. 193].

Problem XII.2.b.iii–1: There is a way of approaching the problem, computationally not the best, but perhaps the most revealing. There is but one state of maximum M, the product of two states each of maximum m values. Check the tables to be sure that the Clebsch-Gordan coefficient for this case is 1. Why should it be? There are two ways of getting the state with $M = M_{max} - 1$; one belongs to J_{max}, the other to the $J_{max} - 1$. The first can be obtained from the state with M_{max} by using the step operator. The state belonging to the next J is orthogonal to it (and to all other states of J_{max}). Hence it can be found by taking a linear combination of the two products with this M value; the coefficients are unknown, but are determined by using the orthogonality condition (and normalizing). Then each of the two products is written as a sum of these two states belonging to different J's; the coefficients are the Clebsch-Gordan coefficients. This is continued for each smaller M, giving the coefficients. Try this for several products; $(0,0), (0,\frac{1}{2}), (0,1), (\frac{1}{2},\frac{1}{2}), (\frac{1}{2},1), (1,1)$, and perhaps a few others. Can a general formula be obtained this way? Although this gives the coefficients, and a computer program can be written to generate them, it is difficult to get a formula expressing them in terms of the j's and m's of the product states, and the J and M of the representation.

Problem XII.2.b.iii–2: There is another way of finding the coefficients. The representation matrices of the product are products of the representation matrices of each term in the product, and we wish to decompose this product matrix into block-diagonal form, each block going with an irreducible representation. This is done by a similarity transformation. Prove that the elements of the matrix of this transformation are the Clebsch-Gordan coefficients giving the sums of states each belonging to an irreducible representation. Work these out using the transformation matrices for SU(2) (sec. X.5.a, p. 284) and SO(3) (eq. X.4.a.ii–2, p. 279), which are the representation matrices for $j = \frac{1}{2}$, and $j = 1$.

Problem XII.2.b.iii–3: The representation matrices for any representation J, is given by the direct product of J of those for $j = \frac{1}{2}$, reduced into block-diagonal form. Verify this for $J = 1$. Find the representation matrices for $J = 3/2$ and $J = 2$. Use these to find the Clebsch-Gordan coefficients reducing the product of $(\frac{3}{2}, \frac{1}{2})$, $(\frac{3}{2}, 1)$, $(2, \frac{1}{2})$ and $(2, 1)$.

XII.2.c Unitarity and reciprocity

The Clebsch-Gordan coefficients are fundamental in physics, and their properties have been thoroughly explored. Because discussions of them are so readily available we do not discuss their properties. But there are a couple of points worth mentioning.

Problem XII.2.c-1: All matrices considered here are (taken to be) unitary. Thus the similarity transformation between unitary matrices is unitary; its entries are the coefficients connecting the states. State explicitly the condition on the coefficients resulting from unitarity.

Problem XII.2.c-2: The product of two states can be written as a sum over states of irreducible representations,

$$|j, m)|j', m') = \sum C(j, m; j', m'; J, M)|J, M). \qquad \text{(XII.2.c-1)}$$

And, solving these equations, a state can be written as a product,

$$|J, M) = \sum D(J, M; j, m; j', m')|j, m)|j', m'). \qquad \text{(XII.2.c-2)}$$

Show that these two sets of coefficients are the same (in what sense, since the indices are differently ordered?). Why? This implies a symmetry of the coefficients [Lyubarskii (1960), p. 234]; they are invariant under permutations of the index sets. What is that symmetry? In fact there are further symmetries, though less obvious [Regge (1958, 1959)].

XII.3 TENSOR OPERATORS AND THE WIGNER-ECKART THEOREM

A fundamental problem of quantum mechanics is the determination of the effect of such objects as fields on a system. How do the energies of an atom change when placed in electric or magnetic fields? What is the probability of a transition from one state to another? To answer questions like these we need the matrix elements of the field operator between states of the system. Having introduced the (say, rotation) group, we now add other objects, such as fields. How do these fit in mathematically; group-theoretically what are they, what restrictions are there on them? How can we combine them with the objects of the group to get a coherent, useful, system?

We expect that for the electric field, the matrix element of a field E_z along z, between two up states along z, is the same as that of the same field E_x along x, between two states up along x; also that there is a relationship between the matrix element of E_z connecting states up along z and that connecting states up along x. Indeed, because of rotational invariance, it is reasonable that knowing the effect of a field on a single state (or perhaps only a few) we can find it on all, and knowing the effect of E_z we can find its effect if it were rotated to be along x. We need to know at least one matrix element — the strength, at least, of the field is not given by group theory — but the invariance relates different matrix elements, so we need not, and cannot, give all arbitrarily.

Here we study what fields are, and whether, when, why, and how matrix elements are related.

Problem XII.3.-1: Although we use a specific example, and terms relating to a specific group, the implication of this discussion is that (the) results apply to all (semisimple) Lie groups. Below we talk about the (quite simple) rotation group. The reader should note whether the statements actually refer to it, or just use a restrictive terminology, and to what extent results (might) hold in general.

XII.3.a Tensor operators

The physical objects that we consider, electric and magnetic fields say, transform as basis states of the relevant group (they have no choice), here the rotation group, but these considerations are broader. Thus we need a terminology that includes objects like fields. A tensor operator of a (semisimple Lie) group is an operator that transforms under an irreducible, linear, representation of the group; it is a set of objects (its components) which are (a realization of) basis vectors of the representation [Barut and Raczka (1986), p. 242; Burns (1977), p. 351; Cornwell (1984), p. 106; Edmonds (1960), p. 68; Inui, Tanabe, and Onodera (1990), p. 142; Jones (1990), p. 113; Joshi (1982), p. 215; Lomont (1961), p. 84; Tinkham (1964), p. 124; Tung (1985), p. 122; Varshalovich, Moskalev and Khersonskii (1988), p. 61; Wigner (1959), p. 244]. A vector (usually) is a tensor of the defining representation, a scalar belongs to the identity representation. The electric and magnetic fields are SO(3) vectors.

So a tensor operator (component), T_i^{α}, transforming as state i of representation α of some group, behaves as

$$T'^{\alpha}_j = R^{-1}T^{\alpha}_j R = \sum D^{\alpha}_{jk}(\theta)T^{\alpha}_k; \qquad \text{(XII.3.a-1)}$$

R is the transformation operator, the D's representation matrices and θ the set of parameters of the transformation. The tensor might be realized, being an operator, as a matrix, so its transformation is a similarity transformation. However it is also a basis state of representation α, and therefore transforms as a basis state as shown by this equation.

We can also take T as a group operator — acting on a basis state, it gives a different basis state. What are the implications of this set of statements for it?

A tensor operator is a tensor — a representation basis state — and an operator, something that acts on a function, specifically on every representation basis state. So the product of a tensor operator acting on another basis state (of the same or different realization, and of the same or different representation) transforms as a basis state of a reducible representation; this can be expanded into basis states of irreducible representations (the Clebsch-Gordan decomposition). And because it itself is regarded as a tensor, there is implicit in its definition the assumption that the function given by the action of the tensor operator on the basis state transforms as the product of two basis states, one the operator and the other the state it acts on, no matter what it is or what it acts on. However, generally the effect of an operator (say a derivative) need not be related to a group representation. Thus for an operator to be a tensor operator, the function it gives when acting on any representation basis state must transform the same whether its transformation properties are found by regarding the operator as a tensor, or by regarding it as an operator. For example, for a derivative to be a tensor operator, the derivative of every state ψ must transform as the product of ψ with a state that transforms as the derivative.

Problem XII.3.a-1: This definition gives the behavior of the operator under a finite transformation; show that it requires that

$$[E_\mu, T_i^\alpha] = \sum (E_\mu)_{ij}^\alpha T_j^\alpha, \qquad \text{(XII.3.a-2)}$$

(and conversely) for all operators E_μ of the algebra of the group; $(E_\mu)_{ij}^\alpha$ is the representation matrix element. This is the Lie-algebra definition of a tensor operator. The E's also are tensor operators (of what representation?). What does this equation say about them?

Problem XII.3.a-2: It should be clear that variables, and derivatives with respect to them, as well as their products, are tensor operators when acting on monomials. Is this also true when they act on multinomials? Polynomials? Are sums of such operators tensor operators? Are there restrictions?

XII.3.b The Wigner-Eckart theorem

The purpose of the Wigner-Eckart theorem is to obtain as much information as possible about the matrix elements of a tensor operator from the knowledge of the state it transforms as, so writing it as fully as possible in terms of functions of the group representations [Burns (1977), p. 355; Chen (1989), p. 107; Cornwell (1984), p. 101; Edmonds

(1960), p. 73; Elliott and Dawber (1987), p. 78; Inui, Tanabe, and Onodera (1990), p. 144, 212; Jones (1990), p. 113; Joshi (1982), p. 224; Schensted (1976), p. 273; Tinkham (1964), p. 131]; many explicit formulas are given in Varshalovich, Moskalev and Khersonskii (1988), chap. 13, p. 475. This reduces to a minimum the knowledge needed about the operator itself (which is the maximum that can be arbitrarily specified) for the determination of matrix elements. It answers the questions: which properties of matrix elements depend on the group and its representations, and which on the operator, and how are these expressed? The theorem gives a matrix element as a product (or for other groups, sums of products, if there is multiplicity), one term determined by how the tensor transforms, and is independent of its realization, the other depends on what it is — how it is realized. The first term is the same for an electric and a magnetic field (or anything else transforming similarly), the second depends on the field.

Tensor operator $T(j,m)$ — belonging to the (semisimple) group representation given by the set of labels j, and the state given by the set m — acts on state $|L,M)$; the resultant state belongs to a reducible representation, and reducing it gives the state of representation K,

$$T(j,m)|L,M) = \sum C(K,f;j,m;L,M)|K,f); \qquad \text{(XII.3.b-1)}$$

the C's are Clebsch-Gordan coefficients. (This can be incomplete. There may be more than one state transforming as $|K,f)$. For example for SU(3), the product of two states of the adjoint representation, the octet, decomposes into a sum of states of different representations, including the octet. But the octet appears twice; in the decomposition of a product of octet states there are two distinct states — two different sums of products — that transform as the same state of the octet. Thus to define fully the Clebsch-Gordan coefficients we introduce multiplicity label μ to distinguish these states and we must resolve the ambiguity in proving and using the Wigner-Eckart theorem. We assume this has been done, although we do not consider how — and there are open questions about the proper way to do it — since it does not occur for the product of two states of the rotation group.)

A matrix element is the scalar product of a state with this reducible-representation state, thus the sum of scalar products of the state with the irreducible-representation states in the decomposition.

Since we wish to find the matrix elements in terms of some one (or a few) using the properties of the group, we write the states as the transforms of some reference states, labeled m_o and M_o, these giving matrix elements in terms of which all others are determined. Then (schematically)

$$T(j,m) = R^j_{mm_o}T(j,m_o), \qquad \text{(XII.3.b-2)}$$

and

$$|L,M) = R^L_{MM_o}|L,M_o), \qquad \text{(XII.3.b-3)}$$

where the R's are the representation matrices of the transformation for the two representations. From these we get

$$\begin{aligned}(L',M'|T(j,m)|L,M) &= \sum C(K,f;j,m;L,M)(L',M'|K,f)\\ &= \sum C(K,f;j,m;L,M)R^{L'}_{M'M'_o}R^K_{ff_o}(L',M'_0|K,f_0)\\ &= C(L',M';j,m;L,M)(L',M'_o|T(j,m_o)|L,M_0), \qquad \text{(XII.3.b-4)}\end{aligned}$$

using the orthogonality of the matrix elements and states; integration over angles is suppressed (sec. VII.5, p. 193). Clebsch-Gordan coefficients being themselves (products of) states, each given by one of the three pairs of labels, are also orthonormal. $(L',M'_o|T(j,m_o)|L,M_0)$ is the scalar product of two basis vectors, so is rotationally invariant, and therefore independent of the choice of the axes so of the reference state labels. Thus we have expressed an arbitrary matrix element of an arbitrary component of the tensor in terms of a single one, and the Clebsch-Gordan coefficients. This gives

$$(L',M'|T(j,m)|L,M) = C(L',M';j,m;L,M)(L'||T||L), \qquad \text{(XII.3.b-5)}$$

defining the reduced matrix element, $(L'||T||L)$. It is independent of the states, m, M and M', being a function only of the representations and the operator, so the reduced matrix element is a scalar. The Clebsch-Gordan coefficient contains the dependence on these labels and it transforms according to the basis states with these labels.

This transfer of the dependence on state labels from matrix elements to Clebsch-Gordan coefficients — which depend only on the group and representations — is the content of the Wigner-Eckart theorem. It gives the state dependence explicitly, in this coefficient; the only property of the operator that enters is how it transforms. The nature of the operator affects the matrix element (the effect of an electric field is different from that of a magnetic field, or a derivative); this is in the reduced matrix element $(L'||T||L)$, only. Given one matrix element, which of course depends on the operator, we know all those of all components of the operator between any two states (of the same two representations). Of course the reduced matrix elements for each pair of representations separately are needed. The effects of a magnetic field on scalars, on spin-$\frac{1}{2}$, on spin-1, objects are different. Likewise the probability of a transition between states of angular momentum representations 1 and 0 is different from that between those of 1 and 1, or between 2 and 0.

(We may need more than one reduced matrix element if there is multiplicity, although far fewer than the total number of matrix elements.)

Problem XII.3.b-1: Work out the algebra explicitly. Prove that the reduced matrix element is independent of M_o and m_o.

Problem XII.3.b-2: One important tensor operator has as its three components the generators of the rotation Lie algebra. For it, find the reduced matrix element, between any pair of representations (which pairs are possible?) [Jones (1990), p. 119, 271]. Does the answer appear correct?

Problem XII.3.b-3: Use the Wigner-Eckart theorem to show that the scalar product of two vectors is actually a scalar. Find it in terms of the reduced matrix elements.

Problem XII.3.b-4: The coordinates form a tensor, as do derivatives with respect to them. For both of these, using the Cartesian and also the spherical forms, find the reduced matrix elements, between arbitrary states. Which representations appear in the states given by their action on an arbitrary one? Does this seem reasonable? Answer these questions for their action on a product of states (including ones from different representations). Repeat this for powers of these operators. Does the Wigner-Eckart theorem hold? Find a formula for the reduced matrix elements for an arbitrary power.

XII.3.c Why does the Wigner-Eckart theorem hold?

The theorem says that the transformation properties of a tensor operator are given by the Clebsch-Gordan coefficients. Why? The answer is in the word tensor. By assuming the operator is a tensor (and the operator need not be one simply because it looks like one) we specify its transformation properties, and these are given by the Clebsch-Gordan coefficients, by their definition. The operator is simply a state of an irreducible representation.

There is however one aspect of a state that is not determined by its transformation properties, its normalization. The tensor-operator matrix elements are undefined up to normalization. Different ones of the same representation are related by Clebsch-Gordan coefficients. All their norms are the same, but unspecified. And the reduced matrix element depends on this (it can also depend on other variables, like position in space, that are unaffected by the group).

XII.4 THE GROUP O(3) AND PARITY

Requiring that the product of vectors be invariant forces the value of the determinant of the transformations to be ± 1. This gives a group, the orthogonal group O(3), with two invariant subgroups, SO(3), for which the determinant is 1, and a two-element group, consisting of the identity and the value of the determinant [Tinkham (1964), p. 139; Wigner (1959), p. 181].

Problem XII.4.-1: Check that this two-element subgroup, as well as SO(3), are invariant. Does this seem physically reasonable? O(3) is thus not simple, being semisimple. However its algebra is simple, since O(3) is a direct product of a Lie and a finite group.

XII.4.a The meaning of parity

What is the physical significance of this extra transformation [Joshi (1982), p. 190, 193; Tung (1985), p. 212]? And how do O(3) representations differ from those of SO(3)? The product of two vectors is invariant if their signs are simultaneously reversed — the signs of all (six) components are changed. This (inversion) is called a parity transformation (the two signs form a pair). The two-element finite group has two representations with transformations (1,1) and (1,-1). Basis functions of the first are unaltered by inversion, for the second they are sent to their negatives; they have even and odd parities.

Representations form pairs mutually inequivalent under O(3), but equivalent for SO(3). Under SO(3), the electric and magnetic fields, E and B, transform the same; under O(3) they differ — these two inequivalent O(3) representations are equivalent under the SO(3) subgroup. From Coulomb's Law $E \sim \hat{r}$, the unit vector (the proportionality constant is a scalar); coordinates change sign under inversion, as does $cos\theta$ ($\theta \rightarrow \theta + \pi$), thus so does the electric field. These quantities are vectors. $B \sim v \times \hat{r}$; the field seen by an observer moving in an electric field is $B \sim v \times E$. Since both v and E, and v and r, change sign under inversion, B does not. It is a pseudovector. B is transformed by the $L = 1$ term in the reduction of the product of two $l = 1$ representations of SO(3).

The parity (the representation of the two-element subgroup) of the spherical harmonics Y_l is $(-1)^l$. It is negative (odd parity) for $l = 1$ (as seen for $cos\theta$) and all odd l, and positive (even parity) for even l (these given by even powers of $cos\theta$, including the scalar). Spherical harmonics are polynomials, and the terms in the sum differ in powers by 2, necessarily as they must all have the same parity.

Besides the parity due to these functions (giving the behavior of the object in space) particles have internal parity. Thus an object in an $l = 1$ state would have odd parity if its internal parity were even — the parity of the total statefunction is even for those objects with odd internal parity in this l state.

Problem XII.4.a-1: Spherical harmonics are functions of both sines and cosines. Check that for all the parity is $(-1)^l$. Why must all terms in the sum have the same parity?

XII.4.b The physical distinction between O(n) and SO(n)

Besides the mathematical definitions of the groups, they can be given physically as sets of transformations leaving certain objects invariant. While it is clear what objects are invariant under O(2) and O(3), circles and spheres, which are unchanged by SO(2) and SO(3), that is by rotations but not by inversions? A current moving in a circle has a direction, clockwise or counterclockwise, and this is changed by inversion through the center of the circle. Thus a current is a physical example of an object invariant under SO(2), but not O(2). A circle with arrows at each point is a geometrical example.

What about three dimensions? A loop of current produces a magnetic field perpendicular to itself; as this is given by a product of two vectors, each sent into its negative by an inversion, it is invariant. Thus a sphere with a current loop at each (geometrical) point, or a spinning particle with all spins the same (clearly quite difficult to draw or produce) with an associated pseudovector, is a physical object invariant under all rotations, SO(3), but not under inversions, so not under O(3). What is a geometrical example? The arrow we use to represent a vector provides a picture of it, though the arrow is not the vector (something we might forget as this picture is so useful). There are in O(3) two objects transforming the same way under SO(3), a vector and a pseudovector. Thus if to each point on a sphere we attach (not an arrow but) a pseudovector, the resultant geometrical object is invariant only under the SO(3) subgroup of O(3). A pseudovector can be represented by a helix with an arrow at the end (or a hand with fingers extended); the length gives the magnitude, the arrow the direction and the sense of the curl labels the objects transforming into each other. Thus to each point on a sphere we attach (but only in thought) an infinitesimally small right or left hand(ed spiral), all with the same handedness. Under a rotation a right hand goes into another, but under an inversion it goes into a left hand. So this gives a geometrical object invariant under SO(3), but not O(3).

Problem XII.4.b-1: This picture of a sphere with a loop attached to each point suggests other models and interpretations. We can put an arrow on each loop and regard its position as a phase. The transformations of this object (which are not symmetries) are given by the rotations of the sphere, and the rotations of each loop. These form (presumably) a group which is infinite dimensional, there being a different transformation at each point (reminiscent of gauge groups, for which the transformations are also space-dependent [Mirman (1995b)]). A sphere can be generated from a point (or a vector to that point) by the application of all rotations. Suppose we did that, with the initial point having a loop. Then the phases at each point would be determined by that at one, so no longer space-dependent. What is the group of transformations now? Is it a symmetry group? Suppose that we took two points, with one loop red, the other blue, with independent phases, and applied all rotations. Would the set of transformations of the resultant object be SU(2)? Why? What would it be?

Problem XII.4.b-2: Can this be generalized to n dimensions? How about spaces of nondefinite signature, like $3 + 1$, say?

XII.4.c The physical effect of parity

The importance of parity appears in selection rules for interactions that conserve parity, as does the electromagnetic, but not the weak, interaction. The (interaction part of the) Hamiltonian acting on a state, $H|l,m\rangle$, belongs to a reducible SO(3) representation. Reducing it gives a sum of terms transforming under different representations each multiplied by a reduced matrix element. SO(3) allows all these terms. However space is invariant under O(3), and (so) the parity of H is even. Thus the parities of initial and final states (including the internal parity) must be the same, else the reduced matrix element is 0. The parity is determined (in part) by the orbital angular momentum, so this gives a selection rule relating the possible final orbital angular momentum to the initial one; thus the allowed final states differ for the electric and magnetic fields, these having opposite parity. This we illustrate below with examples.

XII.4.d The group surfaces of O(3) and SO(3)

As the parameters of a Lie group vary over their range the transformations trace out a (hyper)surface. For SO(n) this surface is connected — it consists of one piece. Each point on it can be reached from any other without leaving the surface (by continuous variation of the transformation parameters). For O(n) it is not connected, it consists of two pieces. From the identity all points on one piece, which is the group surface of SO(n), can be reached by continuous variations of the angles. The other surface, geometrically identical to the first, cannot; it requires a jump, that of the parity transformation.

Problem XII.4.d-1: Show experimentally that any position of the right hand can be reached from any other by a continuous motion, but there are no angles giving a transformation that takes a right to a left hand. And for SO(2), any matrix of the form $\begin{pmatrix} cos\theta & sin\theta \\ -sin\theta & cos\theta \end{pmatrix}$ can be obtained from the identity by continuous variation of θ, but not $\begin{pmatrix} 1 & 0 \\ 0 & -1 \end{pmatrix}$ or $\begin{pmatrix} cos\theta & sin\theta \\ sin\theta & -cos\theta \end{pmatrix}$. Note the effects on the axes of these matrices and the values of their determinants. Diagonalize them and note the difference. Here we have two (matrix) functions, in diagonalized form two complex functions, which give continuous curves over the (part of the) real line labeled by θ, but these curves are distinct. We do not consider that these functions (and curves) can be connected (why?) by letting θ be complex (does it give a group?), but this does not mean it is unimportant [Streater and Wightman (1964), p. 114, 143; Mirman, (1988a,b, 1995a)]. Repeat this for O(3) and SO(3), using their parameterizations. Try it for O(n).

Problem XII.4.d-2: SO(3) is connected, its group surface is a single piece (this use of connected is different from that in "SO(n) is doubly-connected"); O(n) is not connected, its group surface is two — separate, disconnected — pieces. Prove that for any group, the piece containing the identity — for O(3) it is SO(3) — is a subgroup, and invariant [Cornwell (1984), p. 52]. The others are not groups. Why? How are they related?

XII.5 SYSTEMS WITH BROKEN SPHERICAL SYMMETRY

While symmetry is of fundamental importance in physics, broken symmetry is essential. The properties of an atom are quite limited by space being spherically symmetric. But an atom always in such an environment would be forever in its ground state — quite a bore. To lift it to an excited state requires a field, and this has a direction, thus breaks the symmetry. To predict, as we must to understand, the behavior of a system we need to know how the field affects it, so how its behavior is determined by its symmetry and by the (particular) breaking of the symmetry. Consider an object in an electric or magnetic field; if the field is removed the system is (say, spherically) symmetric, which restricts it. Would we not expect that its properties in the field are affected by the symmetry obtained by removing the field? And while properties of a system are strongly limited by its symmetries, many systems do not have complete, but only broken, ones. Still these differ from ones with no symmetry, with no remnant of symmetry at all. What can the fact that a system has a remnant of symmetry tell us about it? How does that, and the remaining symmetry, limit, and determine, and to what extent, its properties?

Here then we consider such breaking of spherical symmetry, though only by electric and magnetic fields, and see what it says about physics, why group theory is relevant and what insight it can provide. What do we wish to know? For a spherically-symmetric system the states are rotation-group basis states. Why; what does "the states" mean? To describe a system we need the eigenstates of the Hamiltonian, the operator giving the time evolution. With symmetry it is rotationally invariant; thus the stationary states, those left invariant by it, are rotation basis states, each of a single representation, and conversely — so we have the states, and their Hamiltonian eigenvalues, their energies. With symmetry breaking we want, again, the eigenstates and eigenvalues of the (total) Hamiltonian. How are these related to the states for the case with symmetry? And what happens to these rotation basis states, are they, or some, no longer Hamiltonian eigenstates, so no longer stationary states? What are they? For those that are, how do their energies now depend on their representation and state labels, j and m?

For spherical symmetry, if the field (the z axis is taken along it) is small, the differences in energy between states of different m are less than the differences in energy between ones of different j; m and j are (reasonably) good quantum numbers — which we take to be the case here. The states of the system are still rotation-group basis states. Now however the effect of rotations on them is more complicated.

Problem XII.5-1: Why with symmetry, are the stationary states rotation basis states, each of a single representation, and conversely?

Problem XII.5-2: In reading this discussion it is useful to consider how it changes, indeed whether any remnant still holds, if the field is strong (which means what?). Are the states of the system (which means what?) still such rotation-group basis states? Why?

XII.5.a The Stark and Zeeman effects

The breaking of spherical symmetry in atomic physics, by electric and magnetic fields, gives the Stark and Zeeman effects, respectively [Cornwell (1984), p. 464; Joshi (1982), p. 202; Tinkham (1964), p. 188; Wherrett (1986), p. 147; Wigner (1959), p. 198]. The interaction Hamiltonians (suppressing constants) are [Schiff (1955), p. 158, 292],

$$H_s = -Ez = -Ercos\theta, \quad \text{(XII.5.a-1)}$$

$$H_z = -BLcos\phi + \zeta B^2r^2sin^2\theta; \quad \text{(XII.5.a-2)}$$

E and B are the electric and magnetic fields (both along z), r, θ and z the particle's coordinates, ζ a numerical factor, L the orbital angular momentum which is at angle ϕ to the z axis (so to B). We take the spin to be 0. The free (noninteraction) part of the Hamiltonian, H_o, is taken spherically symmetric; thus all states with the same m have the same energy if there are no fields.

The problem is to find the Hamiltonian eigenstates and eigenvalues. For spherical symmetry the states are spherical harmonics. That is no longer true with fields.

Problem XII.5.a-1: The last sentence assumes "the states" is defined. Is it?

XII.5.b The Stark effect

To find the Stark Hamiltonian eigenfunctions we write

$$z = rY_{10}(\theta), \qquad \text{(XII.5.b-1)}$$

giving

$$H_s|lm) = -Ez|lm) = -ErY_{l0}|lm) = -Er\sum|l'm)C(l',m;l,m;1,0), \qquad \text{(XII.5.b-2)}$$

where $|lm)$ is state m of representation l, C is the Clebsch-Gordan coefficient, and the sum is over

$$l-1 \leq l' \leq l+1. \qquad \text{(XII.5.b-3)}$$

The m is the same for all states since $m = 0$ for z.

Problem XII.5.b-1: It should be clear that J_z and H_s commute. Show from this that the states have the same m. Justify the range of l'.

Problem XII.5.b-2: The matrix element is $(l'm|H_s|lm)$, but this deceptively simple notation hides an important fact. The matrix element is an integral over angles (it is independent of angle). If $l' = l$ the integral vanishes. Calculate (with what $d\Omega$; why?)

$$I_{cc} = \int d\Omega cos\theta cos^{2l}\theta, \quad I_{sc} = \int d\Omega cos\theta sin^{2l}\theta. \qquad \text{(XII.5.b-4)}$$

A simpler way of doing this is to note that since the integral for the matrix element is over all angles it is unchanged by reflection of the coordinates (or $\theta \to \theta + \pi$). If $l' = l$ the integral changes sign so is 0. Explain what this has to do with conservation of parity. (Might we have guessed that the integral would be 0, and for this reason, if we were unaware of this transformation group?) The symmetry group is O(3), not merely SO(3), because of the realization of the states we (must) use; thus the decomposition of the product of SO(3) states is affected by the O(3) symmetry — by the realization. But the decomposition is purely a property of SO(3), is it not? Would this result obtain for a magnetic field? Why? Could the Hamiltonian be the same? Verify that the only allowed values are

$$l' = l \pm 1. \qquad \text{(XII.5.b-5)}$$

Problem XII.5.b-3: The spherical harmonics are orthonormal. Why is the product of three $(cos\theta, Y_{lm}, Y_{l'm})$ not 0? What information can we get from this product?

Problem XII.5.b-4: Explain why the Stark effect can be regarded as an interaction between the electric field and a permanent electric dipole moment of the atom [Schiff (1955), p. 160]. That moment is 0 for the ground state. If the Hamiltonian conserves parity, show that a nondegenerate eigenstate of it has definite parity — it is even or odd. Why does this fail if there is degeneracy? An electric dipole moment is odd under parity (why?), so its expectation value between two eigenstates is 0. Thus a system described by a nondegenerate eigenstate cannot have an electric dipole moment, if parity is conserved (the converse need not hold; there may be other reasons why it must be 0). Does this mean that all elementary particles have zero electric dipole moments?

XII.5.b.i *The eigenstates*

What is the eigenstate

$$|s) = \sum G_{lm}(E)|lm), \qquad \text{(XII.5.b.i-1)}$$

of the Hamiltonian; $|lm)$ are basis states and G, functions of the field E, coefficients to be determined? With ε the total energy,

$$H|s) = \varepsilon|s) = \varepsilon \sum G_{lm}|l,m) = (H_o + H_s)|s)$$

$$= (H_o + H_s)\sum G_{lm}|l,m) = \sum G_{lm}(\varepsilon_{ol} + H_s)|l,m)$$

$$= \sum G_{lm}\{\varepsilon_{ol}|l,m) - Er\sum |l'm)C(l',m;l,m;1,0)\}. \qquad \text{(XII.5.b.i-2)}$$

States are orthonormal so (schematically, suppressing any integration),

$$\varepsilon G_{lm} = \varepsilon_{ol}G_{lm} - (l-1|Er|l)G_{l-1,m}C(l+1,m;l,m;1,0)$$

$$- (l+1|Er|l)G_{l+1,m}C(l-1,m;l,m;1,0), \qquad \text{(XII.5.b.i-3)}$$

which relates the coefficients of the different states. All this is exact. The problem is finding the coefficients G. These depend on the free

Hamiltonian. This infinite set of equations cannot be solved exactly and perturbation theory is usually used. But this is as far as we go using the rotation group.

XII.5.b.ii *The hydrogen atom*

Consider the hydrogen atom. If the electric field is small, matrix elements between states of different principal quantum number, n, are small and can be disregarded (the field will not — usually — cause transitions between different shells). The lowest shell has only one orbital angular momentum value, so the matrix element of H_s is 0; the field has no effect on an atom in this shell. For $n = 2$, $l = 0, 1$ so only the $m = 0$ element, $(1,0|H_s|0,0)$, and its conjugate are nonzero (as H_s is hermitian the two are equal). So the energy eigenstates are $f_1(r)Y_{1\pm1}$, and with f_0, f_1 functions of r determined by the free Hamiltonian and are unaltered by the field (as is the eigenvalue of H_o), this gives the states,

$$|s_0) = f_0(r)\{|0,0) + G_{10}|1,0)\}, \quad |s_1) = f_1(r)\{|1,0) + G_{00}|0,0)\}. \tag{XII.5.b.ii-1}$$

The coefficients are thus related by

$$\varepsilon G_{10} = \varepsilon_{o1}G_{10} - EG_{00}(0,0|1,0;1,0), \tag{XII.5.b.ii-2}$$

and

$$\varepsilon G_{00} = \varepsilon_{o0}G_{00} - EG_{10}(1,0|0,0;1,0); \tag{XII.5.b.ii-3}$$

for the nonrelativistic hydrogen atom the two energies are the same, but for generality we label them as distinct.

Clebsch-Gordan coefficients are known as are the energies of the free atom. Thus we can now solve these equations for the coefficients and energy ε, given either of the coefficients. Of course it is this point that is the problem. They might be calculated from perturbation theory. But the method of calculation is not relevant here. Once we have the coefficients we know the eigenstates and the energy, solving the problem.

XII.5.b.iii *The role of group theory in studying the Stark effect*

What role did group theory play in this? It provided the eigenstates of the free Hamiltonian, which are essential in finding the effect of the perturbation, though this can be, and often is, disguised. Also E is not arbitrary, but breaks the symmetry in a definite way (along z, but an arbitrary field can be decomposed into a sum of fields along the three axes), which would be a meaningless statement unless there is symmetry to be broken. Then the symmetry-breaking term can be written as a basis state of the invariance group, though invariance no longer holds. The effect of the breaking is then (largely) reduced to a study of the decomposition of products of representations. This done once, using the machinery of group theory, gives much of the result; it is unnecessary to repeat it for each individual problem.

And group theory gives insight, explaining the occurrence of only some splittings and transitions (selection rules), why certain states are unaltered, and why splittings are related, and how.

XII.5.c The Zeeman effect

The Zeeman Hamiltonian (for magnetic field B along z) is

$$H_z = -BJcos\phi + \zeta B^2r^2sin^2\theta, \qquad \text{(XII.5.c-1)}$$

where J is the total angular-momentum quantum number (spin is important), and we start by ignoring the quadratic term.

Problem XII.5.c-1: Check that by the same argument as above $H_z \sim J_z$ does not mix terms with different m. Now J transforms under the $j = 1$ representation of SO(3). How does it transform under O(3)? Under SO(3), J_z and z transform as the same basis state, so this case is somewhat similar to the Stark effect. How do they differ?

XII.5.c.i *Spin-orbit coupling dominant*

Here there are two types of angular momenta, orbital and spin. What is the effect of these? These two interact through spin-orbit coupling and each interacts with the magnetic field. The problem involves the combination of three rotation-group basis states, the orbital and the spin angular momenta, and the magnetic field. Rather than considering this exactly we take first the spin-orbit coupling to be greater than the coupling of either to the magnetic field. Then, ignoring the field, the proper basis states are those of total angular momentum; these are the eigenstates of the Hamiltonian. A measurement of a time-independent state gives its total z-component angular momentum, j_z; the system is not in a state of fixed l_z or s_z, but linear combinations of these.

Problem XII.5.c.i-1: Why are representation labels l and s fixed; the spin-orbit coupling does not mix states of different l or s, but only different z components?

XII.5.c.ii *Calculation of the splitting*

Thus the interaction part of the Hamiltonian is

$$H_z = \zeta(L_z + 2S_z), \quad \zeta = -eB/2mc; \qquad \text{(XII.5.c.ii-1)}$$

e and m are the charge and mass of the electron (ζ has meaning, perhaps partly different, for other systems; in terms of the rotation group it is just a parameter). Ignoring again matrix elements between states of different j, which we (hopefully) expect to be small, we get, using the Wigner-Eckart theorem (sec. XII.3.b, p. 350), since the only nonzero matrix elements are between states of the same j_z,

$$(S,L,J,J_z|H_z|S,L,J,J_z)$$

$$= \zeta(C(J,J_z;1,0;J,J_z)\{(SLJ||L||SLJ) + 2(SLJ||S||SLJ)\}); \tag{XII.5.c.ii-2}$$

C is the Clebsch-Gordan coefficient, $(SLJ||L||SLJ)$ and $(SLJ||S||SLJ)$ the reduced matrix elements. Operator J is defined by

$$J = L + S, \tag{XII.5.c.ii-3}$$

so

$$2J \cdot L = J^2 + L^2 - S^2. \tag{XII.5.c.ii-4}$$

Now

$$2(SLJJ_z|J \cdot L|SLJJ_z) = 2J_z(SLJJ_z|L_z|SLJJ_z)$$
$$= 2J_zC(l,J_z;1,0;lJ_z)(SLJ||L||SLJ), \tag{XII.5.c.ii-5}$$

and similarly for $2(SLJ||S||SLJ)$. Since $J^2 \rightarrow j(j+1)$, and so on, we get, using the value of the Clebsch-Gordan coefficient,

$$H_z = \zeta j_z[\frac{1}{2j(j+1)}\{3j(j+1) - l(l+1) + s(s+1)\}], \tag{XII.5.c.ii-6}$$

for the shift in energy due to the magnetic field. The term in brackets is the Landé g factor [Inui, Tanabe, and Onodera (1990), p. 157; Jones (1990), p. 117; Tinkham (1964), p. 189; Wigner (1959), p. 278].

The Zeeman effect causes the energy of a state to depend on its $m(= j_z)$ value, not surprisingly since the spherical symmetry is broken by the (physical) definition of a z axis, so states degenerate for spherical symmetry become separated. Also the magnetic field causes transitions between states, but to first order in the field, if the matrix elements off-diagonal in j are small, the only effect is splitting of states.

Problem XII.5.c.ii-1: Explain the choice of quantum numbers labeling the states, and in H_z.

Problem XII.5.c.ii-2: Find the selection rules governing the eigenvalues labeling the nonzero matrix elements.

Problem XII.5.c.ii-3: For the case in which the energy due to the field is greater than that due to spin-orbit coupling (the Paschen-Back effect)

[Schiff (1955), p. 294, Tinkham (1964), p. 190], we can consider the shift in energy as due to a shift resulting from L plus a shift resulting from S. Since it is just the sum of these two, it can be found directly. Work out the energy shift in this case. Find the selection rules governing the eigenvalues labeling the nonzero matrix elements. Is group theory useful here?

Problem XII.5.c-4: These give the Zeeman effect in the two limiting cases, weak and strong fields. Discuss it for the intermediate case [Tinkham (1964), p. 191]. Is group theory also relevant here?

XII.5.c.iii *The quadratic Zeeman effect*

The quadratic part of the Zeeman interaction Hamiltonian is

$$H_z^q = \zeta B^2 r^2 sin^2 \theta; \qquad \text{(XII.5.c.iii-1)}$$

it transforms as a basis state of the $l = 2$ representation (the linear part belongs to $l = 1$). Specifically $H_z^q \sim Y_2^{\pm 2}$, so that it changes the value of m (by ± 2), unlike the linear part. This part of the Hamiltonian has nonzero matrix elements for

$$\Delta l = 0, \pm 2, \qquad \text{(XII.5.c.iii-2)}$$

and only the $\Delta l = 0$ gives a diagonal contribution to the energy.

Problem XII.5.c.iii-1: Analyze this similarly to the linear Zeeman effect. How relevant is group theory?

Problem XII.5.c.iii-2: Does the Wigner-Eckart theorem hold for the quadratic Zeeman effect? Does it have the same form as before? To what extent does this theorem depend on the linear nature of quantum mechanics? Does this break down here?

XII.5.d Perturbation theory of the Zeeman effect

It is useful to consider the perturbation expansion of the Zeeman effect, because it shows how group theory is used in perturbation calculations; also it helps clarify the meaning of symmetry breaking. Thus for isotopic spin and unitary symmetry the masses of different states are not quite the same. How can group theory analyze this symmetry breaking? The Zeeman effect introduces some of the concepts. Here we summarize the discussion emphasizing the main points.

A system has a set of states labeled with index k and corresponding energies E_k^o. A (small) perturbation λV is applied and we wish to find the resultant energies E_k. These are given by

$$E_k = E_k^o + \lambda V_{kk} + \lambda^2 \sum \frac{|V_{ik}|^2}{(E_k - E_i)} + \ldots . \qquad \text{(XII.5.d-1)}$$

where V_{ik} is the matrix element of V between states i and k; the sum is taken over all eigenstates of the Hamiltonian, except state k [Wigner (1959), p. 41]. It would be unfortunate if some state i has the same energy as state k.

Now states with different m_z values do have the same energy and if $V \sim H_w$ then the expansion breaks down unless $V_{mm'} = 0$, for different m_w values. This is handled by quantizing along the w axis, the axis along which H lies, which is usually called the z axis and then $H|m_z) \sim |m_z)$. That is, if we write H as an angular momentum eigenstate, its m_z value is 0 (it is along z) so acting on a state it does not change its m value — it only has nonzero eigenvalues between states with the same value of m, thus when the denominator is 0 the numerator is also.

An H transforming as an $l = 1, m_l = 0$ state has matrix elements only between $|L, M)$ and $|L', M')$, where $M' = M$, and $L' = L$, or $L \pm 1$ (in agreement with the vector-addition model). The quadratic term transforms like an $L = 2, m_l = 0$ state, plus a state which is a scalar ($sin^2\theta = 1 - cos^2\theta$). Thus it has matrix elements between states whose L values differ by $\pm 2, \pm 1, 0$, and whose m values are the same.

How is spherical symmetry broken? The breaking term (if only one) must be along z — it defines the z axis. But this is incomplete; we must also tell the representation of the rotation group it transforms under. For the weak-field Zeeman effect, the breaking term transforms as the adjoint representation (which for this group is the defining representation), $L = 1$. For the quadratic case, it transforms as the $L = 2$ representation. Thus the m value of the interaction Hamiltonian is a matter of definition. But the choice of the symmetry-breaking term involves physics. In general it is a sum of terms from different representations (they could be along different axes — have different m values). Which appear, and with what coefficients is a physical assumption. Here we know the answer from electromagnetic theory. But if we were using group theory to help pick out the terms, as we do in other cases, we must determine, in some way, which representations appear, and the coefficients — and even their (set of) state-labeling values if there is more than one term.

Problem XII.5.d-1: For the hydrogen atom states with different l but the same n also have the same energy. Discuss how to handle this.

Problem XII.5.d-2: Take a hydrogen atom in a magnetic field so strong that the energies of a state with principal quantum number n, for some values of l and m, and that of a state with value n', for some l' and m', are equal. Find some examples. This is a case of "accidental degeneracy ". Now with the same magnetic field, see if there are other states whose energies are also "accidentally" equal. If there is a larger symmetry then there is a general rule giving the degeneracy, as with the nonrelativistic hydrogen atom for which the states of all l, for each n, are equal, but in this accidental case, which just happens for a particular value of the magnetic field, this is not true. One, or a few, pairs of states have the same energy but this does not hold generally. Moreover this degeneracy occurs for one exact value of the field. Increasing the experimental precision always shows that it does not really hold. However for rotational invariance of space, giving that all states with the same m values are degenerate, the result is not only general, but there is nothing we can do to get around it (that is there is nothing that theorists can do, but experimenters have to work to achieve it since the environment of an atom is never precisely symmetric, and in reality an increase in experimental precision shows that the symmetry is — apparently — broken).

Problem XII.5.d–3: How was group theory used in these considerations? Is it helpful, essential?

XII.5.e For how long can we use the rotation group?

This discussion is based on the assumption that the system is almost spherically symmetric and the effects of fields are small. But everything we did seemed exact. So why do we need this assumption? Can we not always start with rotation-group basis states?

If the field is very strong then the eigenstates of the Hamiltonian are sums over all states (with the same m; there is cylindrical symmetry) of all representations — over all j (or l). It is in the assumption that we can consider only a single state, and those mixed with it by the field, that rotational invariance enters. But if the fields are too strong, they dominate. Then rotation-group basis states are no longer useful. We can still use them; they form a complete set. But they no longer have anything to do with the system. Any other arbitrary complete set (with the same domain of definition) can also be used; it has the same meaning (or meaninglessness) and the same use (or uselessness).

Chapter XIII

Lie Algebras

XIII.1 ALGEBRAS ON THEIR OWN

A Lie group has a Lie algebra given by a set of operators, realized as differential operators or matrices, say, and their product, commutation. But we can also write a list of symbols and a table of commutators (provided we do so consistently — which is the problem), so defining a Lie algebra independent of its source (if it has one), a Lie group. This can provide insight, freeing us to look at algebras more broadly, and might give more algebras than by starting with groups; we do not (now) know whether every Lie algebra has a Lie group, nor that all their realizations exponentiate to finite functions (of interest). Also it might be easier to study algebras without being concerned (at first) if they came from groups, and what their properties are because of such origin. Here we study Lie algebras without asking why they arise; it is considered a subject whose consistency and value stand on their own. Since we know that Lie groups give Lie algebras we will be aware that they have value in group theory and that this is one reason for considering them; they serve, as we see, as a tool for the study of the groups, one which allows an analysis in many ways easier and more heuristic than attacking the groups directly. We leave open the possibility that they have value beyond group theory [Barut and Raczka (1986); Belinfante and Kolman (1972); Cahn (1984); Chen (1989), chap. 5; Cohn (1965), p. 59; Cornwell (1984), vol. 2; Gilmore (1974); Hamermesh (1962), p. 301; Hausner and Schwartz (1968); Helgason (1964); Hermann (1966); Humphries (1987); Jacobson (1979); Kaplansky (1974), chap. 1; Ludwig and Falter (1988), p. 266; Miller (1968), p. 25; Miller (1972), p. 166, 321; Price (1977), p. 24; Racah (1951, 1964); Samelson (1990); Sattinger and Weaver (1986); Schensted (1976), p. 233; Serre (1987); Varadarajan (1984); Wan (1975);

Wybourne (1974); Zelobenko (1973), p. 238].

Thus we start with definitions and see where they lead.

XIII.2 THE DEFINITION OF A LIE ALGEBRA

A Lie algebra is a set of symbols (operators, generators, elements) X_i, and a rule for their combination

$$[X_i, X_j] = \sum c_{ij}^k X_k, \tag{XIII.2-1}$$

where all X's are members of the set, and we call their product [,] a commutator. This product can be considered abstractly, and is only required to obey the rules for a commutator. The c's are the structure constants of the algebra; they are the constants (independent of the X's) which determine the (structure of the) algebra. An algebra must be closed under commutation — the commutator must equal a sum of the algebra symbols (only).

Different algebras are given by different numbers of X's and different sets of c's. They need not be very different — they could be isomorphic. In distinguishing Lie algebras we ignore unimportant (?) differences like isomorphism. An algebra is given by a set of operators, and the list of their commutators — by the structure constants.

The number of (linearly-independent) operators of an algebra is its order. Here linear independence for the X's is postulated, but in a realization, or a derivation of an algebra from other requirements, the X's are the maximum set of linearly-independent operators (there is no other set with a larger number of such operators); any such set is a basis of the Lie algebra. This discussion is limited to finite Lie algebras — there are only a finite number of (linearly-independent) symbols (a finite number of X's). The maximum number of commuting generators is called the rank of the algebra.

Problem XIII.2-1: Clearly a commutator

$$[a, b] = ab - ba, \tag{XIII.2-2}$$

is antisymmetric,

$$[a, b] = -[b, a]. \tag{XIII.2-3}$$

Show that it satisfies the Jacobi identity

$$[a, [b, c]] + [b, [c, a]] + [c, [a, b]] = 0. \tag{XIII.2-4}$$

Any product with these two properties is an acceptable product for a Lie algebra. We might think that taking different products — with these two properties — would give different algebras. But, abstractly, it does not matter how we interpret the product — different ways give different realizations of the same abstract algebra.

Problem XIII.2.-2: These two conditions on the product give two corresponding conditions on the structure constants [Hamermesh (1962), p. 305]. State them.

Problem XIII.2-3: There are other products, besides commutators, that satisfy. Check that the Poisson bracket of classical mechanics also satisfies [Goldstein (1953), p. 250; Sattinger and Weaver (1986), p. 41; Sudarshan and Mukunda (1983), p. 39]. Does the Lagrange bracket?

Problem XIII.2-4: Is the vector product (cross product) of two vectors, in three-space, a commutator? Does the cross product of unit vectors form a Lie algebra? What are the structure constants? What algebra is it; is this surprising? Can this be generalized to spaces of arbitrary dimension?

Problem XIII.2-5: Show that

$$[M, N_{\pm}] = \pm i N_{\mp}, \quad [N_+, N_-] = 0, \qquad \text{(XIII.2-5)}$$

is a Lie algebra; all Jacobi identities are satisfied [Mirman (1993, 1994, 1995b)].

Problem XIII.2-6: The Heisenberg algebra (also known as the simple harmonic oscillator, SHO, algebra) has commutation relations

$$[p, q] = i, \quad [p, i] = 0, \quad [q, i] = 0. \qquad \text{(XIII.2-6)}$$

Check that it is a Lie algebra [Sattinger and Weaver (1986), p. 25, 48]. What are interpretations of it? Find linear combinations of the generators with useful physical interpretations.

Problem XIII.2-7: Let an "algebra" be given by starting with the symbols

$$X_1 = \frac{d}{d\theta}, \quad X_2 = tan\theta, \qquad \text{(XIII.2-7)}$$

and a commutator defined by their action on arbitrary functions,

$$[X_1, X_2]f(\theta) = X_3 f(\theta) = \frac{d}{d\theta}(f(\theta)tan\theta) - tan\theta\frac{df(\theta)}{d\theta}. \qquad \text{(XIII.2-8)}$$

Does this form a (finite) Lie algebra? Is it closed under commutation? Try this with other trigonometric functions, and also such functions as hyperbolic, and elliptical. Try it also with

$$X_2 = \theta^n, \qquad \text{(XIII.2-9)}$$

for n both positive and negative.

Problem XIII.2-8: An Abelian Lie algebra has all commutators (all c's) zero; else it is non-Abelian. Give examples of Abelian algebras. A group, like an algebra, is Abelian if all its elements commute. Prove that if a Lie group is Abelian so is its algebra. Need the converse hold?

Problem XIII.2-9: Lie algebras form vector spaces; the generators are the coordinate vectors. Check that they obey the conditions on vector spaces (sec. II.4.j, p. 59).

XIII.2.a The nonassociativity of Lie algebras

One of the axioms of group theory is associativity. Lie algebras however are not associative.

Problem XIII.2.a-1: For a Lie algebra

$$a(bc) = (ab)c \text{ means } [a,[b,c]] = [[a,b],c]. \qquad \text{(XIII.2.a-1)}$$

Is this true? How is this related to the Jacobi identity?

XIII.2.b The associative algebra of a Lie algebra

While Lie algebras are not associative, there is an associative algebra associated with them. The algebra operators can be realized using matrices. For matrices there is a product — commutation — which is nonassociative and gives a Lie algebra, and another product — matrix multiplication — which is associative; matrix algebra using this product is associative. For a set of objects, here matrices, there can be more than one algebra. These are related (the definition of the commutator of matrices requires first a definition of matrix multiplication), and we gain information from one using the other. In particular the adjoint (regular) representation matrices of a Lie algebra forms an associative algebra, and this can be used to obtain information about the nonassociative, Lie, algebra.

Problem XIII.2.b-1: Why is matrix multiplication associative?

XIII.2.c Combining the algebra symbols

For a Lie algebra there must be an additional way of combining generators besides the product (the commutator); though not usually explicitly stated, it is essential to all future work (and is needed to be able to interpret [,] as a commutator). This is addition (that means it is commutative). And the generators are required to be linearly independent, but for this to have meaning their sums must be defined. Also we need sums, not only of generators, but of their products. (We should check that in realizations we consider, all these are defined.) Thus given generators T and V, we define $T + V$. Just saying this is trivial. But there is one aspect that is not.

Problem XIII.2.c-1: Given a set of generators T_i forming a Lie algebra, we require

$$[aT + bT', T''] = a[T, T''] + b[T', T''], \qquad \text{(XIII.2.c-1)}$$

the distributive law for commutators; why? Then show

$$[\sum a_i T_i, \sum b_j T_j] = \sum a_i b_j [T_i, T_j]; \qquad \text{(XIII.2.c-2)}$$

$[\sum a_i T_i, \sum b_j T_j]$ acts like a commutator, and arbitrary sums of generators do form a Lie algebra (in particular the set is closed under commutation). There are two rules for combining generators; they must be consistent (so their properties are not arbitrary).

Problem XIII.2.c-2: Do Lie algebras of Lie groups obey the distributive law? Why?

Problem XIII.2.c-3: Simplify $[T_1 T_2 \ldots T_k, T_1 T_2 \ldots T_l]$. To do this, is it necessary to interpret [,] as a commutator? This implies that there is another way of combining generators, that of multiplication. We can regard TT' as just another symbol. But this simplification shows there is more to it than that. What? Are there properties that we must assign to this multiplication for such processes to have meaning and to have the properties that we wish? (If it is not clear that we require both multiplication and addition to have properties, though we may not be explicitly aware of it, this simplification should emphasize that.) For representations of a Lie algebra, the symbols are taken as matrices. Multiplication is just matrix multiplication. Does it have the required properties? The symbols can also be realized as derivatives, as with the angular-momentum algebra. Can multiplication be defined for such cases? How? Does it satisfy the requirements? Thus we have three ways of combining algebra symbols, multiplication, addition, and the algebra product, which we can (but need not) take as commutation (but it must have the properties of commutation). They must be mutually consistent.

Problem XIII.2.c-4: Notice also that we have sneaked in something else, multiplication of an operator by a scalar (a symbol that commutes with all other operators). Have we introduced any requirements on this product? Are all these four operations consistent? Is it possible to think of definitions for which they are not?

XIII.2.d Scalars are also needed to define algebras

Commutators do not completely determine an algebra. More than generators and their products is needed. Given the generators, X_i, we can form sums $\sum a_i X_i$, and the set of all sums and all products (associative and nonassociative) define the algebra. But what are the a's? They are scalars; they commute with all generators. All sums are found by taking all a's of the field, and to specify the algebra we must specify the field: real numbers, complex ones, or something else. (We consider only real or complex a's; unless stated all a's are complex.) Different fields give different algebras; for the same generators and the same product rules, real a's give a different algebra than complex ones [Hamermesh (1962), p. 307; Sattinger and Weaver (1986), p. 153; Wan (1975), p. 210]. Though related these algebras are very different, and have different physical consequences, and relevance.

Problem XIII.2.d-1: The reader might enjoy redoing the whole book with a's that are quaternions [Gilmore (1974), p. 24] (for the more adventurous, octonions — checking whether this is possible). The results here cannot be assumed to hold for these fields — nor even for the real-number field.

Problem XIII.2.d-2: Quaternions (and octonions) can be represented by matrices over complex numbers. So we can ask if algebras defined over quaternions can be recast into algebras over the complex numbers. Can Lie algebras be so recast into Lie algebras? If so how; are the two sets of algebras related? Is it possible to give a rule or algorithm to go from those of one set to those of the other? A more important question is whether algebras over complex numbers can be recast as ones over real numbers, and why not? The reader might think about this while studying (not only groups and Lie algebras).

XIII.2.e Compactness

We have mentioned both compact and noncompact groups — the range of their parameters are finite and infinite, respectively. However even the noncompact Lie groups listed are not "too noncompact". They are locally compact — their algebras are compact. Can a Lie algebra be noncompact?

Essentially, a locally compact space has a finite dimension. So for a locally compact Lie group, a neighborhood of the identity is finite dimensional. If a Lie algebra forms a vector space whose bases are its generators (and the other vectors their sums), and the number of these is finite, the space is finite dimensional — the Lie algebra is compact. We consider only compact Lie algebras.

Problem XIII.2.e-1: For the gauge transformations of electrodynamics

$$A_\mu(x) \to A_\mu(x) + d\delta(x)/dx_\mu, \qquad \text{(XIII.2.e-1)}$$

the space around the identity is not finite dimensional. The reason is that the functions δ are arbitrary and form an infinite-dimensional space. This group is not locally compact [Barut and Raczka (1986), p. 64; Mirman (1995b)]. Find the algebra of this group. Show that it is a Lie algebra. Is it finite dimensional? This is an example of a noncompact Lie algebra.

Problem XIII.2.e-2: In pb. II.4.f-1, p. 55, it was shown that the set of all nonzero, real, rational numbers is a group under multiplication. Show that it is not locally compact. Note that this is not a Lie group. Why?

XIII.2.f Isomorphic algebras

The algebra of SO(3), and of SU(2), is given by the three symbols (proportional to the angular-momentum operators) J_z, J_x, J_y, and also by J_z, J_+, J_- (eq. X.7.c-9, p. 297), with commutation relations

$$[J_x, J_y] = iJ_z, \quad [J_y, J_z] = iJ_x, \quad [J_z, J_x] = iJ_y, \qquad \text{(XIII.2.f-1)}$$

$$[J_+, J_-] = 2iJ_z, \quad [J_z, J_+] = J_+, \quad [J_z, J_-] = -J_-. \qquad \text{(XIII.2.f-2)}$$

These might look like different algebras, but they are not.

This algebra can be written using instead of J's, L's (popular for orbital angular momentum), s's or σ's (for spin), or I's (for isospin). These algebras are isomorphic regarded as abstract symbols. Also these symbols can be realized in different ways, as derivatives on different type spaces for example, or as matrices. States of spin or isospin are usually written as symbols with indices, those for orbital angular momentum as functions of angles. Ignoring these points should not cause too much confusion, but it would be better not to forget them completely. Operators and their eigenstates can have very different physical meanings, but all have the same properties, eigenvalues, and number of states; these are determined by the commutation relations — and the field — and are independent of how the algebra is interpreted. The reason that sets of different physical systems have (much) the same properties is that all are described by the same algebra (isomorphic algebras).

We can now define isomorphic algebras. Two algebras are isomorphic if they have the same structure constants (so the same number of symbols); more generally if the symbols of one are linear combinations of those of the other (with constant coefficients; but we might at some point think about extending the definition to allow them to be functions of other variables, say coordinates). However the word "combinations" is tricky. It means sums with coefficients from the field over which the algebra is defined. Thus algebras may be isomorphic if the field is complex, but not if it is real, as the example of the SU(2) algebra shows.

Isomorphic algebras are not really different (?) so when we talk about an algebra we mean up to isomorphism.

Problem XIII.2.f-1: Using the definition,

$$J_{\pm} = J_x \pm iJ_y, \tag{XIII.2.f-3}$$

show that either set of commutation relations can be obtained from the other (invert this to find J_x and J_y). Notice that the coefficients are complex; this cannot be done for a real field. Check that both are Lie algebras; all Jacobi identities are satisfied for both. Also if they are satisfied for one set, they are for the other.

Problem XIII.2.f-2: Find an algebra isomorphic to that of the J's, but with structure constants real. Can this be done over a real field?

Problem XIII.2.f-3: For order 1 there is a single Lie algebra. How many are there (up to isomorphism) for order 2? Are any non-Abelian? How many isomorphisms are there? Try this for order 3 [Jacobson (1979), p. 11; Kaplansky (1974), p. 5].

XIII.2.g The adjoint representation of a Lie algebra

There is another way of writing a Lie algebra (or in other terms, there is another algebra isomorphic to that of the abstract symbols), the regular (or adjoint) representation of the algebra [Chen (1989), p. 234; Cornwell (1984), p. 415, 488]. For an algebra of order n, the regular representation is a set of n $n \times n$ matrices, one for each algebra generator, acting on a set of basis vectors that are the algebra generators themselves.

Problem XIII.2.g-1: Prove that this is faithful — matrices going with different generators are different (how different?). Is it always, ever, reducible?

XIII.2.h Subalgebras

From an algebra we can obtain other algebras and these are of interest for themselves, for their applications, and for their aid is studying the algebra and its applications. A subalgebra consists of a subset of the operators of the algebra, with the same combination rules, and such that the commutator of two elements of the subalgebra is a sum of elements of that subalgebra. So

$$c^{\gamma}_{\alpha\beta} = 0, \tag{XIII.2.h-1}$$

if α and β are in the subalgebra but γ is not.

Problem XIII.2.h-1: Prove that the algebra of a subgroup is the subalgebra of the group. State and prove the converse (remembering SO(3) and SU(2), and O(3) and U(2)).

XIII.2.i Ideals

An invariant subgroup H contains all its conjugates: $S^{-1}RS$ is in H if R is, for every group element S. An invariant subalgebra is a set of elements Y, which themselves form an algebra, such that the commutator of Y with every X of the algebra is in the subalgebra,

$$[X_i, Y_a] = c^b_{ia} Y_b; \tag{XIII.2.i-1}$$

$$c^d_{ia} = 0, \tag{XIII.2.i-2}$$

if a denotes a Y, while d does not. An invariant subalgebra is also called an ideal. However having an invariant subalgebra is definitely not ideal; it makes things more complicated.

Problem XIII.2.i-1: Show that the subalgebra of an invariant subgroup is an ideal. Why might the converse not hold? Are there examples [Cornwell (1984), p. 398, 815]?

Problem XIII.2.i-2: The sum of two sets (here two ideals) is the set of elements in either one, or in both. The intersection is the set of those elements in both. It can easily be shown that the sum of two ideals is an ideal, as is the intersection. Is the intersection a subalgebra of the sum? Is it an invariant subalgebra?

Problem XIII.2.i-3: Show that an ideal of an ideal is an ideal (a set that is an invariant subalgebra of an invariant subalgebra of algebra L, is itself an invariant subalgebra of L).

Problem XIII.2.i-4: Is there a difference between the regular representations of algebras that have, and do not have, ideals?

Problem XIII.2.i-5: The commutator of two sets (which are not necessarily disjoint) is the set of all elements that equal the commutator of an element of one set with an element of the other (the set of all commutators of an element of one set with an element of the other):

$$[A_i, F_\alpha] = c^\rho_{i\alpha} Z_\rho; \tag{XIII.2.i-3}$$

the set of Z's forms the commutator (set) of the set of A and the set of F. Prove that the commutator of two ideals is an algebra and that it is an ideal (of what?).

Problem XIII.2.i-6: The centralizer (or center) of algebra L is the subset of elements of L that commute with every element of L. Show that it is a subalgebra. Need it be Abelian? Is it an ideal? How does this compare with the definition of centralizer of a group (sec. IV.4, p. 116)? Should they be related?

XIII.3 TYPES OF LIE ALGEBRAS

To understand a set of objects it helps to classify its members. There are two types of Lie algebras, with distinct properties. To study them we need (essentially) consider each separately. We define first solvable algebras, then semisimple ones, in this order because the definition of the latter is negative — these are neither solvable, nor do they contain ideals that are (but their existence is quite positive; they are fundamentally necessary in physics, and elsewhere).

XIII.3.a Solvable algebras, and nilpotent ones

We start with solvable algebras, which has subtypes nilpotent and Abelian [Sattinger and Weaver (1986), p. 27, 119; Wan (1975), p. 12, 21]. Consider Lie algebra L and the commutator $[L, L]$, that is the set of commutators of every pair of elements of the algebra. These commutators are sums of elements of the algebra, but not necessarily of all. Suppose the elements in these sums form set L', symbolically,

$$[L, L] = L'. \qquad \text{(XIII.3.a-1)}$$

Problem XIII.3-1: Prove that L' is itself an algebra and that it is an ideal of L.

XIII.3.a.i *Definition of solvable algebras*

Continuing this process we get a series of ideals (each may have fewer and fewer elements) called the derived series,

$$L^o = L, \qquad \text{(XIII.3.a.i-1)}$$

$$L^1 = [L^o, L^o] = [L, L], \qquad \text{(XIII.3.a.i-2)}$$

$$\dots$$

$$L^{k+1} = [L^k, L^k], \qquad \text{(XIII.3.a.i-3)}$$

$$\dots.$$

At each step we have the commutator of a subalgebra with itself. Suppose that for some j,

$$[L^j, L^j] = 0, \qquad \text{(XIII.3.a.i-4)}$$

that is the final ideal (which may have only a single element) is Abelian. Then L is solvable.

Problem XIII.3.a.i-1: Clearly all ideals in this derived series are also solvable.

Problem XIII.3.a.i-2: Show that for a solvable algebra the matrices of the adjoint representation are all upper (or all lower) triangular; the nonzero matrix elements are all on, or above, the diagonal. Is the converse true?

XIII.3.a.ii *Definition of nilpotent algebras*

Solvable algebras have a subtype. Let

$$L^o = L, \tag{XIII.3.a.ii-1}$$

$$L^1 = [L, L^o] = [L, L], \tag{XIII.3.a.ii-2}$$

$$\dots$$

$$L^{k+1} = [L, L^k], \tag{XIII.3.a.ii-3}$$

$$\dots .$$

Each step is the commutator of algebra L with the subalgebra of the previous step (whereas for a solvable algebra it is the commutator of the subalgebra with itself). If the last commutator is zero, L is nilpotent (as are all subalgebras in this series). So for a solvable algebra we find the commutator of each ideal with itself, finally obtaining an Abelian algebra, while for a nilpotent algebra we find the commutator of each ideal with the original algebra, finally again obtaining an Abelian algebra.

Problem XIII.3.a.ii-1: Show that a nilpotent algebra is also solvable. Find examples of solvable algebras that are not nilpotent.

Problem XIII.3.a.ii-2: State the conditions for solvability and nilpotency in terms of the structure constants.

Problem XIII.3.a.ii-3: Can the regular representation be used to show that an algebra is nilpotent?

XIII.3.a.iii *Abelian algebras*

If all commutators are zero the algebra is Abelian.

Problem XIII.3.a.iii-1: State this in terms of the structure constants. Check that an Abelian algebra is both solvable and nilpotent.

Problem XIII.3.a.iii-2: Can the regular representation be used to show that an algebra is Abelian?

Problem XIII.3.a.iii-3: In these cases, and there are also others, a commutator equals 0. But a commutator must equal a member of the algebra. Is 0 a member of the algebra? Can a commutator equal some other number?

XIII.3.b Semisimple algebras

An algebra that has no solvable ideals is called semisimple. Another definition is that an algebra with no Abelian ideals is semisimple. This makes it sound like semisimple algebras are the ones that are left over. But, at least in physics, these are the most important (apparently). Actually these, semisimple and solvable, are the two classes of Lie algebras; anything else is a combination.

Problem XIII.3.b-1: Explain why the two definitions are the same.

Problem XIII.3.b-2: For semisimple algebra G, is $[G, G] = G$; that is does the set of elements on the right contain all elements of the algebra? Prove that if every element of the algebra appears in one of these sums (so that every element can be expressed as a sum of commutators) then the algebra is semisimple (and conversely?).

Problem XIII.3.b-3: Describe the regular representation of a semisimple algebra.

Problem XIII.3.b-4: For simple algebra L, $[L, L] = L$ (each element of the algebra can be written as a sum of commutators), else there would be a set Z with $c_{XY}^{Z} = 0$, for all X, Y, which would be an ideal contrary to the assumption of simplicity. Justify these statements.

XIII.3.b.i *Simple algebras*

A semisimple algebra can have ideals. Take the algebra consisting of the J's and L's of the rotation group, the total, and the orbital, angular momenta. (Another example is one, or both, of these, and the isospin algebra.) The six operators form a semisimple algebra. But it can be split into two invariant subalgebras (spin and orbital angular momentum commute), and studying these we get complete information about the larger algebra.

If an algebra has no (proper — not including itself) ideals — no invariant (proper) subalgebras — it is simple. If it is simple it is also semisimple (with one exception). Simple algebras are so strongly restricted that we can get much information about them. Hence they are highly popular. But not all physical algebras are simple. Here we concentrate on simple algebras (semisimple ones are trivial extensions).

Problem XIII.3.b.i-1: Find the exception. Show that, except for this, semisimplicity follows from simplicity.

Problem XIII.3.b.i-2: Is the algebra of the L's semisimple? Is the algebra of the J's? Are either of these simple?

Problem XIII.3.b.i-3: For a semisimple algebra all matrices of the adjoint representation cannot be (simultaneously) written as upper (or lower) triangular.

Problem XIII.3.b.i-4: Is there a difference between the regular representations of simple and semisimple algebras? Why?

XIII.3.b.ii *What shall we call simple algebras?*

Having names for things makes it easy to refer to them. So we would like to name the algebras. We cannot do this as we have not found what they are. But we state a general convention and give examples so that we have something to use. Our terminology is (apparently) not quite standard, although the standard terminology is not completely explicit so it is impossible to be sure.

For the rotation group in four dimensions, SO(4), there is a related group, SO(3,1), the Lorentz group. And in general groups, and their algebras, come in sets. There is SL(2), SU(2) and SU(1,1), SO(3) and SO(2,1) and so on. The members of a set are closely related (but definitely different — consider a space of four spatial dimensions, so with transformation group, not including reflections, SO(4), and one with three spatial and one time dimensions, so with group SO(3,1)). These are, however, quite different from other sets. For example each set has the same number of generators.

We therefore give a name to the set, so A_1 (and in general A_l) refers to a set of algebras obtained from each other by putting i's ($= \sqrt{-1}$), and minus signs, in strategic places. For A_1 all algebras have three generators, and similarly for each A_l the members have the same number of generators, and the same number of mutually commuting ones, and closely related commutation relations.

When we refer to the algebra of a specific group of a set we use the same symbol as for the group, except using small, instead of capital, letters. Thus the algebra of SU(2) is su(2), of SU(1,1) it is su(1,1), both belonging to class A_1. The algebras of SO(4) and SO(3,1), so(4) and so(3,1) both belong to class D_2. This notation sometimes means that an algebra has two names. So su(2) and so(3) refer to the same algebra. This is not a defect of the notation but rather reflects a property of the groups: several groups can have the same algebra.

Thus the sets of algebras are labeled by a capital letter from the beginning of the alphabet with a numerical subscript (the set of allowed capital letters has to be worked out). Individual members of a set are denoted by small letters and numbers enclosed in parentheses, the same as the group whose algebra it is, that being given by capital letters from the middle or end of the alphabet.

We list the names of the simple Lie algebras (without showing that the list is complete, and that all do give algebras) [Chen (1989), p. 249; Cornwell (1984), p. 540, 860; Wan (1975), p. 92]. These are A_l, the algebras which include those of the special unitary groups SU($l+1$), B_l, and D_l, the sets which include the algebras of the orthogonal groups for odd-dimensional spaces, SO($2l+1$), and of even-dimensional spaces, SO($2l$), respectively, C_l, those which include the algebras of the symplectic groups, Sp(l) — these four infinite series called the classical algebras (giving the corresponding classical groups). D_2 is only semisimple — over the complex field. There are in addition the exceptional algebras G_2, E_6, E_7, E_8 and F_4.

XIII.4 EXAMPLES OF LIE ALGEBRAS

Different types of algebras have been defined, but are there any actually any satisfying the definitions? Moreover, can we combine algebras of different types? How? Here we give a few examples.

XIII.4.a Simple, semisimple and non-semisimple algebras

The most familiar algebras are probably the semisimple ones. The simplest nontrivial, simple algebra is A_1 (the algebra with one generator is trivial), the algebra of SU(2), of SO(3), and of the isospin group. Its commutation relations are given in sec. XIII.2.f, p. 373, and using symbols L, in eq. X.7.a.ii-6, p. 295.

Problem XIII.4.a-1: Prove that this has no invariant subalgebras by taking arbitrary, linear combinations of its generators and showing there are none that commute with all generators. Does it have any subalgebras?

Problem XIII.4.a-2: The algebra of L(n) (pb. X.4.a.i-3, p. 278), and also of U(n), is given by generators E_i^j, $i, j = 1, \ldots, n$, with commutation relations

$$[E_i^j, E_k^l] = \delta_k^j E_i^l - \delta_i^l E_k^j. \qquad \text{(XIII.4.a-1)}$$

Do these give a Lie algebra? Show that $\sum E_i^i$ commutes with all generators. Removing this gives A_{n-1}, the algebra of the unimodular group, SU(n), and also SL(n). Verify that A_n is simple. Show that all matrices of its adjoint representation have zero trace, and that determinants of all representations equal 1 for a group found (by exponentiating) from an algebra (pb. X.7.a.i-2, p. 293) with

$$\sum E_i^i = 0, \qquad \text{(XIII.4.a-2)}$$

State and show the converse.

Problem XIII.4.a-3: Commutation relations for the orthogonal algebras (classes B_l and D_l) are

$$[E_{ij}, E_{kl}] = \delta_{ik}E_{jl} + \delta_{jl}E_{ik} - \delta_{il}E_{jk} - \delta_{jk}E_{il}, \qquad \text{(XIII.4.a-3)}$$

$$E_{ij} = -E_{ji}. \qquad \text{(XIII.4.a-4)}$$

Check that these give Lie algebras and that these are simple (with one exception). For indices running from 1 to m, how many generators are there? Write the algebra of SO(3) in this form. Show that D_2 is not simple, only semisimple [Cornwell (1984), p. 525, 868]. The orthogonal group is a subgroup of a linear group, so its algebra must be a subalgebra of an algebra of a linear group. Given an algebra B_n or D_n, find the smallest A_k for which it is a subalgebra and find the sums of the generators of the latter algebra that form the orthogonal algebra.

Problem XIII.4.a-4: What are the matrices of the adjoint representation of A_1? Can these be found for arbitrary Lie algebras? Find the adjoint representations of all algebras of orders 1 and 2.

Problem XIII.4.a-5: The subscripts on the symbols for the algebras give their ranks; check that these are correct (for smaller algebras).

Problem XIII.4.a-6: Some of the smaller algebras have more than one name, since there are isomorphisms. Find these [Cornwell (1984), p. 524, 540].

XIII.4.a.i *The direct sum of simple algebras*

The nucleon transforms under both SU(2), when undergoing space rotations, and SU(2), when undergoing isospin transformations. Thus its symmetry algebra is $A_1 \oplus A_1$, the direct sum of two A_1 algebras [Cornwell (1984), p. 419; Wan (1975), p. 11]. This direct sum of two simple algebras is semisimple. The direct sum of algebras is the algebra of all generators of all the algebras, with the requirement that the generators of the different ones commute.

Problem XIII.4.a.i-1: Why is the direct sum of simple algebras always semisimple? What if the algebras in the sum were not all simple?

XIII.4.a.ii *The direct sum of a simple and an Abelian algebra*

There is another symmetry of the nucleon: it has quantum number hypercharge. The corresponding algebra, A_0, has a single generator. This commutes with isospin and space rotations so the symmetry algebra is (that including) $A_0 \oplus A_1 \oplus A_1$, the algebra of group $U(1) \times SU(2) \times SU(2)$. It has an Abelian invariant subalgebra so is not semisimple. Objects whose Hamiltonians have this symmetry form multiplets labeled by their spin, isospin and hypercharge. Thus the nucleon and the Ξ both have spin and isospin $\frac{1}{2}$, while having different hypercharge.

The reason that there is no relationship between the quantum numbers is that the group is not simple. But the algebra su(3) (of class A_2) is simple. For A_2 there is a relationship between isospin and hypercharge. Thus there are two isospin-$\frac{1}{2}$ multiplets (the nucleon and the Ξ), differing in hypercharge, and isospin-1 and isospin-0 multiplets, these part of the adjoint representation, the octet. Given the isospin the hypercharge cannot be picked arbitrarily. In two cases it is fixed. For the other two there are two (allowed) values, one for each of the two multiplets. Thus the quantum numbers are related within an su(3) multiplet because this group is simple.

Problem XIII.4.a.ii–1: From a table of elementary particles [Particle Data Group (1994)] find other $A_0 \oplus A_1 \oplus A_1$ multiplets and notice that there are many that have two quantum numbers the same but differ in the third.

XIII.4.b Inhomogeneous algebras

In describing the universe perhaps the most important algebra is an inhomogeneous one — the Poincaré algebra, for this is the algebra of the group giving the (isometric) transformations of our geometry. An inhomogeneous algebra consists of two parts, a semisimple algebra (the homogeneous part), and an Abelian algebra (the inhomogeneous part) whose operators are the basis vectors of a representation of the homogeneous part. The algebra is a semi-direct sum of these two parts; the Abelian part forms an ideal. So if the semisimple part is S, and the Abelian is P, then the commutation relations are, schematically,

$$[S,S] = S, \quad [S,P] = P, \quad [P,P] = 0. \qquad \text{(XIII.4.b–1)}$$

A Euclidean group is the group of motions in a Euclidean space (for which the distance squared is positive definite) [Elliott and Dawber (1987), p. 326; Sudarshan and Mukunda (1983), p. 251; Talman (1968), p. 189, 215; Tung (1985), p. 152; Wybourne (1974), p. 49]. It's algebra is the semi-direct sum of a rotation algebra (the homogeneous part) and the momenta (or derivatives with respect to coordinates), these transforming as the defining representation of the rotation algebra (in 3-space there are three momentum operators transforming like coordinates). The Poincaré algebra consists of the Lorentz algebra (which acts in a space, usually limited to dimension 3+1, for which the distance squared can be positive, negative or zero) plus four momentum vectors that form its defining representation. Since rotations and Lorentz transformations take momentum vectors into momentum vectors these give an invariant subalgebra.

Problem XIII.4.b-1: Write the commutation relations for the algebras of the inhomogeneous rotation group, and of the Euclidean and Poincaré groups [Cornwell (1984), p. 706; Mirman (1989, 1993, 1994, 1995b); Schweber (1962), p. 36; Sudarshan and Mukunda (1983), p. 254, 274, 357, 439].

Problem XIII.4.b-2: Show that the inhomogeneous part has to be the ideal. Why does it have to be a representation of the homogeneous part? Define an inhomogeneous group (sec. II.3.h, p. 45).

XIII.5 THE CLASSICAL LIE ALGEBRAS

The number of types of semisimple Lie algebras is limited, there are four infinite series, the classical algebras, plus five exceptional algebras [Wan (1975), p. 92]. The members of the infinite series are important for their applications in physics and elsewhere, but also as examples. Here we list the generators and commutation relations of these four series, A_l, the algebra of the unitary groups, B_l and D_l, the algebras of the orthogonal groups in odd and even dimensional spaces respectively, and C_l the algebra of the symplectic groups — properly, one member of each set is the algebra of the group, other members are algebras of the related noncompact groups.

XIII.5.a The general linear and the unitary algebras

The set of all $n \times n$ nonsingular matrices forms a group (pb. X.4.a.i-3, p. 278), the general linear group GL(n). Restrictions of tracelessness and unitarity lead to the special linear group SL(n), the unitary group U(n), and the special unitary group SU(n). The conditions are on the group parameters, so the algebras of the linear and unitary groups belong to the same set (the algebras of which differ by i's), while "special" removes one generator.

An arbitrary $n \times n$ matrix can be written as a sum of n^2 $n \times n$ matrices each having all zero elements except for a single 1, with the position of that 1 differing among the matrices. Thus these n^2 matrices, which we denote by E_i^j, form a basis of the Lie algebra of GL(n), where E_i^j has all 0's except for a 1 at position (i, j).

Problem XIII.5.a-1: Verify that these generators E_i^j, $i, j = 1, \ldots, n$, of the Lie algebra obey the commutation relations of pb. XIII.4.a-2, p. 380. Show that the E_m^m (no sum), $m = 1, \ldots, n$, commute with each other, and no other algebra generator commutes with all; these then form the largest mutually commuting subset. Also $\sum_m E_m^m$ commutes with all generators. For the algebras of the special groups,

$$\sum_m E_m^m = 0. \tag{XIII.5.a-1}$$

Give the number of commuting generators, the rank, of gl(n) and sl(n). Show the former is not semisimple while the latter is.

Problem XIII.5.a-2: The E_m^m are not linearly independent, within su(n), though they are in u(n), therefore are not good for the commuting generators (denoted by H's) of su(n); it is necessary to specify the algebra generators of SU(n), distinguishing them from those for U(n), giving sums of the E_m^m that form proper H's. As $\sum_m E_m^m$ commutes with every generator it is not in the algebra of su(n). For the algebras of unimodular unitary groups [Chen (1989), p. 222], we use $n-1$ independent linear combinations of the n h's [Mirman (1967), appendix],

$$h_m = E_m^m, \tag{XIII.5.a-2}$$

$$H_1 = h_1 - h_2, \tag{XIII.5.a-3}$$

$$\dots$$

$$H_j = \sum_{i=1}^{j} h_i - jh_{j+1}, \quad j = 1, \dots, n-1. \tag{XIII.5.a-4}$$

Check that these are linearly independent, and that the resultant algebra is simple.

Problem XIII.5.a-3: How can these H's be realized as differential operators acting on complex vectors? Find the realization of the group operators giving these differential operators as the realization of the algebra operators. Show that the algebra generators give $z_i \Rightarrow z_j$; state the effect of E_k^l, and H_m on the z's. These operators act by changing indices. What is the effect of a group transformation on the z's?

Problem XIII.5.a-4: Prove that if $exp(i\theta E)$ is unitary then E is hermitian, and conversely. This is the condition that distinguishes the algebras of u(n) and su(n) from that of gl(n). However this is really a condition on the representations (and in particular the adjoint and defining ones), not on the algebra unless the algebra is regarded as given by its list of generators, commutation relations, and this extra condition of hermiticity.

Problem XIII.5.a-5: Find the other algebras of this set by putting i's in everywhere that gives an algebra, not including duplicates. For which are the E's hermitian?

XIII.5.b The orthogonal algebras

Orthogonal group O(n) is given by the nonsingular orthogonal $n \times n$ matrices (pb. X.2.b-1, p. 272), a subset of all nonsingular matrices, so O(n) is a subgroup of GL(n). A subgroup of O(n), SO(n), is that of the subset of these matrices with determinant 1. Since this is the connected part it gives the algebra (varying the angles takes the identity to any SO(n) transformation, but changing the sign of the determinant is a discontinuous jump, so SO(n) is connected, but O(n) is not). The bases of this algebra, clearly a subalgebra of su(n), have matrices forming a subset of the E_i^j.

Problem XIII.5.b-1: Is "nonsingular" redundant?

Problem XIII.5.b-2: Linear groups can have any determinant; to get the special groups they are divided by an arbitrary factor to give matrices with determinant 1 — for orthogonal groups determinants are only ± 1, so to find SO(n) we take a subset of group operators. It should be obvious that the so(n) matrices are a subset of those of su(n). What is this subset? If $exp(i\theta R)$ is orthogonal, show that R is antisymmetric. Is the converse true? Another way of doing this is to take transformations

$$x' = exp(i\theta R)x, \qquad \text{(XIII.5.b-1)}$$

leaving x^2 invariant, where x is a real vector. Then R is antisymmetric. Thus so(n) is the set of $n \times n$ antisymmetric matrices with real entries. A basis is clearly

$$R_{ij} = E_i^j - E_j^i, \qquad \text{(XIII.5.b-2)}$$

where the E's are given in pb. XIII.4.a-2, p. 380, and

$$R_{ij} = -R_{ji}. \qquad \text{(XIII.5.b-3)}$$

Find the commutation relations of the R's. Groups SO(n) and O(n) are different, but algebras so(n) and o(n) are not.

Problem XIII.5.b-3: Commutation relations can be found using a differential realization (sec. X.7.c, p. 295), or matrices realizing the R's (from the matrix realization of sec. XIII.5.a, p. 383, for the E's), or the commutation relations for the E's [Chen (1989), p. 224]. Hopefully all answers are the same:

$$[R_{ij}, R_{kl}] = -i(\delta_{ik}R_{jl} + \delta_{jl}R_{ik} - \delta_{il}R_{jk} - \delta_{jk}R_{il}), \qquad \text{(XIII.5.b-4)}$$

(pb. XIII.4.a-3, p. 381); i is put in to give hermitian generators. Why? If corresponding indices on the R's, match the sign is +; if opposite indices match, the sign is −. Indices on the R's on the right side are in the same order as those on the left.

Problem XIII.5.b–4: Find all (distinct) algebras obtained by insertion of i's (or minus signs, if there is a difference).

XIII.5.c The symplectic algebras

A $2n \times 2n$ symplectic matrix M satisfies

$$M^t JM = J, \tag{XIII.5.c–1}$$

where M^t is the transpose of M,

$$M^t_{ij} = M_{ji}, \tag{XIII.5.c–2}$$

and

$$J = \begin{pmatrix} 0 & I_n \\ -I_n & 0 \end{pmatrix}; \tag{XIII.5.c–3}$$

I_n is the $n \times n$ unit matrix.

Problem XIII.5.c–1: Show that these matrices form a group, symplectic group Sp(n) [Cahn (1984), p. 55; Chen (1989), p. 226; Cornwell (1984), p. 869; Gilmore (1974), p. 187; Hamermesh (1962), p. 403; Miller (1972), p. 174, 344]. These matrices are nonsingular, of course, so this is a subgroup of GL(2n); its algebra, sp(n), is a subalgebra of gl(n). Prove that Sp(n) is semisimple, and simple, and that its rank is n. Demonstrate that this requires the algebra generators S (with real entries) to satisfy

$$S^t J = -JS. \tag{XIII.5.c–4}$$

This means that any algebra generator can be written in the form, with U and V $n \times n$ matrices,

$$S = \begin{pmatrix} U & V \\ W & -U^t \end{pmatrix}, \tag{XIII.5.c–5}$$

where V and W are symmetric. Check that these requirements mean that Sp(n) exists only for n even.

Problem XIII.5.c–2: With E^j_i the su(n) generators of sec. XIII.5.a, p. 383, and

$$E^j_{i+} = E^j_i + E^i_j, \quad E^j_{i-} = E^j_i - E^i_j, \tag{XIII.5.c–6}$$

show that a basis for the symplectic algebra is [Cornwell (1984), p. 869]

$$\begin{pmatrix} iE^j_{i+} & 0 \\ 0 & -iE^j_{i+} \end{pmatrix}, \begin{pmatrix} E^j_{i-} & 0 \\ 0 & E^j_{i-} \end{pmatrix}, \begin{pmatrix} 0 & iE^j_{i+} \\ iE^j_{i+} & 0 \end{pmatrix}, \begin{pmatrix} 0 & E^j_{i+} \\ -E^j_{i+} & 0 \end{pmatrix}. \tag{XIII.5.c–7}$$

Problem XIII.5.c-3: Another form [Gould (1989)] uses a metric (which however is not the algebra metric defined below) $g_{ij} = -g_{ji}$, which is antisymmetric as might be expected for the symplectic algebra, and the generators are written as

$$S_{ij} = g_{ik}E_k^j + g_{jk}E_k^i \quad \text{(XIII.5.c-8)}$$

These are symmetric. The commutation relations should then be

$$[S_{ij}, S_{kl}] = g_{kj}S_{il} + g_{lj}S_{ik} - g_{il}S_{kj} - g_{ik}S_{lj}. \quad \text{(XIII.5.c-9)}$$

The metric is taken as

$$g_{ij} = \delta_{j,i+1},\ i \text{ odd};\ g_{ij} = -\delta_{j,i-1},\ i \text{ even};\ 1 \leq i, j \leq n. \quad \text{(XIII.5.c-10)}$$

This is similar to the commutation relations for the orthogonal algebra, except for signs. But signs are important.

Problem XIII.5.c-4: Find all algebras obtained by insertion of i's (and minus signs).

XIII.6 THE CONDITION THAT AN ALGEBRA BE SEMISIMPLE

Lie algebras are strongly restricted (fortunately), limiting the kinds of algebras, and the forms they can take, allowing us to analyze, classify, understand and use them, and (perhaps) bound physics. Most strongly restricted are the semisimple, really the simple, algebras. Here we start to find how the requirements that the product be a commutator, and of semisimplicity, limit, and determine, these algebras.

But first how do we recognize a semisimple algebra? A Lie algebra is specified (given the number of generators and the field over which it is defined) by its structure constants. If an algebra is semisimple there are conditions on these. What are they? How can we tell if an algebra is semisimple?

XIII.6.a The metric

The quantity that provides the answer to these questions is the metric tensor of an algebra

$$g_{\alpha\beta} = \sum c^{\sigma}_{\alpha\rho} c^{\rho}_{\beta\sigma}. \quad \text{(XIII.6.a-1)}$$

Problem XIII.6.a-1: The adjoint representation is the algebra itself so it contains all information. How can we use it to learn about the structure? The Killing form $K(M,N)$, is the trace (sum of the diagonal elements), Tr, of the matrix that is the product of the adjoint-representation matrices of M and N, $Ad(M), Ad(N)$,

$$K(M,N) = Tr\{Ad(M)Ad(N)\}. \tag{XIII.6.a-2}$$

Show that this is the same as the metric, and that it satisfies the conditions for a metric [Cahn (1984), p. 17; Cornwell (1984), p. 485; Jacobson (1979), p. 69; Kaplansky (1974), p. 35; Wan (1971), p. 16; Wybourne (1974), p. 46]. This version is sometimes more useful.

XIII.6.b The metric gives the condition for semisimplicity

The necessary and sufficient condition for an algebra to be semisimple, the Cartan criterion, is

$$det|g| \neq 0; \tag{XIII.6.b-1}$$

$|g|$ is the determinant of the metric (matrix whose elements are $g_{\alpha\beta}$) [Cornwell (1984), p. 487, 823; Racah (1964) p. 8; Sattinger and Weaver (1986), p. 119; Wan (1971), p. 19, 44, 47].

Problem XIII.6.b-1: Use the criterion to see if the algebra of pb. XIII.2-5, p. 369, is semisimple. What does it say about the Heisenberg algebra (pb. XIII.2-6, p. 369)?

Problem XIII.6.b-2: It should be clear that the criterion is not satisfied by an Abelian algebra, and thus the determinant of that part of the metric referring to an Abelian subalgebra is 0. Apply this to the algebra so(3) and test whether it is semisimple. Try it for su(3). Is there a difference for so(2,1) and su(2,1)? Is u(1) semisimple? How about u(n), su(n)? Check it for o(n) and Sp(2n).

Problem XIII.6.b-3: To verify the rule we first consider algebra Λ with a solvable ideal. Denoting the elements of the invariant subalgebra by Latin letters, and the other elements by initial Greek ones, check that

$$c^{\alpha}_{\mu k} = 0. \tag{XIII.6.b-2}$$

Then note that

$$g_{\mu k} = c^{\sigma}_{\mu\rho} c^{\rho}_{k\sigma} = 0. \tag{XIII.6.b-3}$$

Show that $|g|$ is zero. How does the ideal being solvable enter? It remains to demonstrate the converse: if $|g|$ vanishes, Λ has a solvable ideal, hence if it does not, Λ is semisimple. Now if

$$|g| = 0, \text{ then } g_{\rho\sigma} y^{\sigma} = 0 \tag{XIII.6.b-4}$$

has nonvanishing solutions. Why? For each set of solutions we form $\sum y^\sigma X_\sigma$; the X's are the elements of Λ. Prove that these form a solvable subalgebra, and an invariant one — an ideal of Λ. Why does this proof work for solvable ideals, but not for simple ones — why does it allow the algebra to be semisimple, and not require it to be simple? So whether the determinant of the metric is, or is not, 0, tells if an algebra is not, or is, semisimple.

Problem XIII.6.b-4: There is another way of stating the criterion for semisimplicity. A Lie algebra L is solvable if

$$K(M,N) = \mathrm{Tr}\{Ad(M)Ad(N)\} = 0, \qquad \text{(XIII.6.b-5)}$$

for all M and N of L', where $L' = [L,L]$, and is semisimple if

$$\det K(M,N) \neq 0. \qquad \text{(XIII.6.b-6)}$$

Prove this, and show that it is equivalent to the metric condition.

Problem XIII.6.b-5: Explain why, if an algebra is semisimple all its ideals are also, and any algebra to which it is homomorphic is semisimple. Is this surprising?

Problem XIII.6.b-6: Show that for a semisimple algebra there is a reciprocal tensor $g^{\mu\nu}$, such that

$$g_{\mu\nu}g^{\mu\zeta} = \delta_\nu^\zeta. \qquad \text{(XIII.6.b-7)}$$

Is semisimplicity both necessary as well as sufficient? Can $g^{\mu\nu}$ be calculated in terms of $g_{\mu\nu}$? So there is another set related to the structure constants,

$$c_{\alpha\beta\gamma} = c_{\alpha\beta}^{\rho}g_{\rho\gamma}. \qquad \text{(XIII.6.b-8)}$$

These would not be useful if we could not get $c_{\alpha\beta}^{\gamma}$ back from them,

$$c_{\alpha\beta}^{\gamma} = g^{\gamma\rho}c_{\alpha\beta\rho}. \qquad \text{(XIII.6.b-9)}$$

This is possible only because g is nonsingular — the algebra is semisimple. Show that $c_{\alpha\beta\gamma}$ is totally antisymmetric.

Problem XIII.6.b-7: Find $c_{\alpha\beta\gamma} = 0$ for the algebra of pb. XIII.2-5, p. 369, and for the SHO algebra (pb. XIII.2-6, p. 369).

Problem XIII.6.b-8: Find $c_{\alpha\beta\gamma}$ for the su(2) and so(3) algebras. Repeat for su(3). Does this work for su(1,1), so(2,1) and su(2,1)? Are there differences among these algebras?

Problem XIII.6.b-9: Algebras can be combined in various ways (although not many). One way is a direct sum — a set consisting of algebras that mutually commute and share no elements. So a semisimple algebra is a direct sum of simple algebras. Why?

XIII.7 THE BEST WAY OF WRITING SIMPLE LIE ALGEBRAS

Given a set of generators, an infinite number of other sets can be found by taking linear combinations. We might expect that there is one set, or perhaps a few, with simple properties, which is not only easiest to use (and understand) but is closely related to the structure of the algebra, so providing information about that structure. How do we find this canonical set of generators and commutation relations, if there is one [Cahn (1984), p. 25; Chen (1989), p. 244; Cornwell (1984), p. 497, 528; Racah (1951), p. 16; Racah (1964), p. 9; Sattinger and Weaver (1986), p. 130; Wan (1975), p. 48; Wybourne (1974), p. 57; Zelobenko (1973), p. 264]? And why should there be one? Might these canonical generators be the (only?) ones with physical interpretation? Why? Is there a relationship between the answers to these questions and the mathematical definition of canonical? (In writing the commutation relations we have in mind simple algebras; semisimple ones are direct sums, so their commutation relations follow immediately from those of the simple algebras.)

For so(3), there are two standard, simple, ways of writing the algebra, two sets of operators, J_x, J_y, J_z and $J_z, J_\pm$, so two sets of c's. Which is better? J_x, J_y, J_z have symmetrical commutation relations. However to label states we need diagonal operators whose eigenvalues distinguish the states. This can be any of the three — there is no distinction between x, y, z — but it is conventional, for some reason, to take J_z as diagonal. Further we need operators to change the state label; these are $J_\pm$. The less symmetrical form has the advantage that it distinguishes diagonal, J_z, and step, $J_\pm$, operators, and it is on this that we model our discussion. However this is a good choice for additional, perhaps more important, reasons.

Note this interesting point about so(3). The operators come in pairs, J_+ and J_-; these have the same commutation relations with J_z, except for a sign. And $[J_+, J_-] \sim J_z$. Does something like this hold true in general? We denote the operators of the algebra by H_i (the diagonal ones), and E_r (the step operators). No E commutes with all H's.

Problem XIII.7.-1: Why?

XIII.7.a The Cartan subalgebra

Choose any maximal set of mutually commuting, linearly independent, generators of the algebra, denoted by H_i, $i = 1, \dots, r$. (We do not prove that r is independent of the choice of the diagonal generators.) This Abelian subalgebra is called the Cartan subalgebra [Kaplansky (1974), p. 39; Wan (1975), p. 31]; its order, the number r of its operators, is the rank of the Lie algebra (the order of that is the total number of algebra operators). For so(3) there is one H (say J_z).

XIII.7.b The canonical form of the commutation relations

The canonical commutation relations are

$$[H_i, H_j] = 0, \qquad \text{(XIII.7.b-1)}$$

$$[H_i, E_{\pm r}] = \pm r_i E_{\pm r}, \qquad \text{(XIII.7.b-2)}$$

$$[E_r, E_{-r}] = \sum r^i H_i, \qquad \text{(XIII.7.b-3)}$$

$$[E_r, E_s] = N_{rs} E_{r+s}, \qquad \text{(XIII.7.b-4)}$$

where the set of vectors r_i are called the roots (it is from these that the algebra grows); we call r^i the adjoint roots. For each operator E_r there is another, E_{-r}, having a root of opposite sign (like E_+ and E_-). Then E_{r-r} belongs to the set of H's so the last equation becomes the same as the one before it and corresponds to $[J_+, J_-]$, except now, since there are several H's, there may be a sum of terms. The notation implies that we regard E_0 as a sum of H's. To avoid redundancy then, we only use the last equation for $s \neq -r$.

This holds for all semisimple Lie algebras (any set of commutators, no matter how messy, can be brought into this form), with the different ones distinguished (up to insertion of i's) by their orders, ranks, and roots (the r's — which give the N's).

Problem XIII.7.b-1: One of these equations should be reminiscent of an eigenvalue equation for the H's. Prove that in the regular representation it is in such a form, so it states that the E's are eigenvectors of the H's.

XIII.7.b.i *The symmetric form of the generators*

While the canonical form is the usual way of writing Lie algebras, and useful for proving theorems, it is not the only one — the most familiar form of the angular momentum algebra uses J_x, J_y, J_z. In general, we can write the step operators, up to possible insertion of i's, as

$$E_{\alpha\pm} = \frac{1}{2}(E_\alpha \pm E_{-\alpha}), \qquad \text{(XIII.7.b.i-1)}$$

giving the algebra in symmetric form. This is also useful and we consider it at times.

Problem XIII.7.b.i–1: Derive the commutation relations for the symmetric form from the canonical one. Note that (some) conditions for the canonical form are needed for there to be a symmetric one.

Problem XIII.7.b.i–2: Check the so(3) algebra to see if these two sets of generators agree with standard usage. In particular it may be necessary to introduce i's. Why?

Problem XIII.7.b.i–3: If the canonical step operators with opposite roots are hermitian conjugates, what about the operators in symmetric form?

XIII.7.b.ii *What has to be proved?*

To justify these commutation relations what has to be shown? That the H's commute follows from their definition (such a set, even if only of a single operator, can always be chosen). We must show that the commutator of an H and an E is a sum of E's (no H's). Given this we can always choose linear combinations such that the commutator is proportional to the same E — the main point on which the proof of this form rests. Further we have to show that the E's come in pairs, with equal and opposite roots, and that their commutators are sums of H's, only, with coefficients which are (essentially) the same as the roots. Finally the commutator of two E's is a sum of E's only, no H's, with only a single E in the sum, and its root is the sum of the roots of the two E's in the commutator.

Problem XIII.7.b.ii–1: From the commutation relations in sec. XIII.5.a, p. 383, write the commutation relations of the algebra A_l in canonical form, identifying the H's and pair the step operators with equal and opposite roots. Find these roots. Repeat for the orthogonal and the symplectic algebras (sec. XIII.5.c, p. 386). Are the algebras in canonical form? Determine their metrics. Compare all results.

XIII.7.c The simple roots

The set of roots is overcomplete — the number of roots is greater than the dimension of the root space (the rank of the algebra), so are not all independent. This complicates things. Thus we pick a minimal set of independent roots; for a rank l algebra, whose root space is therefore l-dimensional, this consists of (any) l roots that are linearly independent — none can be written as a sum over the others.

These are called the simple roots. All other roots can be written as a sum over the simple roots, and the coefficients are integers, as we will see.

XIII.8 PROOF OF THE CANONICAL COMMUTATION RELATIONS

To show that all simple algebras can be written in canonical form (it then follows that semisimple algebras can) we have to prove that the step generators, the E's, are all eigenvectors, in a sense, of the Cartan generators, the H's. Using this, the Jacobi identities, and the Cartan criterion ($|g| \neq 0$), we get the canonical form, also known as the Cartan-Weyl basis.

XIII.8.a The step generators as eigenvectors

The canonical commutation relations imply that the regular representation matrices for the H's are diagonal. And not every matrix is diagonalizable. Hence we have to show that for a simple algebra there is (at least) one diagonal matrix (in the regular representation) and then show that the E's are eigenvectors of it, using commutation as the product. If one member of a maximal set of mutually commuting semisimple-algebra generators is diagonalizable, all are.

Problem XIII.8.a-1: Why do these commutation relations imply this diagonalizability?

Problem XIII.8.a-2: Can the Cartan subalgebra matrices be diagonalized? It must be shown that if they cannot be reduced to diagonal form the algebra has a solvable ideal so is not semisimple. Any matrix M can either be diagonalized or written in Jordan canonical form, with nonzero elements only on the diagonal and the superdiagonal [Cullen (1990), p. 195; Gel'fand (1989), p. 132],

$$M_{ij} = 0 \text{ unless } j = i, \text{ or } j = i + 1; \;\; M_{i,i+1} = 1 \text{ or } 0. \qquad \text{(XIII.8.a-1)}$$

If no element of the Cartan subalgebra were diagonalizable, any one could be brought to this form. (We note that if all matrices of a set commute, and one is in Jordan canonical form, all are upper triangular). Thus either the H's are diagonalizable, as we want, or are all upper triangular. Show that the eigenvalues of a matrix in Jordan canonical form are its diagonal elements. A matrix with all distinct eigenvalues can be diagonalized [Gel'fand (1989), p. 85]. So if the Cartan subalgebra could not be, every H would have at least two equal diagonal elements. Any matrix can be reduced to block-diagonal form, each block with all the same diagonal elements. We bring each block in turn into Jordan Canonical form. However the set of commutators with this matrix forms a solvable ideal. Why? So for semisimple algebras, the Cartan subalgebra regular-representation matrices are diagonalizable, and we assume them diagonal.

Problem XIII.8.a-3: Check that if the regular-representation matrix of H is diagonal, then for all E,

$$[H, E_\alpha] = r_H E_\alpha, \tag{XIII.8.a-2}$$

where r_H is the corresponding eigenvalue of H.

XIII.8.b The Jacobi identities and the roots

The main tool for proving the canonical form of the commutation relations is now the Jacobi identities. First we note that if one H is diagonal, all H_i are. Then

$$[H, [H_i, E_\alpha]] + [H_i, [E_\alpha, H]] + [E_\alpha, [H, H_i]] = 0, \tag{XIII.8.b-1}$$

where the H's are any members of the Cartan subalgebra, gives

$$[H, [H_i, E_\alpha]] = r_H [H_i, E_\alpha], \tag{XIII.8.b-2}$$

so that $[H_i, E_\alpha]$ is, in this sense, an eigenvector of H. But

$$[H_i, E_\alpha] = r_i E_\alpha, \tag{XIII.8.b-3}$$

so, for all H,

$$[H, E_\alpha] = r_H E_\alpha; \tag{XIII.8.b-4}$$

the commutator of every H with an E is proportional to that E — all E's are eigenvectors of the H's.

From the next Jacobi identity,

$$[H, [E_\alpha, E_\beta]] + [E_\alpha, [E_\beta, H]] + [E_\beta, [H, E_\alpha]] = 0, \tag{XIII.8.b-5}$$

$$[H, [E_\alpha, E_\beta]] = (\alpha_H + \beta_H)[E_\alpha, E_\beta]. \tag{XIII.8.b-6}$$

A commutator of two E's is also an eigenvector of the H's, thus is an E (we cannot have

$$[E_\alpha, E_\beta] = E_\gamma + c_{\alpha\beta i} H_i). \tag{XIII.8.b-7}$$

With $E_\gamma \sim [E_\alpha, E_\beta]$, the root of E_γ is

$$\gamma = (\alpha + \beta). \tag{XIII.8.b-8}$$

The commutator of two E's is equal to that (single) E whose root is the sum of the roots.

Problem XIII.8.b-1: If $\beta = -\alpha$ then the only term the commutator can equal is a sum of H's. Check that a Jacobi identity shows this directly.

Problem XIII.8.b-2: From these results, $g_{\alpha\beta} = 0$, unless $\beta = -\alpha$. From this show that $det|g| \neq 0$ requires that for every root α there is a root $-\alpha$. Check this for all algebras with three generators. That the roots are paired is a necessary consequence of simplicity (and conversely?).

Problem XIII.8.b-3: Prove that not all components of a root can be 0; $[H_i, E_\alpha] = 0$ for all i, is impossible.

Problem XIII.8.b-4: Show, summing over all roots α, that, with s, t referring to the H's,

$$g_{st} = \sum \alpha_s \alpha_t. \tag{XIII.8.b-9}$$

Problem XIII.8.b-5: Using the antisymmetry of c_{ijk}, check that (up to signs, which we consider next)

$$\alpha^i = c^i_{\alpha-\alpha} = g^{ik}\alpha_k. \tag{XIII.8.b-10}$$

Thus

$$[E_\alpha, E_{-\alpha}] = \sum \alpha^i H_i = \sum g^{ik}\alpha_k H_i. \tag{XIII.8.b-11}$$

This completes the proof that all Lie algebras can be brought into the standard (canonical) form.

Problem XIII.8.b-6: There is some freedom in the choice of the structure constants; prove that we can resolve this by choosing

$$g_{\alpha-\alpha} = 1, \tag{XIII.8.b-12}$$

for all α, as we do. Demonstrate that it is also possible to choose

$$g_{\alpha-\alpha} = -1, \tag{XIII.8.b-13}$$

and find the commutation relations for A_1 and A_2 with this choice. Also show that $g_{\alpha-\alpha} = 1$, for some roots, -1 for others, can be chosen, and find the commutation relations for some such choice. The commutation relations, and algebras, are determined up to signs. There are different sets of signs, giving related, but non-isomorphic, algebras. For each algebra considered, there is a set, different members differing in sign choices. It is this (mere?) arbitrariness of sign that allows this richness, and so the existence of noncompact algebras (thus spaces that include time dimensions, permitting the universe to exist) and infinite-dimensional multiplets. Finding the algebras has two aspects, considered separately, determining the sets (the structure constants with the signs undetermined), and finding the members of each (determining the signs) — which we do not do here, studying only algebras with all signs positive.

XIII.8.c Conditions on the roots

The commutation relations in canonical form express certain structure constants as roots. To get these we used Jacobi identities. However we did not use all, implying conditions on the roots [Racah (1951), p. 21; Racah (1964), p. 12]. These can be regarded as vectors in some space (root space) and in it we define a scalar product

$$(\alpha, \beta) = \sum \alpha_i \beta^i = \sum \alpha^i \beta_i = \sum g_{ij} \alpha^i \beta^j = \sum g^{ij} \alpha_i \beta_j. \qquad \text{(XIII.8.c-1)}$$

Problem XIII.8.c–1: Prove that this product is really a scalar — it is invariant under rotations of the axes, that is the replacement of the H's by (all?) linear combinations of H's.

Problem XIII.8.c–2: Now what are the restrictions on the roots? First $2(\alpha, \beta)/(\alpha, \alpha)$ is an integer (as is $2(\alpha, \beta)/(\beta, \beta)$), for roots α and β. And $\beta - 2\alpha(\alpha, \beta)/(\beta, \beta)$ is a root. Start the proof by verifying that there is a γ for every α such that $\alpha + \gamma$ is not a root. Then

$$[E_{-\alpha}, E_\gamma] = N_{-\alpha\gamma} E_{\gamma-\alpha} = E'_{\gamma-\alpha}, \qquad \text{(XIII.8.c-2)}$$

$$[E_{-\alpha}, E_{\gamma-\alpha}] = (NN) E_{\gamma-2\alpha} = E'_{\gamma-2\alpha}, \qquad \text{(XIII.8.c-3)}$$

$$\ldots$$

$$[E_{-\alpha}, E_{\gamma-j\alpha}] = (N \ldots N) E_{\gamma-(j+1)\alpha} = E'_{\gamma-(j+1)\alpha}, \qquad \text{(XIII.8.c-4)}$$

where the product of N's has been indicated schematically and the primes indicate that they have been ignored. Check that this is acceptable. Show that for some g,

$$E_{\gamma-(g+1)\alpha} = 0. \qquad \text{(XIII.8.c-5)}$$

Define the coefficient μ, which depends on root γ,

$$[E_\alpha, E'_{\gamma-(j+1)\alpha}] = \mu_{j+1} E'_{\gamma-j\alpha}. \qquad \text{(XIII.8.c-6)}$$

Then we get the recursion relation

$$\mu_{j+1} = (\alpha, \gamma) - j(\alpha, \alpha) + \mu_j, \qquad \text{(XIII.8.c-7)}$$

from the Jacobi identity

$$[E_\alpha, [E_{-\alpha}, E'_{\gamma-j\alpha}]] + \ldots = 0. \qquad \text{(XIII.8.c-8)}$$

We can choose $\mu_0 = 0$. Check that then

$$\mu_j = j(\alpha, \gamma) - \frac{1}{2} j(j-1)(\alpha, \alpha), \qquad \text{(XIII.8.c-9)}$$

$$(\alpha, \gamma) = \frac{1}{2} g(\alpha, \alpha), \qquad \text{(XIII.8.c-10)}$$

where g is the value for which the E is zero, for this pair of roots α and γ. Further

$$\mu_j = \frac{1}{2}j(g - j + 1)(\alpha, \alpha). \qquad \text{(XIII.8.c-11)}$$

Is it possible that $(\alpha, \alpha) = 0$? There is an integer j such that $\beta + j\alpha$ is a root, but $\beta + (j + 1)\alpha$ is not. Why? Show that

$$(\alpha, \beta) = \frac{1}{2}(g - 2j)(\alpha, \alpha). \qquad \text{(XIII.8.c-12)}$$

So this gives the result that the ratio of the scalar products is an integer. Thus

$$g = 2(\alpha, \gamma)/(\alpha, \alpha). \qquad \text{(XIII.8.c-13)}$$

Verify that, for $\alpha + \gamma$ not a root, this also shows that there is a string of roots $\gamma, \gamma - \alpha, \gamma - 2\alpha, \ldots, \gamma - g\alpha$, which is invariant under reflection in the hyperplane perpendicular to α, and that every root β belongs to one of these strings.

Problem XIII.8.c-3: This establishes the result. What result?

Problem XIII.8.c-4: Demonstrate that these root conditions have a geometrical interpretation: reflecting β in a (hyper)plane through the origin perpendicular to α gives a new root vector, and the difference between β and this new vector is an integer multiple of α.

Problem XIII.8.c-5: Any root can be written as a sum of simple roots with integer coefficients. Show that it follows from $2(\alpha, \beta)/(\alpha, \alpha)$ being an integer that the coefficients are also.

Problem XIII.8.c-6: It should be easy to establish that the only multiples of α that are roots are α, 0, and $-\alpha$. What does the 0 mean? This gives the number of roots, three, in this string. How about other strings? The maximum number is four. Suppose there were five, which we label $\beta - 2\alpha, \beta - \alpha, \beta, \beta + \alpha, \beta + 2\alpha$. Why is this labeling general? Since neither $2\alpha = (\beta + 2\alpha) - \beta$, nor $2(\beta + \alpha) = (\beta + 2\alpha) + \beta$ can be roots (why?), and similarly $(\beta - 2\alpha) \mp \beta$ are not roots, so

$$(\beta \pm 2\alpha, \beta) = 0, \qquad \text{(XIII.8.c-14)}$$

giving $(\beta, \beta) = 0$, a contradiction. If α, β and $\alpha + \beta$ are roots, then why does $[E_\alpha, E_\beta] \neq 0$? The E's are labeled by their roots but this assumes that there is a single E for each root. Why must this be so? Show that $2(\alpha, \beta)/(\alpha, \alpha)$ being an integer means that [Wybourne (1974), p. 64],

$$(\alpha, \beta) = \frac{1}{2}m(\alpha, \alpha) = \frac{1}{2}n(\beta, \beta), \qquad \text{(XIII.8.c-15)}$$

where m and n are integers. Then show that if ϕ is the angle between roots α and β,

$$cos^2\phi = \frac{1}{4}mn, \qquad \text{(XIII.8.c-16)}$$

so the only values of ϕ are

$$\phi = 0, \frac{\pi}{12}, \frac{\pi}{8}, \frac{\pi}{6} \text{ and } \frac{\pi}{4}. \tag{XIII.8.c-17}$$

Does it matter if ϕ is in other than the first quadrant? Can it be? Also the ratio of the magnitudes of two roots are respectively, ?, $\sqrt{3}$, $\sqrt{2}$, $\sqrt{1}$, and ?.

Problem XIII.8.c-7: This determines not $cos\phi$, but $cos^2\phi$, so that these restrictions give not a value of ϕ, but a pair. Given the set of values m and n, there are several algebras allowed. We note, but do not discuss, this possibility here. Was it mentioned before?

Problem XIII.8.c-8: How many different root lengths are there? We might expect for algebras of large dimension there can be many, giving very complicated systems. Fortunately there are only two. If there were three lengths, prove that two would have ratios $\sqrt{3/2}$, which is not possible. These roots are called, not surprisingly, long and short. If they all have the same lengths they are all called long.

XIII.8.d The values of the N's

The last structure constants that we have to find are the N's. We can give each product of two N's, and that is all that is determined. The only question is how are the signs related [Cornwell (1984), p. 511; Gilmore (1974), p. 280].

Problem XIII.8.d-1: Taking $\beta = \gamma - j\alpha$, show that [Wybourne (1974), p. 65],

$$\mu_j E_{\alpha+\beta} = [E_\alpha, [E_{-\alpha}, E_{\alpha+\beta}]] = N_{-\alpha,\alpha+\beta}[E_\alpha, E_\beta] = N_{-\alpha,\alpha+\beta}N_{\alpha,\beta}E_{\alpha+\beta} \tag{XIII.8.d-1}$$

So

$$N_{-\alpha,\alpha+\beta}N_{\alpha,\beta} = \mu_j = \frac{1}{2}j(g - j + 1)(\alpha, \alpha). \tag{XIII.8.d-2}$$

So $N_{\alpha\beta} \neq 0$ if $\alpha + \beta$ is a root. And

$$N_{\alpha\alpha} = 0, \tag{XIII.8.d-3}$$

so 2α is not a root (because E_α commutes with itself). Next show that

$$N_{\alpha\beta} = \pm\{\frac{1}{2}(m + 1)n(\alpha, \alpha)\}^{\frac{1}{2}}, \tag{XIII.8.d-4}$$

where m and n are the integers relating (α, β) to (α, α) and (β, β). Prove that the signs of the N's are related by

$$N_{\alpha\beta} = -N_{\beta\alpha} = -N_{-\alpha,-\beta} = N_{-\beta,-\alpha}. \tag{XIII.8.d-5}$$

How are they related to $N_{\beta,-\alpha-\beta}$ and $N_{-\alpha-\beta,\alpha}$?

Problem XIII.8.d-2: Consider the commutators (say for the matrices of the regular representation $[E_\alpha, E_\beta], [E_{-\alpha}, E_{-\beta}]$. Take their hermitian conjugates (why are these matrices hermitian?) and use the requirement that the structure constants be real (why?) to find all relations between the signs [Gilmore (1974), p. 266]). Note that these relations are based on an assumption, hermiticity.

XIII.8.e Diagonalization of the metric

As g is symmetric it can be diagonalized; its determinant is nonzero, so it has no zero eigenvalues. Thus

$$g_{ii} = c^{\beta}_{i\alpha} c^{\alpha}_{i\beta} \neq 0, \tag{XIII.8.e-1}$$

so the matrices c_i with elements $c^{\beta}_{i\alpha}$ have inverses, for all diagonal generators i of the algebra.

Problem XIII.8.e-1: Why does this show that the adjoint representation (for the algebra in canonical form) is completely reduced, and also faithful? Are either or both these statements true for non-semisimple algebras? Can they be?

XIII.8.f Why semisimple algebras have this canonical form

The canonical form is basic to the study and use of simple algebras. Why do they have this form? The key is that a matrix that does not leave a subspace invariant can be diagonalized, and a simple algebra has no invariant subspace (subalgebra), hence there is one regular-representation matrix that is diagonalizable (semisimple algebras can be block-diagonalized so add nothing new). Thus there is a set of these matrices that can be taken diagonal, and with what eigenvectors? Only the E's, for the H's mutually commute. This means that the commutation relations of H's and E's are of this eigenvector form, they state the nonexistence of an invariant subspace — that the algebra is simple. And the commutator of two E's equals an E.

The structure constants can be regarded as matrices. If there is an ideal, rows and columns would be 0, giving a zero determinant of the metric, and conversely. Thus, simplicity and the nonzero value of this determinant are equivalent. But this nonzero value also means that the metric tensor cannot have rows or columns 0. So the roots come in pairs, equal and opposite. And these provide the basic features of the canonical form of the algebra.

XIII.8.g The Cartan matrix

The roots give the algebra so it would be useful to have a concise description of them, thus of the algebra. This is given by the Cartan matrix, an $r \times r$ matrix for a rank r algebra, defined by

$$A_{ij} = 2\frac{(\alpha_i, \alpha_j)}{(\alpha_j, \alpha_j)}, \tag{XIII.8.g-1}$$

where the α 's are the simple roots [Cahn (1984), p. 43; Chen (1989), p. 250]. So it is the matrix of products of the simple roots, divided by the square of their magnitudes (or if we want to be picturesque, the square of their lengths). It need not be symmetric.

Problem XIII.8.g-1: Show that the scalar product of two positive roots is nonpositive. Thus all diagonal elements of the Cartan matrix are 2, and the off-diagonal ones are 0, -1, -2, or -3. Using the Schwarz inequality [Arfken (1970), p. 445] for the scalar product of two vectors,

$$(V \cdot W)^2 \leq |V|^2|W|^2, \tag{XIII.8.g-2}$$

prove that

$$A_{ij}A_{ji} < 4, \text{ for } i \neq j. \tag{XIII.8.g-3}$$

So if $A_{ij} = -2$, or -3, then $A_{ji} = -1$.

XIII.8.h The simplest simple algebras

The simplest algebras to use in illustrating these results are those of A_1 and A_2, though the first is perhaps a little too simple.

Problem XIII.8.h-1: For A_1, using the expressions for $cos\phi$ and N, find the roots; check all results.

Problem XIII.8.h-2: A_2 is somewhat more interesting. For this, all angles between pairs of neighboring root vectors are equal. What is this angle? What is the ratio of the lengths of root vectors? Labeling the roots consecutively, show that $\alpha_1 + \alpha_3$ is a root and $\alpha_1 + 2\alpha_3$, $\alpha_1 - \alpha_3$ are not. Then

$$n = 1, \quad m = 0. \tag{XIII.8.h-1}$$

Take

$$N_{12} = \frac{1}{\sqrt{6}}; \tag{XIII.8.h-2}$$

find the other N's. Why is this value chosen? Compute the root vectors to show

$$\alpha_1 = \frac{1}{\sqrt{3}}(1,0), \tag{XIII.8.h-3}$$

$$\alpha_2 = \frac{1}{2\sqrt{3}}(1, \sqrt{3}), \tag{XIII.8.h-4}$$

$$\alpha_3 = \frac{1}{2\sqrt{3}}(-1, \sqrt{3}), \tag{XIII.8.h-5}$$

and find the magnitude of each. Give the commutation relations.

Problem XIII.8.h-3: Prove that there are two other rank-two simple algebras, and derive the corresponding results for them.

Problem XIII.8.h-4: For su(3), that the simple roots are (2,0), $(-1, \sqrt{3})$, with magnitudes 4 and product -2, and that the Cartan matrix is

$$A = \begin{pmatrix} 2 & -1 \\ -1 & 2 \end{pmatrix}. \tag{XIII.8.h-6}$$

Verify these.

Problem XIII.8.h-5: Compute the Cartan matrix for su(2), su(4), so(3), so(4) and so(5). Try it for the symplectic algebras.

Chapter XIV

Representations of Lie Algebras

XIV.1 WHAT IS A REPRESENTATION OF AN ALGEBRA?

The concept of a Lie algebra representation is essentially the same as that for Lie groups (not surprisingly). A Lie algebra is a set of symbols X_α with law of combination

$$[X_\alpha, X_\beta] = c^\gamma_{\alpha\beta} X_\gamma. \qquad \text{(XIV.1-1)}$$

A representation of it is a set of matrices with the same commutation relations. The basic problem of representation theory is to find all sets of matrices — there can (must?) be infinitely many — with the same commutation relations as an algebra. Replacing "commutation" by "multiplication", we get the statement for groups. How then do these types of representations, and the processes for finding them, differ? And why go to the trouble of developing two distinct (?) procedures; is not one enough?

That an algebra is semisimple strongly restricts the properties of its representations (fortunately). All are equivalent to unitary ones for compact, (semi)simple groups (all we consider here, so our concepts and results may not be general). These, so those of the corresponding algebras, are all decomposable and all finite dimensional (so their matrix elements are discrete). We restrict the representations to be unitary — ones that exponentiate to unitary representations of the group; the algebra matrices are hermitian or pairs of hermitian conjugates.

So we want for each semisimple algebra, all sets of (irreducible) hermitian matrices (up to equivalence) satisfying the algebra commutation relations — all unitary irreducible representations. For this we need names — labels — for the matrices, and their elements, so for the basis states, plus an understanding of how these matrices and labels, and their properties, are related to, and determined by, the algebra. After finding representations we face the question (which we avoid here) of whether all can be exponentiated to give (unique) representations of the group.

If an algebra is realized by derivatives, as we did for the rotation algebra, then we would want representation basis states that are functions of the variables (with respect to which the derivatives are taken), and the effect of the operators on these. How this gives the representation matrices is another question to be answered.

Here then we consider representations of semisimple algebras.

XIV.1.a Why these are algebra representations

The concepts introduced here apply to groups and algebras. However we limit the discussion to representations of the algebras in the sense that the operators used are from the algebra; also we do not try to find functions of the group parameters (like generalizations of spherical harmonics).

We start to determine how to generate representations [Cahn (1984); Cornwell(1984), chap. 15,16; Mirman(1989); Racah(1951);Racah(1964); Sattinger and Weaver (1986), p. 233; Wan (1975), chap. 9-14; Zelobenko (1973)], beginning with their labels and those of the states. This leads to two concepts introduced in the discussion of the rotation group, but here made explicit — invariants and weights — and to a study of how these provide the labels, and how the two types of representation labels that they give are related. This chapter concentrates on terminology, concepts, and properties of the representations. Actual calculations are, as might be expected, harder, and are not considered. Occasionally some examples are mentioned, these being "well-known"; some relevant information is given in the appendix.

These are representations by matrices with discrete elements. This is not the only possibility; momentum operators id/dx have eigenfunctions $exp(ikx)$, for all real values of k, giving representations of the translation algebra (which is not semisimple).

Problem XIV.1-1: It is worthwhile in reading about representations to guess which assertions apply to algebras, which to groups, and which to both, and why, how assertions about algebras and groups are related, and what the significance of such relations might be, and why. Also it is useful to consider to what extent these questions are answered here, or even considered.

XIV.1.b Completeness of the representation

The representations that we consider are finite-dimensional. Are these all representations? Unfortunately we do not discuss here the Peter-Weyl theorem (sec. XI.4.a.ii, p. 317) which says that all (well--behaved) functions of the group parameters (defined over the same range) can be expanded in terms of sums over the basis states of these representations — the basis states of these representations are complete (here we are claiming it only for compact, semisimple groups) [Barut and Raczka (1986), p. 172, 426; Bröcker and tom Dieck (1985), p. 133; Chevalley (1962), p. 203; Miller (1968), p. 77; Miller (1972), p. 215; Talman (1968), p. 92; Weyl (1946), p. 189; Zelobenko (1973), p. 70]. (And we do not consider the exact definition of "expanded" and "complete".)

Problem XIV.1-1: In reading this it is useful to notice why the representations are finite-dimensional. Also explain, though, why the Peter-Weyl theorem shows that these are all the (inequivalent, independent) irreducible representations. How is this related to Fourier's theorem? How is it related to the result for finite groups that the decomposition of the regular representation contains all irreducible representations (sec. VI.1.b, p. 171)?

Problem XIV.1-2: The states of an irreducible representation are generated by applying the generators of the algebra a sufficient number of times to any state. Why? Would there be a difference if the representation were reducible? What would happen if a state were the sum of two from different representations (say describing a particle that can be found in several orbital angular momentum states)? Try examples.

XIV.1.c The defining representation

One representation is particularly important, the defining representation (pb. XI.4.b.i-5, p. 320), the one whose transformations define the group (or algebra). So SU(n) is the group that leaves invariant scalar products of a set of n complex vectors. These n complex vectors form the basis space of the SU(n) defining representation; the matrices transforming them are the representation matrices of it. Likewise SO(n) is defined over n real vectors, these forming the space of its defining representation. The unitary, orthogonal and symplectic groups and algebras are those that leave invariant certain products in the relevant n-dimensional spaces. The coordinate vectors in these spaces are the basis vectors of their defining representations and the representation matrix elements are found from the action of the algebra on these vectors.

Problem XIV.1.c-1: This discussion is for the classical algebras. Try extending it to the exceptional algebras [Cornwell (1984), p. 850, 880; Wan (1975), p. 160]. In particular what, for such algebras, are the (equivalent of the) defining representations?

XIV.1.d The adjoint representation

As with a group there is one more representation, again called the adjoint, or regular, representation, of especial interest: it is the algebra as a representation of itself. Commutation relations

$$[X_\alpha, X_\beta] = c^\gamma_{\alpha\beta} X_\gamma, \qquad \text{(XIV.1.d-1)}$$

can be regarded as operations by operators $[X_\alpha, \,]$ on matrices X_β giving the sum of matrices $c^\gamma_{\alpha\beta} X_\gamma$. Thus we can regard the X_α as a set of matrices acting on matrices X_β. These X_α are the matrices of the regular representation.

Problem XIV.1.d-1: Explain how an operation that is commutation is converted into one that is matrix multiplication, and why these matrices actually do give a representation of the algebra.

Problem XIV.1.d-2: What are these matrices? Write them for the canonical and symmetric forms of an algebra, starting with su(2), then for all A_1 algebras, then in general. How are these related to the matrices of the regular representation of a Lie group?

Problem XIV.1.d-3: There are (now) two representations that we have decided are important enough to be given names, the defining and adjoint representations. These are different. However there is one (important) case in which they are the same. Why is this an orthogonal algebra?

Problem XIV.1.d-4: Representations of different degrees are inequivalent, but representations of the same degree may, or may not, be equivalent. Would you expect that there could be other inequivalent representations of the same degree as the defining representation; as the adjoint representation?

Problem XIV.1.d-5: Is there a general relationship between the adjoint and the defining representations? Why?

XIV.2 HOW TO GIVE A REPRESENTATION

To be able to find and deal with representations each, and each of its states and each element of each of its matrices, must be given a "name".

These names are, of course, numbers. We need a set of numbers that determine a representation, and a set determining the states, so labeling the matrix elements. This, familiar from quantum mechanics, is the state-labeling problem. How is it dealt with for (compact, simple) Lie algebras? Again, as in quantum mechanics, the standard way is to find a (complete) set of (commuting) operators whose eigenvalues label representations and states.

Problem XIV.2.-1: Why do the states label the matrix elements? What is the relationship between their labels?

XIV.2.a The enveloping algebra

Besides generators of a Lie algebra, we also need all polynomials in them. This set is the enveloping algebra of the Lie algebra [Barut and Raczka (1986), p. 249; Humphries (1987), p. 89; Wan (1975), p. 180]. It provides representation labels and its members are physically important.

Problem XIV.2.a-1: What are the elements of this algebra? What operations are needed to define it? Are there any beyond those needed for the Lie algebra itself?

XIV.2.b Invariants — Casimir operators

An operator whose eigenvalues label representations is invariant under the algebra — it commutes with all generators (so it is called an invariant). Otherwise if $XI \neq IX$ then I's eigenvalue for states $|\rangle$ and $X|\rangle$ would differ — it would not be the same for all states — so would not have a single value for a representation. What operators commute with all of the algebra, and what is a (or the) minimum (independent) set? Invariants are not generators themselves — the algebra is simple.

For so(3), J^2, the square of the total angular momentum, belongs to the enveloping algebra and its commutators with all rotation-algebra operators are zero, giving the well-known physical result that the total angular momentum is rotationally invariant; it is constant under rotation — which is fortunate physically (pb. X.7.c-3, p. 296).

Hopefully (all) required representation-labeling operators are provided by the enveloping algebra. For algebras of all compact, simple, groups, there are in fact enveloping-algebra members whose set of eigenvalues do uniquely determine the representations (two nonequivalent representations differ in at least one eigenvalue). These are called Casimir operators, though sometimes only the lowest (second) order one is so named [Barut and Raczka (1986), p. 254; Cahn (1984), p. 84; Cornwell (1984), p. 592; Elliott and Dawber (1987), p. 148, 248;

Racah (1951), p. 44; Wan (1975), p. 122, 134; Wybourne (1974), p. 139; Zelobenko (1973), p. 156, 368]. They provide one, but not the only, way of labeling representations of simple groups.

Problem XIV.2.b-1: Why, if $XI \neq IX$, does I's eigenvalue for $|)$ and $X|)$ differ?

Problem XIV.2.b-2: By Schur's Lemma(s) matrices that commute with all those of a group representation are used to show irreducibility; this is also true — less explicitly — here (why?). State and prove Schur's Lemma(s) for Lie algebras [Wan (1975), p. 119]. Why do these hold for both groups and algebras? How are they used?

Problem XIV.2.b-3: The enveloping algebra consists, of course, of an infinite number of operators. Which of these might provide (a complete set of independent) labels? Check that, for every semisimple algebra, the commutators of

$$C_s = \sum H_i^2 + \sum E_\alpha E_{-\alpha}, \qquad \text{(XIV.2.b-1)}$$

with all generators, are zero, where the sums are over all H's and all roots. Thus C_s is an invariant, a Casimir operator. Also [Racah (1951), p. 45], with the algebra operators denoted by X_α, and

$$X^\alpha = g^{\alpha\beta} X_\beta, \qquad \text{(XIV.2.b-2)}$$

the operator

$$I = \sum d^{\beta_1}_{\alpha_1\beta_n} d^{\beta_2}_{\alpha_2\beta_3} \ldots X^{\alpha_1} \ldots X^{\alpha_n}, \qquad \text{(XIV.2.b-3)}$$

commutes with all generators, with sums over dummy indices, where the d's are coefficients, for example the structure constants (for second order, writing the X's as H's and E's, this is the same as C_s). A saturated index is one that appears twice, once as a covariant (lower) index, the other as a contravariant (upper) index, and is summed over. Show that the necessary and sufficient condition that an operator be invariant is that all its indices are saturated. Verify that it (also) commutes with all elements of the enveloping algebra (or is this redundant?).

Problem XIV.2.b-4: Check this for the A_1 and A_2 algebras. For each there is more than one algebra, related by multiplication of certain generators by i's. How does this affect the invariants?

Problem XIV.2.b-5: Prove that there is a degree d invariant (one that is a product of d operators) that can be written

$$C_s^d = \sum g^{ik} \ldots H_i H_k \ldots + \sum g^{ij} \ldots H_i \ldots E_\alpha E_\beta \ldots; \qquad \text{(XIV.2.b-4)}$$

both sums have d generators, the sum of the roots equals 0 (why?), and we can place all negative (and also all positive) roots on the right. Show that for degree 2 this gives (as before?)

$$C_s^2 = \sum g^{ij} H_i H_j + \sum E_\alpha E_{-\alpha}. \tag{XIV.2.b-5}$$

Problem XIV.2.b-6: While taking the structure constants as the coefficients in I gives an invariant, it does not prove that the only invariants are of this form; there might be other tensors, so other ways of constructing invariants. Show that the difference between two invariants differing only in the ordering of generators is an invariant of lower degree (fewer X's). Thus we can construct a set of invariants using $g^{\mu\nu}$, starting with the lowest possible degree. What is the lowest degree; why? However while there are invariants of each degree, not all, in fact very few, are independent — if we know the eigenvalues of the set of independent invariants, for a representation, we know the eigenvalues for all. For example, if C_s is invariant, clearly C_s^2 is also, but not independent. Why is it clear? The question is what is the smallest set of invariants, all of whose eigenvalues must be known to determine (uniquely) a representation? How many sets are there? How do we choose the best? Determining all needed labels, say by finding a complete set of invariants (if these give a complete set of labels), and their eigenvalues and eigenvectors, is the first step in obtaining the representations.

Problem XIV.2.b-7: For so(3), obviously $(J^2)^2$ is not independent of J^2. Write an invariant of third degree and show that it also is not independent of J^2 [Wybourne (1974), p. 141]. Other algebras have further, independent, invariants (A_1 has only one, but it is very simple).

XIV.2.c Semisimplicity, conservation laws and invariants

Invariants are used to label the representations. Finding them and their eigenvalues is part of the standard state-labeling problem. But their existence is required by the semisimplicity of these algebras describing nature. Thus conservation laws like that of angular momentum are related to this semisimplicity. Our universe would be very different, and would be described very differently (if it were still possible) if the algebras giving its laws did not include ones that are semisimple.

XIV.3 WEIGHTS

Lie algebra generators have a maximal commuting set, the H's (which are taken diagonal), the Cartan subalgebra. Representation basis states are eigenstates of these; their eigenvalues are thus state labels, so are important enough to be given a name: weights [Cahn (1984), p. 31;

Cornwell (1984), p. 561; Humphries (1987), p. 67, 107; Miller (1972), p. 323; Samelson (1990), p. 94; Sattinger and Weaver (1986), p. 170; Serre (1987), p. 56; Wan (1975), p. 126; Wybourne (1974), p. 97]. Each state has a weight, a set of numbers (regarded as components of a vector) that are the eigenvalues of the H's for that state,

$$H_i|w) = w_i|w). \tag{XIV.3-1}$$

Different weights label different states, but the converse need not be true. Several states can have the same weight; a weight may not fully determine a state. The baryon octet (see the appendix) of the eightfold way [Ludwig and Falter (1988), p. 311; Sattinger and Weaver (1986), p. 163], whose statefunctions are basis states of the adjoint representation of su(3), has eight states; there are two with weight (0,0). The two labeling operators are H_1 (the z component of the isospin) and H_2 (the hypercharge). The two nucleons have weights $(\pm\frac{1}{2}, 1)$, the two Ξ's weights $(\pm\frac{1}{2}, -1)$, the charged Σ's have weights $(\pm 1, 0)$, all these are simple (there is only one particle — state — with any weight), but the (isospin-1) Σ^o and (isospin-0) Λ both have weight (0,0). This weight is not simple (there are two states with it, distinguished by their isospin). It appears twice (its multiplicity is 2). All other weights are simple and mutually equivalent. The Λ and Σ^o are thus distinguished by their isospin (given by the Casimir operator of the isospin subalgebra, implying a rule that holds more generally).

The weights provide (some) labels for the states, and labels are essential; also a subset of weights label representations. Thus we have to find how to obtain them, see how they depend on the algebra and representation, study their properties (whatever this means) and relate these to properties of the states (if states have properties independent of their weights).

XIV.3.a Ordering weights

Weights provide (in part) a means of giving states. But how do we give weights? Neither the components of a vector, nor a vector in a set, can be specified without ordering conventions, and, for the weights, both of these require that the H's be ordered. Obviously we take the H we call H_1 as first, and so on; trivial unless H's are distinguished, as they sometimes are (as with isospin and hypercharge). This orders the components.

The ordering of the weights themselves follows from that of the components. For two weights, that with the larger first component comes first, if this component is the same, the one with the larger second component comes first, and so on. Thus

$$w < w' \text{ if } w_1 > w'_1, \qquad \text{(XIV.3.a-1)}$$

and if $w_1 = w'_1$ then

$$w < w' \text{ if } w_2 > w'_2, \qquad \text{(XIV.3.a-2)}$$

and so on (note the difference in direction of the inequality signs). The first weight is said to be higher then the second (or sometimes, more consistently, heavier). The weights then are ordered, from highest to lowest. The highest plays a special role, as we see (by convention; the lowest can be chosen).

Weights are equivalent if they can be transformed into each other by reordering H's. In a set of equivalent weights the highest is called the dominant weight. The highest weight of a representation is not unique; there is a set of weights any of which can be taken, by proper ordering of the H's, to be the highest. We assume an ordering convention has been decided, so the highest weight is fixed.

A weight is positive if its first nonvanishing component is.

Problem XIV.3.a-1: What are the highest weights of the so(3) representations?

Problem XIV.3.a-2: Check that the weights are completely ordered: there are no weights whose order is undecided. Prove that they are totally ordered: if weight 1 < weight 2 and weight 2 < weight 3, then weight 1 < weight 3. Clearly the components are.

Problem XIV.3.a-3: Why are the roots the weights of the adjoint representation?

Problem XIV.3.a-4: Show that a simple root (sec. XIII.7.c, p. 392) cannot be written as a sum of two positive roots. Does this hold for weights of other representations?

XIV.3.b Simple and nonsimple weights

For each root there is a single algebra element (adjoint-representation state). However, in general, for each weight there is more than one state. A weight is simple if there is only one state of the representation with that weight. For the su(3) octet — the su(3) adjoint representation — all weights are simple except (0,0), and all, except (0,0), are equivalent to each other. And this must be so since this algebra has more than one H (these have weight (0,0)). There is no such excuse for other representations, but most have nonsimple weights — ones with more than one state.

Notice a difference between the way the terms weight and root are often used. Weight (0,0) of the su(3) adjoint representation, which corresponds to the H's, is not considered a root. The other weights of this representation (six for su(3)) correspond to roots. So there is a difference in terminology with weight being more inclusive.

Problem XIV.3.b-1: For the su(3) octet, show how each simple weight can be taken as highest, but (0,0) cannot be. Show that they are all equivalent.

Problem XIV.3.b-2: For the 27-dimensional representation of A_2, find the weights and their multiplicity; note which weights are simple and that there are others besides (0,0) with multiplicity greater than 1.

Problem XIV.3.b-3: For a representation of degree N (the matrices are $N \times N$) there are at most N distinct weights; such a matrix can have at most N linearly independent eigenvectors (and each has a single weight). Check this. Show that it can have fewer. Since not all weights are simple, in general, there have to be fewer than N distinct ones.

Problem XIV.3.b-4: Would you expect the defining representation of an algebra to have nonsimple weights? Are all weights equivalent?

XIV.3.c Extending properties of roots to weights

The roots determine an algebra and label its step generators. To what extent do they determine the representations and the weights? And how fully do the weights determine a representation and the states? We start the analysis of these questions by studying the relationship between weights and roots.

There are relations among the roots. Might there be similar ones for the weights? The difference between weights and roots is that each root is a simple weight. So we have to consider how to extend the results we found for roots, if possible, to nonsimple weights, or see whether for them anything analogous holds.

Problem XIV.3.c-1: What would go wrong with an algebra if there were several states of the adjoint representation with the same nonzero weight (root)?

Problem XIV.3.c-2: For every root there is one that is its negative. Thus for the adjoint representation, weights come in pairs, the members being equal and opposite, except for zero weights. Is this a peculiar property of the adjoint representation, or is it true in general? The trace of a commutator is 0,

$$tr[A,B] = \sum(A_{ij}B_{ji} - B_{ij}A_{ij}) = 0. \qquad \text{(XIV.3.c-1)}$$

Thus every matrix that equals a commutator, as does every representation matrix of a semisimple algebra (why?) (pb. XIII.8.a-4, p. 378), has zero trace. The H's are diagonal, so their traces are the sums of components of the weights. Thus all, except zero, weights of every representation of a semisimple algebra, come in pairs with equal and opposite values. Check this for all representations of the rotation group that are listed in the nearest quantum mechanics book, and also for any other convenient representations, say those of SU(3). There is a slight weakness in this argument; make it rigorous.

XIV.3.d Weights as vectors in root space

The weights form a set of vectors, whose components are eigenvalues of the Cartan generators, the H's, just as for the root vectors. Thus in root space we can draw the weight vectors, with the same axes labeled by the H's. This allows statements about the relationships between weights and roots and between weights and weights to be expressed geometrically. The weights of the adjoint representation are the roots, so we expect that the results that we have found for this representation would generalize in a reasonable manner.

XIV.3.d.i *Scalar products*

If we consider roots and weights as sets of vectors (in the same space) we can ask if we can take (meaningful) products, as we expect for vectors. We have considered products of roots; can we extend this to weights and roots, and weights and weights? Why would we want to? As we have seen for roots, there are restrictions on such scalar products. If there are similar restrictions involving weights, then they would be limited, and this would help find them, so find the representations. And limitations would help clarify why representations have the properties that they do.

Problem XIV.3.d.i-1: The scalar product of two roots was defined as an ordinary dot product in root space (sec. XIII.8.c, p. 395). In this space we can also draw the weight vectors (for each representation), these having the same number of components as the roots (why?), and define similarly a scalar product between a weight and a root,

$$(w, \alpha) = \sum w_i \alpha^i = \sum g^{ij} w_i \alpha_j. \qquad \text{(XIV.3.d.i-1)}$$

Show that this has the properties of a scalar product, in particular it is invariant under rotations of the axes (what is the significance of these rotations?). Repeat for products of weights.

Problem XIV.3.d.i-2: Show that states of different weight are linearly independent.

XIV.3.d.ii *Restrictions on products and implications*

Is the value of the scalar product of a weight and a root, (w, α), limited? We know a restriction for the adjoint representation, and it does hold in general:

$$p_w = 2\frac{(w, \alpha)}{(\alpha, \alpha)}, \tag{XIV.3.d.ii-1}$$

for any weight w and any root α, is an integer and $w - 2\frac{(w,\alpha)}{(\alpha,\alpha)}\alpha$ is a weight — results analogous to those for roots (sec. XIII.8.c, p. 395). The proof is similar [Cornwell (1984), p. 563; Miller (1972), p. 326; Racah (1951), p. 35; Wan (1975), p. 128]. To show this for any α, we define

$$H_\alpha = \sum \frac{\alpha^i H_i}{(\alpha, \alpha)}, \quad E'_{\pm\alpha} = \frac{1}{\sqrt{2(\alpha, \alpha)}} E_{\pm\alpha} \tag{XIV.3.d.ii-2}$$

giving an su(2) subalgebra (denoted by $su(2)_\alpha$), with half-integral matrix elements for H_α. Thus the component of any weight, along a line on which a root lies, is a half-integer. This is a consequence of the A_1 algebra having only integer or half-integer representations — an interesting connection between the weights of representations of (semisimple) algebras and the fact that angular momentum is integral or half-integral. The reason, which is similar to that of the corresponding result for the roots, is that step operator E_α applied to a state increases (and $E_{-\alpha}$ decreases) its weight by α, and gives zero applied enough times. Thus adding α to, or subtracting it from, w, an integral number of times gives zero.

Also, with w_h the highest weight, $E_{-\alpha}$ acting on the state with weight

$$w = w_h - p_w\alpha = w_h - \frac{2(w_h, \alpha)}{(\alpha, \alpha)}\alpha, \tag{XIV.3.d.ii-3}$$

gives zero. Applying lowering operator $E_{-\alpha}$ to $|w_h)$, $p_w + 1$ times, gives zero since $\frac{(w_h,\alpha)}{(\alpha,\alpha)}\alpha$ is the component of w_h in the $su(2)_\alpha$ subalgebra, so is the highest weight of one of its representations. And the result is true for this subalgebra.

Problem XIV.3.d.ii-1: Use this last statement to give a slightly different argument for the result.

Problem XIV.3.d.ii-2: Similarly, weights w and $w - 2\frac{(w\alpha)}{(\alpha\alpha)}\alpha$ have the same multiplicity. So equivalent weights have the same multiplicity.

XIV.3.e The action of step generators

Applying to a state of definite weight (an eigenstate of all H's), step generator E_α — which "steps" up or down the weight, we get a basis state with a different weight. What is its action, what state does it give, of what weight, and with what coefficients? And might some of these be 0?

Problem XIV.3.e-1: Show that if $|w)$ has weight w, then $E_\alpha|w)$ is (unless zero) a state of definite weight, $w + \alpha$. The weight of the resultant state is a sum of the weight of the original plus the root labeling the step generator. There is a multiplicative factor that depends on the normalization of the two states; also weight $w + \alpha$ could be multiple, so the state may be a sum of basis states. The weights of a representation are found from each other by adding, or subtracting, roots; any weight equals any other plus a sum of roots, with integer coefficients. Explain how this gives a relationship between weights and roots. We have considered one set of weights, the roots. Show that this holds for the roots. To find explicitly the effect of E_α, start with state $|0)$ of weight w such that $w + \alpha$ is not a weight, define

$$|1) = E_{-\alpha}|0), \quad |2) = E_{-\alpha}|1), \ldots, \qquad \text{(XIV.3.e-1)}$$

and show that

$$E_\alpha|j+1) = \mu_{j+1}|j), \qquad \text{(XIV.3.e-2)}$$

by induction, using

$$E_\alpha|j+1) = E_\alpha E_{-\alpha}|j) = \{(\alpha, w) - j(\alpha, \alpha)\}|j) + \mu_j|j), \qquad \text{(XIV.3.e-3)}$$

analogous to the results of sec. XIII.8.c, p. 395, where $\mu_0 = 0$ since $w + \alpha$ is not a weight. What is the corresponding statement for the roots? Is multiplicity accounted for?

XIV.3.f Expressing weights in terms of simple roots

For the roots we can choose a subset, the simple roots (sec. XIII.7.c, p. 392), and write every root as a sum of these, with integer coefficients. The situation is not quite so favorable for weights, but we can still express them in terms of the simple roots,

$$w = \sum \mu_i \alpha_i, \qquad \text{(XIV.3.f-1)}$$

for every weight, where the sum is over all simple roots, and the μ 's are real (of course, the H's are hermitian), and rational. Taking the product of this with root α, we get

$$2\frac{(w, \alpha)}{(\alpha, \alpha)} = \sum \mu_j 2\frac{(\alpha_j, \alpha)}{(\alpha, \alpha)}, \qquad \text{(XIV.3.f-2)}$$

the left side is integral, as is $2\frac{(\alpha_j,\alpha)}{(\alpha,\alpha)}$, thus each μ is rational.

Problem XIV.3.f-1: Why can a weight be written as a sum over simple roots?

Problem XIV.3.f-2: Given the weights of a representation, is it possible to find the algebra whose representation it is? If not, what else need be known? Would the answer be different for the weights of the adjoint representation?

XIV.4 USING WEIGHTS TO LABEL REPRESENTATIONS

One way to label representations is to use the eigenvalues of the invariants. For su(2), the representations are labeled by j, the angular-momentum quantum number, which is the maximum value of the z component, the highest weight. And the eigenvalue of Casimir invariant J^2 is $j(j+1)$. Here the highest weight and the eigenvalue determine each other, so either can be used to specify the representations. Is this true in general?

Representations of every simple algebra are, in fact, labeled by their highest weight — two irreducible representations are equivalent if their highest weights are equal, else not. A weight, of algebra of rank l, has l components, so the eigenvalues of l (independent) invariants are needed to distinguish representations. Thus we have to show that the highest weights do label the representations, that there are (exactly) l independent invariants, find these and also the relationship between their eigenvalues and the components of the highest weight. This we start here.

Problem XIV.4-1: Consider the state $|h)$ having the highest weight, w_h, and all states of the form $\dots E_{-\beta}E_{-\alpha}|h)$, using only step-down generators. For an irreducible representation, why do these states span the representation space (any state of the representation can be written as a sum of these)? Why is "irreducible" required? Would they span the representation space if $|h)$ did not have highest weight? Suppose step-up generators were included? Can the minimum set of the states of this form be found? Check this for the su(2) representations, and a few of those of su(3).

Problem XIV.4-2: For some su(3) representations, check that inequivalent ones have different highest weights. This is obvious for ones of different dimension. Is it true for ones of the same dimension? Can you prove inequivalence?

XIV.4.a Simplicity of the highest weight

The highest weight of an irreducible representation labels the representation; it determines the eigenvalues of the invariants and vice versa — representations are equivalent if they have the same highest weight, else are inequivalent. And all states of a representation can be generated from that of the highest weight. This would be impossible if there were several states with that weight. Fortunately the highest weight is simple [Cornwell (1984), p. 837; Miller (1972), p. 328; Racah (1951), p. 39; Racah (1964), p. 18; Sattinger and Weaver (1986), p. 172; Zelobenko (1973), p. 128].

Problem XIV.4.a-1: Check this for the $3, 6, 8, 10, 10^*$ and 27 dimensional representations of A_2.

Problem XIV.4.a-2: To prove that the highest weight, w_h, of an irreducible representation is simple, consider state $|h)$ with this weight. Then simplicity is established (why?) if it is shown that every state with weight w_h of the form $\ldots E_\delta E_\gamma E_\beta E_\alpha |h)$ equals $k|h)$, where k is a constant — applying any product of generators gives the same state. Check that applying (all) such products to either su(3)-octet state of weight (0,0) does not give the same state. What does it give? Now explain why

$$\ldots + \delta + \gamma + \beta + \alpha = 0. \tag{XIV.4.a-1}$$

From this it follows that one root must be positive. Taking γ as the first positive root (from the right), moving E_γ to the right, using

$$E_\gamma |h) = 0, \tag{XIV.4.a-2}$$

(why?) we get a sum of terms acting on $|h)$, with the same weight, but with fewer E's. Write out the sum. Eliminating in this way all E's with positive roots gives a sum of products of H's acting on $|h)$. Why? What happened to the E's with negative roots? And this gives $|h)$ multiplied by a polynomial in the components of the weight w_h. How are the components of this weight obtained? Thus all states with the highest weight are proportional, so the weight is simple. It would be unfortunate in various ways if the highest weight were not simple. Why are we so lucky that it does turn out to be so? To go from one state to another we (must) apply the step operators. But for a highest-weight state only step-down operators are available, and these change the weight.

Problem XIV.4.a-3: How does this proof break down if the weight is not a highest weight?

Problem XIV.4.a-4: Does it break down if the representation is reducible? Why?

XIV.4.b Representations are labeled by their highest weights

To show that representations R, R', of the same highest weight are equivalent, but not if they differ in this weight, consider their highest-weight basis vectors. For each weight of the representation, there is a state given by a product of E's acting on the highest-weight state. Thus the weights of the two representations are in one-to-one correspondence. This almost gives the result. How different could the representations be? There is a slight difficulty that other weights need not be simple and we have to consider states of weights with multiplicity greater than 1. It is conceivable that starting with the highest-weight states of the two representations, there could be states of the same weight, but different. How could such states differ? The weights might have different multiplicity. Also we know that there are states that must also be labeled by operators of the enveloping algebra of a subalgebra (such as those giving their subalgebra representations, like isospin for the su(3) octet), these being invariants of subalgebras (though not, of course, of the algebra). These labeling operators, distinguishing states with the same weight, might have eigenvalues differing for the two representations. Thus a step operator applied to a state might give a sum of states, all with the same weight, but with the terms in the sum different for different representation. Could this happen?

Problem XIV.4.b-1: The 8 and 27 dimensional su(3) representations contain different su(2) representations. They might have states, of the same weight, on which an E_α gives a sum containing states of different su(2) representations. Are these such states?

Problem XIV.4.b-2: Thus we have to show that if the highest weights are the same, then so are the multiplicities of all states [Wan (1975), p. 128], and all eigenvalues of all labeling operators, for all states, so the representations are equivalent. First, clearly if the highest weights are the same, then so are all weights, although a rigorous proof would be nice, as well as an explanation of why this is not true for other weights. Now consider two purported nonequivalent representations with the same highest weight [Hausner and Schwartz (1968), p. 147; Miller (1972), p. 328; Racah (1951), p. 38; Racah (1964), p. 19; Serre (1987), p. 58; Wan (1975), p. 130; Zelobenko (1973), p. 312]. The state-labeling operators, the weight operators (the H's) and those of the enveloping algebra, have all eigenvalues the same for the highest-weight states $|h)$, $|h)'$, with the highest weight w_h (a vector in root space). Why do the enveloping-algebra operators have the same eigenvalues on the two highest-weight states? Apply step-down operators (these change eigenvalues of the labeling operators) until reaching the weight giving subspaces in which the representations differ. In such a subspace, with weight w, the states are given by

$$|i) = \ldots E_{-\beta}E_{-\alpha}|h), \quad |i)' = \ldots E_{-\beta}E_{-\alpha}|h)', \qquad \text{(XIV.4.b-1)}$$

with all roots negative, and the full set of states is given by all sums of negative roots so

$$w = w_h - \alpha - \beta - \cdots. \qquad \text{(XIV.4.b-2)}$$

Take a maximal set of independent states in this subspace, for one representation, and assume that the dimension of the subspace for the other is less, so that for the first,

$$|1) + \cdots + |k) = 0, \qquad \text{(XIV.4.b-3)}$$

but, for the other,

$$|1)' + \cdots + |k)' = |V), \qquad \text{(XIV.4.b-4)}$$

where $|V)$ is some state. Applying the H's to these equations, since all terms have the same weight, we get $kw = 0 \neq 0$, a contradiction. Thus the number of independent states in the subspace is the same — the multiplicity is the same for all weight vectors. Explain why the same argument shows that if two states are orthogonal for one representation, they are for the other. Thus for a maximal set of linearly independent (orthogonal) vectors in one space (a coordinate system), the corresponding vectors for the other representation also form such a coordinate system. Further if we take any state in this space and expand it in terms of the coordinate vectors, the expansion of the corresponding state (obtained with the same set of E's from the highest-weight state) is the same — the coefficients are the same. The coefficients are found by taking the product of each basis state and its conjugate, which equals 1, so the coefficients depend on the algebra, just the roots. Justify. What about the effect of a subalgebra enveloping-algebra operator on these states? Applying one to $\ldots E_{\alpha}|h)$, check that we can move all E's with positive roots to act on $|h)$, giving zero, leaving then a sum of terms with H's and E's with negative roots; each term in the sum gives a state with the same weight. But each term in the sum is a state (acted on by the H's) of a subspace of the same weight, and each such state has an expansion, the same for both representations, in terms of the subspace coordinate vectors. And the H's have the same eigenvalues on all states in the subspace. Hence the action of any operator, on corresponding states of the representations, gives the same expansion, thus the matrix elements of the operator are the same. From these it follows that representations with the same highest weight are equivalent. Why? This argument is probably somewhat redundant. Give the most compact one. Comment on the relationship of this to those (different forms?) in the references. Which is the best? Why? Give a simple summary of why representations are determined by their highest weight — up to equivalence. This last word leaves some flexibility. What is its source?

Problem XIV.4.b-3: Suppose two representations had different highest weights. Why could they not be equivalent?

XIV.4.c Are there labeling operators?

This discussion on labeling representations and states by a set of operators is based on our faith that there is such a set. It seems reasonable, for these algebras, that there might be different sets of matrices, obeying the commutation relations of the algebra symbols, that cannot be converted into each other by a similarity transformation, but for which there is no operator having different eigenvalues for them. How do we know that for semisimple algebras this cannot happen? How do we know that for these algebras we can find a complete, independent, set of labels; might there be inequivalent representations that have the same set of eigenvalues for all invariants? Are the invariants sufficient; might there be insufficient — independent — ones? Labeling operators are (taken) diagonal. The diagonal elements for each row determine that row (and likewise column). Might there be representation matrices having two rows all labeled by the same set of diagonal elements? In general, all these unfortunate things do occur. But, for all compact, semisimple algebras there are always enough operators to label every representation, and every state, completely.

That invariants (which so label the representations) do exist is shown above (sec. XIV.2.b, p. 406). That each representation has a state from which it can be generated, and which labels it, that of highest weight, and that the number of weight components is finite, means that all representations can be distinguished by a finite number of labels. Invariants are functions of the highest weights, so a finite number of these, equal to the number of weight components, provides sufficient labels.

Are there operators, and a sufficient set, to label, uniquely, all states of all representations of all semisimple groups? As we have seen, there (fortunately) are. However, might the states be different functions of the transformation variables (like angles), though having the same eigenvalues of the labels? Starting with a realization of the highest-weight state, which is thus completely specified, as are the algebra operators, all states generated are also completely determined. The action of the step operators changes the eigenvalues of the labeling operators, and this change depends on the commutation relations, not on the realization of the states. If two states are different, say of different functional form, there must be algebra operators going from one to the other. But these also change the eigenvalues of the labeling operators, because it is always possible to find ones that do not commute with a given step operator. Hence, if the functional forms (within a realization) differ, the labels for the two states must differ. States and representations are labeled — but not realizations.

Problem XIV.4.c–1: Are these labeling invariants independent (notice, the modifier "linearly" is not included); might there be (polynomial) relations? There are a finite number of invariants needed, and an infinite number of representations. Explain why there can be no polynomial relations. How about arbitrary functions; is something like $I_1 = exp(I_2)$, relating invariants I_1 and I_2, possible?

Problem XIV.4.c–2: Prove that there cannot be representation matrices with two rows, or columns, given by the same eigenvalues of all labeling operators. How is this related to the states being completely labeled? Explain why, and how, labeling operators also label the representation matrix elements.

Problem XIV.4.c–3: The generators of the algebra, and those of its enveloping algebra, thus label the states and representations, these being completely labeled, and there are no further operators to provide additional labels. This is true for semisimple algebras — of compact groups. The Poincaré algebra shows that it cannot be extended beyond these [Mirman (1995a)]. There is a difference between the algebra representations for compact, and noncompact, semisimple groups. The unitary representations of compact algebras are finite dimensional, those of noncompact ones are infinite dimensional. (This also is relevant to the rows being completely labeled by operator eigenvalues; the states are, and it would be uncomfortable if the number of rows of the transformation matrices differed from the number of states. For infinite-dimensional matrices more care is needed.) We restrict consideration to compact cases; whether the results hold elsewhere must be determined in each instance. But for the case considered here, compact, semisimple algebras, there are (fortunately) all needed labeling operators.

Problem XIV.4.c–4: It is interesting to examine the extent to which this can be applied to noncompact algebras. Can highest (or perhaps lowest) weights be defined for these? If not, is there an analog? What, if anything, is the defining representation? Are representations determined by their highest weights (or the equivalent)? Are they, and the states (completely) labeled by the operators from the algebra, and the enveloping algebra? How about inhomogeneous algebras?

XIV.5 DECOMPOSABILITY OF THE REPRESENTATIONS

Of the properties of representations, perhaps most important are decomposability, and closely linked, unitarity. Similarity transformations take the representation matrices into equivalent block-diagonal ones. A representation is reduced if it is in block-diagonal form — all blocks below (or above) the diagonal are zero for all matrices. If this is the best that can be done — the matrices are all upper (or lower) triangular — the representation is reducible, but not completely reducible. It is completely reduced (decomposed) if all off-diagonal blocks are zero. It is completely reducible (decomposable) if it is equivalent to one with all blocks — for all matrices — on the diagonal, a completely reduced (decomposed) representation. This applies equally to groups and algebras. But can Lie algebra representations be decomposed? And is there a relationship between decomposability for groups and (their) algebras?

For semisimple algebras (like semisimple groups), only, all representations are decomposable. This we consider next, studying how it is related to the semisimplicity, and its implications.

XIV.5.a Complete reducibility for a group and for its algebra

First let us see the effect of a representation that is not completely reduced. A nondecomposed representation leaves a subspace invariant, clearly a consequence of matrices being upper triangular. Does this contradict the requirement that each element of a group have an inverse? There is a difference in the effect of non-complete reducibility for a group and for its algebra.

Problem XIV.5.a-1: Show that matrices $\begin{pmatrix} 1 & k\alpha \\ 0 & 1 \end{pmatrix}$, $\begin{pmatrix} 1 & -k\alpha \\ 0 & 1 \end{pmatrix}$, form a group, the two matrices are inverses and that its algebra generator is $\begin{pmatrix} 0 & 1 \\ 0 & 0 \end{pmatrix}$. Check that they both leave a subspace invariant. This is not an impressive algebra but it does illustrate the (what?) point.

Problem XIV.5.a-2: Find the group generated by $\begin{pmatrix} 1 & \alpha & \beta \\ 0 & 1 & \gamma \\ 0 & 0 & 1 \end{pmatrix}$, and its inverse (which is ?); check that the algebra is given by the three matrices $\begin{pmatrix} 0 & 1 & 0 \\ 0 & 0 & 0 \\ 0 & 0 & 0 \end{pmatrix}$, $\begin{pmatrix} 0 & 0 & 1 \\ 0 & 0 & 0 \\ 0 & 0 & 0 \end{pmatrix}$, $\begin{pmatrix} 0 & 0 & 0 \\ 0 & 0 & 1 \\ 0 & 0 & 0 \end{pmatrix}$. Is it semisimple? Which subspace is invariant?

Problem XIV.5.a-3: Both the group and the algebra leave a subspace invariant. By an inverse operator we mean one that applied after the operator, gives the original vector. A group operator applied to a vector with a component not in the subspace gives one having a component in the subspace. Its inverse sets this component to 0, giving the original vector. For the algebra however, a vector outside the subspace is moved into it, and there is then no way of getting back out. Thus for the group there is a product of group elements that applied to any vector gives the same vector. For the algebra the original vector cannot be regained. Check this for these two groups.

Problem XIV.5.a-4: For the angular-momentum algebra, consider any completely-reduced representation (say the two or three-dimensional ones) using both J_z, J_x, J_y, and also J_z, J_+, and J_-. Is it possible to return to every vector?

Problem XIV.5.a-5: The action of the algebra does not look interesting as given. A generator applied to a vector gives only one other vector. Starting from a vector it is impossible to reach all of the representation space. Of course this is not how we treat the algebra. We define (usually without saying so) another product (besides commutation) so that $X_1X_2\ldots X_k$ means (as expected) that operator X_k acts on a vector, then operator X_{k-1} acts on the resultant, ..., then on the resulting vector X_2 acts, then X_1. Check that this is the effect of matrix multiplication. Using this we traverse the entire representation space. Why? Show that for a completely-reducible representation we can traverse it starting from any (set of?) vector(s), for an indecomposable one there is one vector we must start from to get the entire space [Wan (1975), p. 247]. (That it is possible to cover this whole space for any representation using all powers of the operators is essentially the import of the Poincaré-Birkhoff-Witt theorem [Barut and Raczka (1986), p. 249; Humphries (1987), p. 91; Jacobson (1979), p. 156; Varadarajan (1984), p. 166; Wan (1975), p. 182; Zelobenko (1973), p. 158].) The complete set of (the infinite number of) these products of generators, plus all their sums, is the enveloping algebra of the Lie algebra.

Problem XIV.5.a-6: What must the original vector be for the three-dimensional representation of pb. XIV.5.a-2?

XIV.5.b All representations of semisimple algebras are decomposable

As seen from these examples it is not (always?) possible to decompose representations of a solvable algebra. But for a semisimple algebra all representations can be completely reduced [Barut and Raczka (1986), p. 204; Humphries (1987), p. 28; Jacobson (1979), p. 75; Wan (1975), p. 134; Zelobenko (1973), p. 166]. Why? There is one representation of a simple algebra that is always decomposable, the adjoint representation.

Problem XIV.5.b-1: What is the relationship between semisimplicity and the decomposability of the adjoint representation?

Problem XIV.5.b-2: How does the complete reduciblity of all other representations follow from the algebra being simple? Suppose the representation matrices are all upper diagonal. Then their commutators give matrices that are nilpotent and these would form a nilpotent ideal contradicting the simplicity of the algebra. Why? This is suggestive, but not conclusive. Prove that if matrices are not completely reducible but upper diagonal with elements that are not numbers, but blocks, their commutators need not be nilpotent matrices.

XIV.5.b.i *Why the representations are decomposable*

So it remains to show complete reducibility in general. To do this, we use the fact that simple algebras have invariants. Suppose that a representation is not completely reducible. Then all matrices are upper (or lower) block-diagonal.

Problem XIV.5.b.i-1: Demonstrate that the product, so the commutator, of two upper block-diagonal matrices is upper block-diagonal. Any invariant would then also be upper block-diagonal, so not proportional to the unit matrix. But since it is an invariant it has the same value on all states. Consider $\begin{pmatrix} 1 \\ 0 \end{pmatrix}$ with 1,0 now matrices; this the invariant leaves unchanged while it does not for $\begin{pmatrix} 0 \\ 1 \end{pmatrix}$. Hence it, so the representation matrices, cannot be upper block-diagonal. They must be block-diagonalizable; the representation is completely reduced (or reducible). Expand this example and check that in general, using the Jordan canonical form (pb. XIII.8.a-2, p. 393), an invariant that is upper block-diagonal is not an invariant. Also an invariant cannot be zero for any representation except the scalar. Thus the decomposability follows from the existence of invariants, which follows from semisimplicity (sec. XIV.2.b, p. 406).

Problem XIV.5.b.i-2: Decomposibility for any algebra also follows from that for all su(2)-algebra representations. Start with any su(2) subalgebra. Take the (hermitian) transpose of the three commutators. Then the identification

$$H \Rightarrow H^t, \quad E_{\pm} \Rightarrow E^t_{\mp}, \tag{XIV.5.b.i-1}$$

gives the same commutators; we obtain a representation from another by taking the transpose and interchanging the two step operators. Now any matrix can be brought into either diagonal or Jordan canonical form (pb. XIII.8.a-2, p. 393) — this has nonzero elements only on the diagonal except for ones on the superdiagonal (only H_{jj}, and

$$H_{i,i+1} = 1, \tag{XIV.5.b.i-2}$$

can be nonzero elements). Suppose it is the Jordan form; take the $(n,1)$ matrix element, of the n-dimensional matrix,

$$[H, E_+] = E_+ \tag{XIV.5.b.i-3}$$

giving

$$(H_{11} - H_{nn}) = 1, \text{ unless } E_{n1} = 0. \tag{XIV.5.b.i-4}$$

From

$$[H, E_-] = -E_-, \quad (H_{11} - H_{nn}) = -1 \Rightarrow E_{n1} = 0. \tag{XIV.5.b.i-5}$$

Next taking the $(n-1,1)$ matrix element we get

$$E_{n-1,1} = 0, \tag{XIV.5.b.i-6}$$

and so on. Continuing, we find that all matrix elements below the diagonal are 0. Check this. If all three matrices have all elements below the diagonal 0, then the representation is not completely reducible; but we know the su(2) representations are. Thus the H of this subalgebra, so all H's since they commute, are diagonal (or can be diagonalized; the alternative, the Jordan form, being impossible) — the diagonalizability of the H's follows from the complete reducibility of the subalgebra representations. Then it clearly follows from the canonical form that if E_α is upper triangular (can it have nonzero diagonal elements?) $E_{-\alpha}$ is lower triangular, and so in the symmetric form they both have elements above and below the diagonal — implying that the representation is decomposed. How is this argument related to that of the previous problem? Is this completely rigorous; if necessary can it be made so?

Problem XIV.5.b.i-3: The decomposibility of simple-algebra representations follows from diagonalizability of the H's, which can be seen another way. Suppose that the representation matrices are all upper triangular. Explain why these form a solvable algebra (a clue that representation matrices of semisimple algebras are completely reducible; we would not expect a semisimple algebra to be represented by matrices forming a solvable algebra). Now $[H, E_\alpha] \sim E_\alpha$, $[E_\alpha, E_{-\alpha}] \sim H$ (any H can be written as a sum of such commutators); thus show that the commutator of two upper triangular matrices is strictly upper triangular (all diagonal elements are 0), so all E's and H's would be nilpotent matrices (obviously?). Applied to a vector these give one with fewer nonzero components. Finally there would be one vector on which the nilpotent matrices give zero. Invariant C acting on this vector would be 0, and so be 0 for all vectors; $C = 0$ for all non-decomposable representations. Thus if $C \neq 0$ for a representation, not all matrices are upper triangular; the Jordan canonical form gives one diagonal (but not a multiple of the unit matrix; why?). Then all matrices that commute with it, the H's, are diagonal. Also if E_α increases a weight component, $E_{-\alpha}$ decreases it by the same amount. So it is always possible to go from one weight of the space to any other by using a product of E's, implying (?) that the representation is completely reduced. What goes wrong with the last sentence if the representation is reducible? To use this argument the H's have to be diagonal, and this is the key to showing complete reducibility. Are these arguments rigorous? If not make them so. Do they include all possible cases?

XIV.5.b.ii *Complete reducibility and unitarity*

By the same argument as for (finite) groups (sec. VII.2, p. 181), if a representation is finite-dimensional and completely reduced, it is equivalent to a unitary representation. Then the H's are hermitian and can be diagonalized, emphasizing again this connection. Also if H has real eigenvalues (giving a unitary representation), then $[E_\alpha, E_{-\alpha}] \sim H$ means E_α, $E_{-\alpha}$ are hermitian conjugates. So all terms in invariant C are positive and $C = 0$ is possible only if all operators are zero — if the representation is the scalar.

Problem XIV.5.b.ii-1: Show this rigorously, for both Lie groups and Lie algebras.

XIV.5.b.iii *The physics of complete reducibility*

To see the importance of complete reduciblity consider SO(3). If it were not completely reducible, J_z would not be diagonalizable. States might be labeled by their expectation values (although nilpotent matrices have no diagonal elements) if these were all different. Otherwise it would be difficult to distinguish states. However there would be no labeling operator that left them invariant (except for multiplication by a constant, the eigenvalue). Thus a magnetic field would always cause transitions — there would never be stable eigenstates. Also rotation operators would not be hermitian (complete reducibility and unitarity are closely related), so a rotation would change the probability of finding a particle.

This is not to say that nondecomposable and nonunitary representations do not have physical significance (as for mass-zero representations of the Poincaré group [Mirman (1993, 1994, 1995b)]). But if physics were described by algebras whose representations were not completely reducible the universe would be very different, and very strange, probably uninhabitable and possibly impossible. It seems quite fortunate that our universe is described by semisimple groups and these have representations which are (all) decomposable, so unitary.

Problem XIV.5.b.iii-1: Describe what scattering would be like if the representations of the isospin algebra or su(3) were not decomposable.

XIV.5.c When, and why, are representations finite dimensional?

If an algebra is compact its representations are finite, and equivalent to unitary ones; all unitary representations of semisimple, compact algebras are finite dimensional. (For a noncompact semisimple algebra the finite-dimensional representations are nonunitary, the unitary ones are infinite dimensional.) For unitary representations (whose Casimir invariants are real numbers — they must be for these), E_α, $E_{-\alpha}$ are hermitian conjugates, and E_α increases the eigenvalue of the H's, so there has to be a vector on which it gives zero (and another on which $E_{-\alpha}$ does), for

$$C = \sum H^2 + \sum E_\alpha E_{-\alpha}, \qquad \text{(XIV.5.c-1)}$$

and each term is positive definite; thus the representation is finite-dimensional. The converse is just been shown. This is the argument giving the representations of SO(3), and holds generally.

To get a noncompact group some E_α's are multiplied by i giving

$$C = \sum H^2 + \sum E_\alpha E_{-\alpha} - \sum E_\alpha E_{-\alpha}, \qquad \text{(XIV.5.c-2)}$$

where the first sum is over the E's that have not been multiplied by i, the second over those that have, and now, by the same argument, the last term is negative definite, so for C to be constant the effect of this term must be counterbalanced by increases in the values of the H's, thus the representation is infinite dimensional.

Problem XIV.5.c-1: Why must the eigenvalue of the Casimir invariant be real for unitary representations?

Appendix A

SU(3) and States of Some of its Representations

Here we list the su(3) commutation relations [Cahn (1967), p. 9], and states of a few smaller representations.

The generators Y, T_z, $T_\pm$, T_z, $U_\pm$, U_z, $V_\pm$, have commutation relations [Elliott and Dawber (1987), p. 234],

$$[T_+,\ T_-] = 2T_z,\ [T_z,\ T_\pm] = \pm T_\pm,\ [T_z,\ Y] = 0,\ [T_\pm,\ Y] = 0, \tag{A-1}$$

$$[U_+,\ U_-] = 2U_z,\ \ [U_z,\ U_\pm] = \pm U_\pm,\ \ [Y,\ U_\pm] = \pm U_\pm, \tag{A-2}$$

$$[V_+,\ V_-] = 2V_z,\ \ [V_z,\ V_\pm] = \pm V_\pm,\ \ [Y,\ V_\pm] = \mp V_\pm, \tag{A-3}$$

$$[T_+,\ U_+] = V_-,\ \ [T_-,\ U_-] = -V+, \tag{A-4}$$

$$[T_-,\ V_-] = U_+,\ \ [T_+,\ V_+] = -U_-, \tag{A-5}$$

$$[T_z,\ U_\pm] = \mp U_\pm/2,\ \ [T_z,\ V_\pm] = \mp V_\pm/2, \tag{A-6}$$

$$[U_-,\ V_-] = -T_+,\ \ [U_+,\ V_+] = T_-, \tag{A-7}$$

$$[T_+,\ U_-] = [T_-,\ \ U_+] = [T_+,\ V_-] = 0, \tag{A-8}$$

$$[T_-,\ \ V_+] = [U_+,\ V_-] = [U_-,\ \ V_+] = 0, \tag{A-9}$$

where

$$U_z = 3Y/4 - T_z/2,\ \ V_z = -3Y/4 - T_z/2; \tag{A-10}$$

any other commutators are zero.

Matrices of the Defining Representation

The defining representation of the su(3) algebra is

$$T_z = \begin{pmatrix} \frac{1}{2} & 0 & 0 \\ 0 & -\frac{1}{2} & 0 \\ 0 & 0 & 0 \end{pmatrix}, \quad T_+ = \begin{pmatrix} 0 & 1 & 0 \\ 0 & 0 & 0 \\ 0 & 0 & 0 \end{pmatrix}, \quad T_- = \begin{pmatrix} 0 & 0 & 0 \\ 1 & 0 & 0 \\ 0 & 0 & 0 \end{pmatrix},$$

$$Y = \begin{pmatrix} \frac{1}{3} & 0 & 0 \\ 0 & \frac{1}{3} & 0 \\ 0 & 0 & -\frac{2}{3}, \end{pmatrix} \quad U_+ = \begin{pmatrix} 0 & 0 & 0 \\ 0 & 0 & 1 \\ 0 & 0 & 0 \end{pmatrix}, \quad U_- = \begin{pmatrix} 0 & 0 & 0 \\ 0 & 0 & 0 \\ 0 & 1 & 0 \end{pmatrix},$$

$$V_+ = \begin{pmatrix} 0 & 0 & 0 \\ 0 & 0 & 0 \\ 1 & 0 & 0 \end{pmatrix}, \quad V_- = \begin{pmatrix} 0 & 0 & 1 \\ 0 & 0 & 0 \\ 0 & 0 & 0 \end{pmatrix}. \qquad \text{(A-11)}$$

Representations

In the table, for each state of a representation [Cornwell (1984), p. 738] we give its su(2) representation, denoted by its highest weight I, the state labels I_z, and hypercharge Y, listed as (I, I_z, Y). The representations are labeled by their dimensions, and Young frames; for two of the same dimension, one is labeled by *. The dimension of a representation is denoted by d. It is interesting to notice how the frames, and also the states, of conjugate representations, those with the same dimensions, are related, and to guess, in terms of the patterns of both representation and state labels, why those that are self-conjugate are so.

Repres-entation	*d*	*States*
	1	$(0,0,0)$
	3	$(\frac{1}{2},\frac{1}{2},\frac{1}{3}),(\frac{1}{2},-\frac{1}{2},\frac{1}{3}),(0,0,-\frac{2}{3})$
	3*	$(\frac{1}{2},-\frac{1}{2},-\frac{1}{3}),(\frac{1}{2},\frac{1}{2},-\frac{1}{3}),(0,0,\frac{2}{3})$
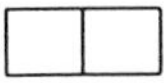	6	$(1,1,\frac{2}{3}),(1,0,\frac{2}{3}),(1,-1,\frac{2}{3})$ $(\frac{1}{2},\frac{1}{2},-\frac{1}{3}),(\frac{1}{2},-\frac{1}{2},-\frac{1}{3}),(0,0,-\frac{4}{3})$
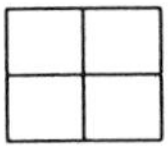	6*	$(1,-1,-\frac{2}{3}),(1,0,-\frac{2}{3}),(1,1,-\frac{2}{3}),$ $(\frac{1}{2},-\frac{1}{2},\frac{1}{3}),(\frac{1}{2},\frac{1}{2},\frac{1}{3}),(0,0,\frac{4}{3})$
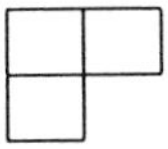	8	$(\frac{1}{2},\frac{1}{2},1),(\frac{1}{2},-\frac{1}{2},1),$ $(1,1,0),(1,0,0),(1,-1,0)(0,0,0),$ $(\frac{1}{2},\frac{1}{2},-1),(\frac{1}{2},-\frac{1}{2},-1)$
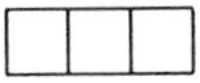	10	$(\frac{3}{2},\frac{3}{2},1),(\frac{3}{2},\frac{1}{2},1),(\frac{3}{2},-\frac{1}{2},1),(\frac{3}{2},-\frac{3}{2},1),$ $(1,1,0),(1,0,0),(1,-1,0),$ $(\frac{1}{2},\frac{1}{2},-1),(\frac{1}{2},-\frac{1}{2},-1)$ $(0,0,-2)$
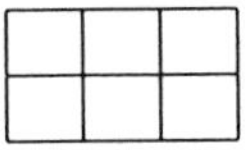	10*	$(\frac{3}{2},-\frac{3}{2},-1),(\frac{3}{2},-\frac{1}{2},-1),(\frac{3}{2},\frac{1}{2},-1),(\frac{3}{2},\frac{3}{2},-1),$ $(1,-1,0),(1,0,0),(1,1,0),$ $(\frac{1}{2},-\frac{1}{2},1),(\frac{1}{2},\frac{1}{2},1)$ $(0,0,2)$
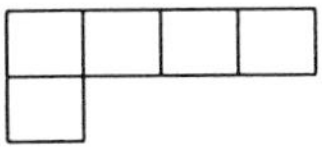	27	$(1,1,2),(1,0,2),(1,-1,2),(\frac{1}{2},\frac{1}{2},1),(\frac{1}{2},-\frac{1}{2},1),$ $(\frac{3}{2},\frac{3}{2},1),(\frac{3}{2},\frac{1}{2},1),(\frac{3}{2},-\frac{1}{2},1),(\frac{3}{2},-\frac{3}{2},1),$ $(2,2,0),(2,1,0),(2,0,0),(2,-1,0),(2,-2,0),$ $(1,1,0),(1,0,0),(1,-1,0),(0,0,0),$ $(\frac{3}{2},\frac{3}{2},-1),(\frac{3}{2},\frac{1}{2},-1),(\frac{3}{2},-\frac{1}{2},-1),(\frac{3}{2},-\frac{3}{2},-1),$ $(\frac{1}{2},\frac{1}{2},-1),(\frac{1}{2},-\frac{1}{2},-1),$ $(1,1,-2),(1,0,-2),(1,-1,-2)$

References

Arfken, George (1970), Mathematical Methods for Physicists, second ed. (New York: Academic Press).

Armstrong, M. A. (1988), Groups and Symmetry (New York: Springer-Verlag).

Barut, A. O. and Raczka R. (1986), Theory of Group Representations and Applications (Singapore: World Scientific Publishing Co.).

Baumslag, B. and Chandler B. (1968), Theory and Problems of Group Theory (New York: McGraw-Hill Book Co.; Schaum's Outline Series in Mathematics).

Belinfante, Johan G. F. and Kolman, Bernard (1972), A Survey of Lie Groups and Lie Algebras with Applications and Computational Methods (Philadelphia: SIAM).

Bhagavantam, S. and Venkatarayudu, T. (1951), Theory of Groups and its Application to Physical Problems, second ed. (Waltair, India: Andhra University).

Biedenharn, L. C. and Van Dam, H., eds. (1965), Quantum Theory of Angular Momentum (New York: Academic Press).

Blichfeldt, H. F. (1917), Finite Collineation Groups (Chicago: University of Chicago Press).

Boerner, Hermann (1963), Representations of Groups (Amsterdam: North Holland Publishing Co.).

Bröcker, Theodor and tom Dieck, Tammo (1985), Representations of Compact Lie Groups (New York: Springer-Verlag).

Burn, R. P. (1991), Groups, A Path to Geometry (Cambridge: Cambridge University Press).

Burns, Gerald (1977), Introduction to Group Theory with Applications (New York: Academic Press).

Burnside, W. (1955), Theory of Groups of Finite Order, second ed. (New York: Dover Publications).

Burrow, Martin (1993), Representation Theory of Finite Groups (New York: Dover Publications).

Cahn, Robert N. (1984), Semi-Simple Lie Algebras and Their Representations (Menlo Park, Ca.: Benjamin/Cummings Publishing Co.).

Cartan, Élie (1981), The Theory of Spinors (New York: Dover Publications).

Chen, Jin-Quan (1989), Group Representation Theory for Physicists (Singapore: World Scientific Publishing Co.).

Chevalley, Claude (1962), Theory of Lie Groups (Princeton, NJ: Princeton University Press).

Cohn, P. M. (1965), Lie Groups (Cambridge: Cambridge University Press).

Cornwell, J. F. (1984), Group Theory in Physics (London: Academic Press).

Courant, R. and Hilbert, D (1955), Methods of Mathematical Physics, vol. I (New York: Interscience Publishers, Inc.).

Coxeter, H. S. M. and Moser, W. O. J. (1980), Generators and Relations for Discrete Groups, fourth ed. (Berlin: Springer-Verlag).

Cullen, Charles G. (1990), Matrices and Linear Transformations, second ed. (New York: Dover Publications).

Dixon, John D. (1973), Problems in Group Theory (New York: Dover Publications).

Duffey, George H. (1992), Applied Group Theory for Physicists and Chemists (Englewood Cliffs, NJ: Prentice Hall).

Edmonds, A. R. (1960), Angular Momentum in Quantum Mechanics (Princeton, NJ: Princeton University Press).

Edwards, H. (1984), Galois Theory, Graduate Texts in Mathematics, #101 (New York: Springer-Verlag).

Elliott, J. P. and Dawber, P. G. (1987), Symmetry in Physics, vol. 1, principles and simple applications (Houndsmills: Macmillan Publishers).

Elliott, J. P. and Dawber, P. G. (1986), Symmetry in Physics, vol. 2, further applications (New York: Oxford University Press).

Falicov, L. M. (1966), Group Theory and its Physical Applications (Chicago: The University of Chicago Press).

Fässler, A. and E. Stiefel (1992), Group Theoretical Methods and Their Applications (Cambridge, MA: Birkhauser Boston).

Fejes Toth, L. (1964), Regular Figures (Oxford: Pergamon Press).

Gel'fand, I. M. (1989), Lectures on Linear Algebra (New York: Dover Publications).

Gel'fand, I. M., Minlos, R. A. and Shapiro, Z. Ya. (1963), Representations of the Rotation and Lorentz Groups and their Applications (New York: The Macmillan Company).

Ghyka, Matila (1977), The Geometry of Art and Life (New York: Dover Publications).

Gilmore, Robert (1974), Lie Groups, Lie Algebras, and Some of Their Applications (New York: John Wiley and Sons, Inc.).

Goldstein, Herbert (1953), Classical Mechanics (Cambridge, MA: Addison-Wesley Publishing Company).

Gorenstein, Daniel (1982), Finite Simple Groups: An Introduction to their Classification (New York: Plenum Press).

Gorenstein, Daniel (1983), The Classification of Finite Simple Groups (New York: Plenum Press).

Gould, M. D. (1989), Representation Theory of the Symplectic Groups. I, *J. Math. Phys.* **30**, #6, 1205-1218.

Grossman, Israel and Magnus, Wilhelm (1992), Groups and their Graphs (Washington, DC: The Mathematical Association of America, New Mathematical Library).

Grove, L. C. and Benson, C. T. (1985), Finite Reflection Groups, second ed. (New York: Springer-Verlag).

Gursey, F., ed. (1964), Group Theoretical Concepts and Methods in Elementary Particle Physics (New York: Gordon and Breach).

Hall, Marshall, Jr. (1959), The Theory of Groups (New York: Macmillan).

Hamermesh, M. (1962), Group Theory and its Application to Physical Problems (Reading MA: Addison-Wesley).

Hargittai, Istvan and Hargittai, Magdolna (1987), Symmetry through the Eyes of a Chemist (New York: VCH Publishers).

Hausner, Melvin and Schwartz, Jacob T. (1968), Lie Groups; Lie Algebras (New York: Gordon and Breach).

Heine, Volker (1993), Group Theory in Quantum Mechanics (New York: Dover Publications).

Helgason, Sigurdur (1964), Differential Geometry and Symmetric Spaces (New York: Academic Press).

Hermann, Robert (1966), Lie Groups for Physicists (New York: W. A. Benjamin, Inc.).

Höhler, G. ed. (1965), Ergeb. d. Exakt. Naturw., vol. 37, Tracts in Modern Physics (Berlin: Springer-Verlag).

Humphries, James E. (1987), Introduction to Lie Algebras and Representation Theory (New York: Springer- Verlag).

Inui, T., Tanabe, Y. and Onodera, Y. (1990), Group Theory and its Applications in Physics (Berlin: Springer-Verlag).

Jacobson, Nathan (1979), Lie Algebras (New York: Dover Publications).

James, Gordon and Liebeck, Martin (1993), Representations and Characters of Groups (Cambridge: Cambridge University Press).

Jansen, Laurens and Boon, Michael (1967), Theory of Finite Groups. Applications in Physics (Amsterdam: North-Holland Publishing Co.).

Janssen, T. (1973), Crystallographic Groups (Amsterdam: North-Holland Publishing Co.).

Jones, H. F. (1990), Groups, Representations and Physics (Bristol: Adam Hilger, IOP Publishing).

Joshi, A. W. (1982), Elements of Group Theory for Physicists, third ed. (New York: John Wiley and Sons, Inc.).

Kaplansky, Irving (1974), Lie Algebras and Locally Compact Groups (Chicago: University of Chicago Press).

Kurosh, A. G. (1960a,b), The Theory of Groups, vol. 1, 2 (New York: Chelsea Publishing Co.).

Landin, Joseph (1989), An Introduction to Algebraic Structures (New York: Dover Publications).

Lax, Melvin (1974), Symmetry Principles in Solid State and Molecular Physics (New York: John Wiley and Sons, Inc.).

Ledermann, Walter (1953), Introduction to the Theory of Finite Groups (Edinburgh and London: Oliver and Boyd).

Ledermann, Walter (1987), Introduction to Group Characters, second ed. (Cambridge: Cambridge University Press).

Littlewood, Dudley E. (1958), The Theory of Group Characters and Matrix Representations of Groups, second ed. (London: Oxford University Press, at the Clarendon Press).

Lockwood, E. H. and Macmillan, R. H. (1978), Geometric Symmetry (Cambridge: Cambridge University Press).

Lomont, J. S. (1961), Applications of Finite Groups (New York: Academic Press).

Ludwig, W. and Falter, C. (1988), Symmetries in Physics, Group Theory Applied to Physical Problems (Berlin: Springer-Verlag).

Lyapin, E. S., Aizenshtat, A. Ya. and Lesokhin, M. M. (1972), Exercises in Group Theory (New York: Plenum Press).

Lyndon, Roger C. (1989), Groups and Geometry, London Mathematical Society, Lectures Notes Series 101 (Cambridge: Cambridge University Press).

Lyubarskii, G. Ya. (1960), The Application of Group Theory in Physics (Oxford: Pergamon Press).

Magnus, Wilhelm, Karrass, Abraham and Solitar, Donald (1966), Combinatorial Group Theory: Presentations of Groups in Terms of Generators and Relations (New York: John Wiley and Sons, Inc.).

Margenau, Henry and Murphey, George Mosely (1955), The Mathematics of Physics and Chemistry (New York: D. Van Nostrand Co., Inc.).

Martin, George E. (1987), Transformation Geometry, An Introduction to Symmetry (New York: Springer-Verlag).

Maxfield, John and Maxfield, Margaret (1992), Abstract Algebra and Solution by Radicals (New York: Dover Publications).

Miller, D. W. (1958), On a Theorem of Hölder, *Am. Math. Monthly*, **65**, 252-254.

Miller, Willard, Jr (1968), Lie Theory and Special Functions (New York: Academic Press).

Miller, Willard, Jr (1972), Symmetry Groups and their Applications (New York: Academic Press).

Mirman, R. (1967), Representations of Inhomogeneous SU(n) Groups with a Mixture of Homogeneous and Inhomogeneous Operators Diagonal, *J. Math. Phys.* **8**, #1, 57-62.

Mirman, R. (1968a), Invariants and Scalars of Compact Unitary Algebras, *J. Math. Phys.* **9**, #1, 39-46.

Mirman, R. (1968b), Number of Polynomial Invariants of Adjoint and Fundamental Compact Inhomogeneous Unitary Algebras, *J. Math. Phys.* **9**, #1, 47-49.

Mirman, R. (1973), Experimental Meaning of the Concept of Identical Particles, *Nuovo Cimento* **B18**, #1, 110-121. Reprinted in Mirman (1995a), p. 191-202.

Mirman, R. (1979), Nonexistence of Superselection Rules; Definition of Term "Frame of Reference", *Found. Phys.*, **9**, #3/4, 283-299. Reprinted in Mirman (1995a), p. 203-219.

Mirman, R. (1986), Quantum Mechanics Determines the Dimension of Space, *Annals of the New York Academy of Sciences*, **480** (New Techniques and Ideas in Quantum Measurement Theory, D. M. Greenberger, Ed.) 601-603.

Mirman, R. (1987a), Expansion of Symmetric Group Products and States, *Can. J. Phys.*, **65**, 185-192.

Mirman, R. (1987b), Tensors of Symmetric and Unitary Groups, *Can. J. Phys.*, **65**, 193-197.

Mirman, R. (1988a), Complex Groups, Quantum Mechanics, and the Dimension and Reality of Space, *Helv. Phys. Acta*, **61**, 966-978. Reprinted in Mirman (1995a), p. 221-233.

Mirman, R. (1988b), The Reality and Dimension of Space and the Complexity of Quantum Mechanics, *Int. J. Theor. Phys.*, **27**, #10, 1257-1276. Reprinted in Mirman (1995a), p. 235-254.

Mirman, R. (1989), Unitary Group Canonical Representations, *Can. J. Phys.*, **67**, 766-773.

Mirman, R. (1991a), Poincaré Group Gives Quantum Mechanical Equations, *Bull. Am. Phys. Soc.*, **36** #1 GF1, P. 103 (January 1991); *AAPT Announcer* (December 1990), P. 105.

Mirman, R. (1991b), Interactions Give Quantum Mechanics and 3 + 1 Space, *Bull. Am. Phys. Soc.*, **36** #1 GF5, P. 103 (January 1991); *AAPT Announcer* (December 1990), P. 106.

Mirman, R. (1991c), Quantum Mechanics Describes Reduction of Statefunctions, *Bull. Am. Phys. Soc.*, **36** #3 B22 8, P. 413.

Mirman, R. (1991d), Quantum Linearity is Group Theoretical, *Bull. Am. Phys. Soc.*, **36** #3 B22 10, P. 413.

Mirman, R. (1991e), Transpositions and Their Diagonalization, *Rep. Math. Phys.*, **29**, #1, 1-16.

Mirman, R. (1992), How Nonlinearity Determines the Laws of Nature, *Physics Essays* **5**, #1, 97-114.

Mirman, R. (1993), Poincaré Representations of the Massless Class, *Rep. Math. Phys.*, **32**, #2, 251-265.

Mirman, R. (1994), Poincaré Mass-Zero Representations, *Int. J. Mod. Phys.*, **A9**, #1, 127-156.

Mirman, R. (1995a), Group Theoretical Foundations of Quantum Mechanics (Commack, NY: Nova Science Publishers, Inc.).

Mirman, R. (1995b), Massless Representations of the Poincaré Group, with applications to electromagnetism and gravitation (Commack, NY: Nova Science Publishers, Inc.).

Murnaghan, Francis D. (1962), The Unitary and Rotation Groups, Lectures on Applied Mathematics, Vol III (Washington, D. C.: Spartan Books).

Murnaghan, Francis D. (1963), The Theory of Group Representations (New York: Dover Publications).

Naimark, M. A. (1964), Linear Representations of the Lorentz Group (New York: The Macmillan Company).

Oyibo, Gabriel A. (1993), New Group Theory for Mathematical Physics, Gas Dynamics and Turbulence (Commack, NY: Nova Science Publishers, Inc.).

Particle Data Group (1994), Review of Particle Properties, Physical Review **D50**, no. 3.

Patterson, C. W. and Harter, W. G. (1976), Canonical symmetrization for unitary bases. I. Canonical Weyl bases, *J. Math. Phys.* **17**, 1125-1136.

Price, John F. (1977), Lie Groups and Compact Groups (Cambridge: Cambridge University Press).

Racah, G. (1951), Group Theory and Spectroscopy (Princeton, NJ: Lecture Notes, Institute for Advanced Study); reprinted in Höhler (1965), p. 28-84.

Racah, G. (1964), Lectures on Lie Groups, in Gursey (1964), p. 1-36.

Regge T. (1958), Symmetry Properties of Clebsch-Gordan's Coefficients, *Nuovo Cimento* [10] **10**, #3, 544-545, reprinted in Biedenharn and Van Dam (1965), p. 296.

Regge T. (1959), Symmetry Properties of Racah's Coefficients, *Nuovo Cimento* [10] **11**, #1, 116-117, reprinted in Biedenharn and Van Dam (1965), p. 298.

Robinson, G. de B. (1961), Representation Theory of the Symmetric Group (Toronto: University of Toronto Press).

Rutherford, Daniel Edwin (1948), Substitutional Analysis (Edinburgh: at the University Press).

Sagan, Bruce E. (1991), The Symmetric Group; Representations, Combinatorial Algorithms, and Symmetric Functions (Pacific Grove, CA: Brooks/Cole Publishing Co.).

Samelson, Hans (1990), Notes on Lie Algebras (New York: Springer-Verlag).

Sattinger, D. H. and Weaver, O. L. (1986), Lie Groups and Algebras with Applications to Physics, Geometry, and Mechanics (New York: Springer-Verlag).

Schattschneider, Doris (1990), Visions of Symmetry: Notebooks, Periodic Drawings, and Related Work of M. C. Escher (New York: W. H. Freeman and Co.).

Schensted, Irene Verona (1976), A Course on the Application of Group Theory to Quantum Mechanics (Peaks Island, ME 04108: NEO Press).

Schiff, L. I. (1955), Quantum Mechanics, second ed. (New York: McGraw-Hill Book Co.).

Schindler, Susan and Mirman, R. (1977a), The Decomposition of the Tensor Product of Representations of the Symmetric Group, *J. Math. Phys.*, **18**, #8, 1678-1696.

Schindler, Susan and Mirman, R. (1977b), The Clebsch-Gordan Coefficients of S_n, *J. Math. Phys.*, **18**, #8, 1697-1704; Addendum **19**, 2665 (1978).

Schindler, Susan and Mirman, R. (1977c), The Clebsch-Gordan Decomposition and the Coeficients for the Symmetric Group, in Sharp and Kolman (1977), pp. 661-668.

Schindler, Susan and Mirman, R. (1978a), Generation of the Clebsch-Gordan Coefficients for S_n, *Comput. Phys. Commun.*, **15**, #1/2, 131-145.

Schindler, Susan and Mirman, R. (1978b), Functions on Tableaux and Frames of the Symmetric Group, *Comput. Phys. Commun.*, **15**, #1/2, 147-152.

Schweber, Silvan S. (1962), An Introduction to Relativistic Quantum Field Theory (New York: Harper and Row).

Scott, W. R. (1987), Group Theory (New York: Dover Publications).

Segal, I. E. (1940), On the Automorphisms of the Symmetric Group, *Bull. Am. Math. Soc.*, **46**, 565.

Senechal, Marjorie (1990), Crystalline Symmetries, An Informal Mathematical Introduction (Bristol: Adam Hilger, IOP Publishing).

Serre, J-P. (1987), Complex Semisimple Lie Algebras (New York: Springer-Verlag).

Sharp, Robert T. and Kolman, Bernard, eds. (1977), Group Theoretical Methods in Physics. Proceedings of the Fifth International Colloquium (New York: Academic Press).

Shubnikov, A. V. and Koptsik, V. A. (1974), Symmetry in Science and Art (New York: Plenum Press).

Soto, M. F., Jr. and Mirman, R. (1981a), Objects for the Symmetric Group, *J. Math. Phys.*, **22**, #6, 1144-1148; Addendum **22**, 3010 (1981).

Soto, M. F., Jr. and Mirman, R. (1981b), Computation of Group Tables for the Symmetric Group, *Comput. Phys. Commun.*, **23**, 81-93.

Soto, M. F., Jr. and Mirman, R. (1981c), Construction of Symmetric Group Representation Matrices and States, *Comput. Phys. Commun.*, **23**, 95-107.

Soto, M. F., Jr. and Mirman, R. (1982), Number of Representations and Maximum Dimensions for S(N), *Comput. Phys. Commun.*, **27**, 57-64.

Stephenson, G. (1986), An Introduction to Matrices, Sets and Groups for Science Students (New York: Dover Publications).

Streater, R. F., and Wightman, A. S. (1964), PCT, SPIN AND STATISTICS, AND ALL THAT (New York: W. A. Benjamin, Inc.).

Sudarshan, E. C. G. and Mukunda, N. (1983), Classical Dynamics: A Modern Perspective (Malabar, FL: Robert E. Krieger Publishing Co.).

Talman, James D.(1968), Special Functions, A Group Theoretic Approach (New York: W. A. Benjamin, Inc.).

Tinkham, Michael (1964), Group Theory and Quantum Mechanics (New York: McGraw-Hill Book Co.).

Tung, Wu-Ki (1985), Group Theory in Physics (Singapore: World Scientific Publishing Co.). Aivazis, Michael (1991), Group Theory in Physics, Problems and Solutions (Singapore: World Scientific Publishing Co.).

Varadarajan, V. S. (1984), Lie Groups, Lie Algebras, and Their Representations, Graduate Texts in Mathematics, #102 (New York: Springer-Verlag).

Varshalovich, D. A., Moskalev, A. N. and Khersonskii, V. K. (1988), Quantum Theory of Angular Momentum: Irreducible Tensors, Spherical

Harmonics, Vector Coupling Coefficients, 3nj Symbols (Singapore: World Scientific Publishing Co.).

Wan, Zhe-Xian (1975), Lie Algebras (Oxford: Pergamon Press).

Weyl, Hermann (1931), The Theory of Groups and Quantum Mechanics (New York: Dover Publications).

Weyl, Hermann (1946), The Classical Groups, Their Invariants and Representations (Princeton, NJ: Princeton University Press).

Weyl, Hermann (1989), Symmetry (Princeton, NJ: Princeton University Press).

Wherrett, Brian S. (1986), Group Theory for Atoms, Molecules and Solids (Englewood Cliffs, NJ: Prentice Hall).

Wigner, E. P. (1959), Group Theory, and its Application to Quantum Mechanics of Atomic Spectra (New York: Academic Press).

Wilson, E. Bright, Jr., Decius, J. C., and Cross, Paul C. (1980), Molecular Vibrations (New York: Dover Publications).

Wybourne, Brian G. (1970), Symmetry Principles and Atomic Sectroscopy (New York: John Wiley and Sons, Inc.).

Wybourne, Brian G. (1974), Classical Groups for Physicists (New York: John Wiley and Sons, Inc.).

Yale, Paul B. (1988), Geometry and Symmetry (New York: Dover Publications).

Zelobenko, D. P. (1973), Compact Lie Groups and their Representations, Translations of Mathematical Monographs, vol 40, (Providence, RI: American Mathematical Society).

Index

Definitions are listed first, in bold; other references are italicized.